Lecture Notes in Electrical Engineering

Volume 580

The book series *Lecture Notes in Electrical Engineering* (LNEE) publishes the latest developments in Electrical Engineering - quickly, informally and in high quality. While original research reported in proceedings and monographs has traditionally formed the core of LNEE, we also encourage authors to submit books devoted to supporting student education and professional training in the various fields and applications areas of electrical engineering. The series cover classical and emerging topics concerning:

- Communication Engineering, Information Theory and Networks
- Electronics Engineering and Microelectronics
- Signal, Image and Speech Processing
- Wireless and Mobile Communication
- Circuits and Systems
- Energy Systems, Power Electronics and Electrical Machines
- Electro-optical Engineering
- Instrumentation Engineering
- Avionics Engineering
- Control Systems
- Internet-of-Things and Cybersecurity
- Biomedical Devices, MEMS and NEMS

For general information about this book series, comments or suggestions, please contact leontina. dicecco@springer.com.

To submit a proposal or request further information, please contact the Publishing Editor in your country:

China

Jasmine Dou, Associate Editor (jasmine.dou@springer.com)

India

Aninda Bose, Senior Editor (aninda.bose@springer.com)

Japan

Takeyuki Yonezawa, Editorial Director (takeyuki.yonezawa@springer.com)

South Korea

Smith (Ahram) Chae, Editor (smith.chae@springer.com)

Southeast Asia

Ramesh Nath Premnath, Editor (ramesh.premnath@springer.com)

USA, Canada:

Michael Luby, Senior Editor (michael.luby@springer.com)

All other Countries:

Leontina Di Cecco, Senior Editor (leontina.dicecco@springer.com)

**** Indexing: The books of this series are submitted to ISI Proceedings, EI-Compendex, SCOPUS, MetaPress, Web of Science and Springerlink ****

More information about this series at http://www.springer.com/series/7818

Reji Kumar Pillai · Girish Ghatikar ·
Ravi Seethapathy · Vijay L. Sonavane ·
S. A. Khaparde · Pradeep Kumar Yemula ·
Samir Chaudhuri
Editors

ISGW 2018 Compendium of Technical Papers

4th International Conference and Exhibition on Smart Grids and Smart Cities

Springer

Editors
Reji Kumar Pillai
India Smart Grid Forum (ISGF)
New Delhi, Delhi, India

Girish Ghatikar
Electric Power Research Institute (EPRI)
Palo Alto, CA, USA

Ravi Seethapathy
Biosirus
Markham, ON, Canada

Vijay L. Sonavane
Former Member (Tech)
MERC
Mumbai, Maharashtra, India

S. A. Khaparde
Department of Electrical Engineering
Indian Institute of Technology Bombay
Mumbai, Maharashtra, India

Pradeep Kumar Yemula
Department of Electrical Engineering
Indian Institute of Technology Hyderabad
Hyderabad, Andhra Pradesh, India

Samir Chaudhuri
Chair Working Group 7
India Smart Grid Forum (ISGF)
New Delhi, Delhi, India

ISSN 1876-1100 ISSN 1876-1119 (electronic)
Lecture Notes in Electrical Engineering
ISBN 978-981-32-9118-8 ISBN 978-981-32-9119-5 (eBook)
https://doi.org/10.1007/978-981-32-9119-5

This Springer imprint is published by the registered company Springer Nature Singapore Pte Ltd.
The registered company address is: 152 Beach Road, #21-01/04 Gateway East, Singapore 189721, Singapore

Contents

Smart Grid Technology

Advanced Distribution Management System: An Integrated System

Tarun Bhardwaj, Aakash Verma and Virender Mittal

Abstract Tata Power-DDL has been the frontrunner in implementing power distribution reforms in the capital city and is acknowledged for its consumer friendly practices. The company has implemented high-tech automated systems for its entire distribution network. In the world of digitalization, TATA POWER-DDL adopted smart technologies like SCADA-DMS and OMS to improve reliability and customer outages. These were implemented at different times, from different vendors and operated in different environments resulting in inefficient outage durations, non-optimal costs, and customer dissatisfaction. The flow of information from DMS to OMS happens through ICCP, but no reverse flow of information from OMS to DMS happens. Consequently, the DMS dispatchers have no clue as to the impact of outages on customers and OMS dispatchers have little clue about the cause of outages. This often results in delayed and to some extent, inaccurate information being conveyed to the customers, and no benefit accruing to reliability from customer calls. For better efficiency, digitalization of network and outage management, accurate reporting integration of the two systems was required. However, technological integration was beyond question due to the incompatibilities in technologies, vendors and data systems. So Advance Distribution Management System (ADMS) has been implemented to manage all levels of outages efficiently. This system acts as an integrator between all IT & OT systems, which is integrated with SAP-CRM, GIS, MWM/FFA,ICCP, Power manager portfolio, SAP BW-BO, RTU/DCU/FRTU/IED, Social media etc. In the near future we are going to integrate it with EV charging Solar Roof top, Battery storage, micro grids, Energy Audit, DPS (CYME, ETAP), HES, MDMS, ADR etc. Thus, ADMS is going to be the biggest integrator in the utility sector which will lead us towards the digitalization of our utility. This will also lead to desired improvements in

T. Bhardwaj (✉) · A. Verma · V. Mittal
Tata Power Delhi Distribution Ltd., New Delhi, India
e-mail: tarun.bh@tatapower-ddl.com

A. Verma
e-mail: aakash.verma@tatapower-ddl.com

V. Mittal
e-mail: virendra.mittal@tatapower-ddl.com

R. K. Pillai et al. (eds.), *ISGW 2018 Compendium of Technical Papers*,
Lecture Notes in Electrical Engineering 580,
https://doi.org/10.1007/978-981-32-9119-5_1

reliability and customer outage services. This paper attempts to outline and analyze the issues in the old system and explain how the new system has addressed those issues and added value to the utility by improving reliability and customer services simultaneously.

Keywords SCADA-DMS · OMS · ADMS · RTU · DCU · FRTU · IED · DPS · MDMS · HES · ADR · MWM & ICCP

Abbreviations

Tata Power-DDL	Tata power Delhi distribution Limited
ADMS	Advanced distribution management system
SCADA	Supervisory control and data acquisition
EMS	Energy management system
DMS	Distribution management system
OMS	Outage management system
GIS	Geographical information system
ERP	Enterprise resource planning
GSAS	Grid sub-station automation system
DA	Distribution automation
AT& C	Aggregate technical and commercial
SAIDI	System average interruption duration indices
SAIFI	System average interruption frequency indices
CMI	Consumer minutes interruption
ADR	Automatic demand response
DER	Distributed energy resource
BESS	Battery energy storage system
FFA	Field force automation
AMI	Advanced metering infrastructure
APRS	Automated power restoration system
SA	Switching advisor
IVVC	Integrated Volt/VAR control
OFR	Optimal feeder reconfiguration
NOC	Normal operating condition
AVL	Automated vehicle locator
CIM	Common information model

1 Introduction

Tata Power Delhi Distribution Limited is a Power Distribution Utility, distributes electricity in North and North- West part of Delhi, serving about 1.5 million consumers. Reliability and Aggregate Technical & Commercial (AT&C) Losses are the most important criteria for Regulators for measuring performance of any Power Utility. TATA POWER-DDL has been forerunner in implementing power distribution reforms in its licensee area to reduce AT& C Loss and improve reliability and consumer delight.

In present era of industrial and technological development, purpose of electricity is not only to supply power to bulb and fan but to life support systems, essential lifesaving services, water treatment plants, commutation and communication infrastructure, continuous process industries, cooling and heating appliances etc. In this aspect, responsibility of power utility is not just provide power availability to their consumer but also with quality and services. Hence the definition of Reliability is not limited to subside power outages but to provide quality of power with services which improve customer satisfaction. In this context reliability is basically measure of overall effectiveness of distribution utility which includes:

- Availability of Power to meet demand of all consumers at all time.
- Secured network to transmit power to consumer premises in all contingencies.
- Deliver power within accepted standards desired for delicate electrical appliances and gadgets.
- Provide consumer friendly services to meet their expectation and satisfaction.

To improve the reliability TATA POWER-DDL had already taken so many conventional measure like augmentation of existing network, segregation of long feeders, providing protective device with proper coordination at optimized location, adding sectionalizing and load break switch at proper location, improving lightening protection, introducing covered conductor and bird guard etc. In addition to these conventional method, some progressive technology has been adopted to enhance reliability and consumer satisfaction.

- Supervisory Control and Data Acquisition (SCADA) along with Grid Sub-station Automation System (GSAS): to monitor and control Subtransmission Network.
- Distribution Management System (DMS) along with Distribution Automation (DA): to supervise and control 11 kV distribution network.
- Outage Management System (OMS): for complaint handling, outage management and crew management.
- Customer Information Service (SAP-CIS): unified juncture for consumer interaction and convey necessary information.

Due to all these continuous initiatives and technology adaptation, there is incessant improvement in Reliability Indices (SAIDI, SAIFI) since its inception in 2002. *SAIDI decreased from 8.6–2.68 h whereas SAIFI decreased from 6.5 to 2.5 in last 10 years* (Fig. 1).

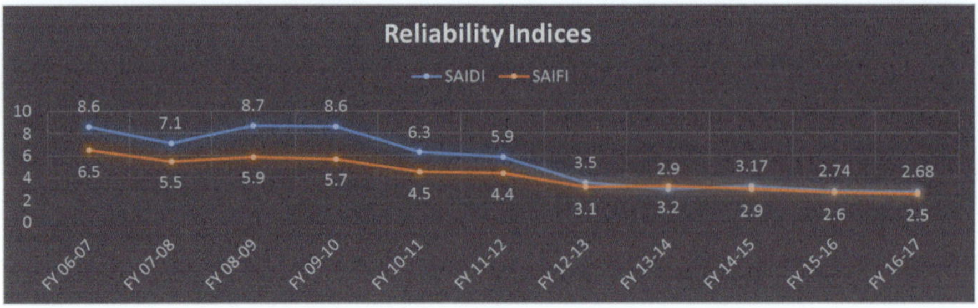

Fig. 1 Graph of reliability indices (SAIDI and SAIFI) of TATA POWER-DDL

2 Tata Power-DDL Challenge

Although TATA POWER-DDL's performance is going high, but due to increase in delicate electrical appliances, inclusion of Distributed Energy Resource like roof top solar plant and growth in Electrical Vehicles, consumer as well as regulators raise their bar of expectation. To meet their further expectation TATA POWER-DDL has stride toward Smart Grid technology like Smart Meters, Automatic Demand Response (ADR), Battery Energy Storage System (BESS) and Field Force Automation (FFA). These evolving technologies increase volume of data to be managed and visualized. Existing Technology like SCADA, DMS and OMS can handle and analyze limited data and unable to respond on new and relevant data coming from both grid sensors and consumers smart meters effectively. These applications are not capable of providing required level of reliability and safety to this emerging smart electric network. Thus, requirement of more advanced tool arises which are capable to organizes and analyses the gigantic volumes of proximate real-time complex data and integrate existing technologies into a unified platform for better network, outage and crew management. This study will describe how reliability and consumer satisfaction will improve in reference to the above stated challenges through Advanced Distribution Management System (ADMS).

Tata Power Delhi Distribution Limited is the first utility in the world to implement ADMS at such an extent with total integration of all the system in the utility, i.e. GIS, SAP, MWM/FFA, AMI etc. (Fig. 2).

3 Methodology

ADMS is an integrated platform with advanced monitoring application for better analysis and improved supervision and control of enormous real time electrical data (Fig. 3).

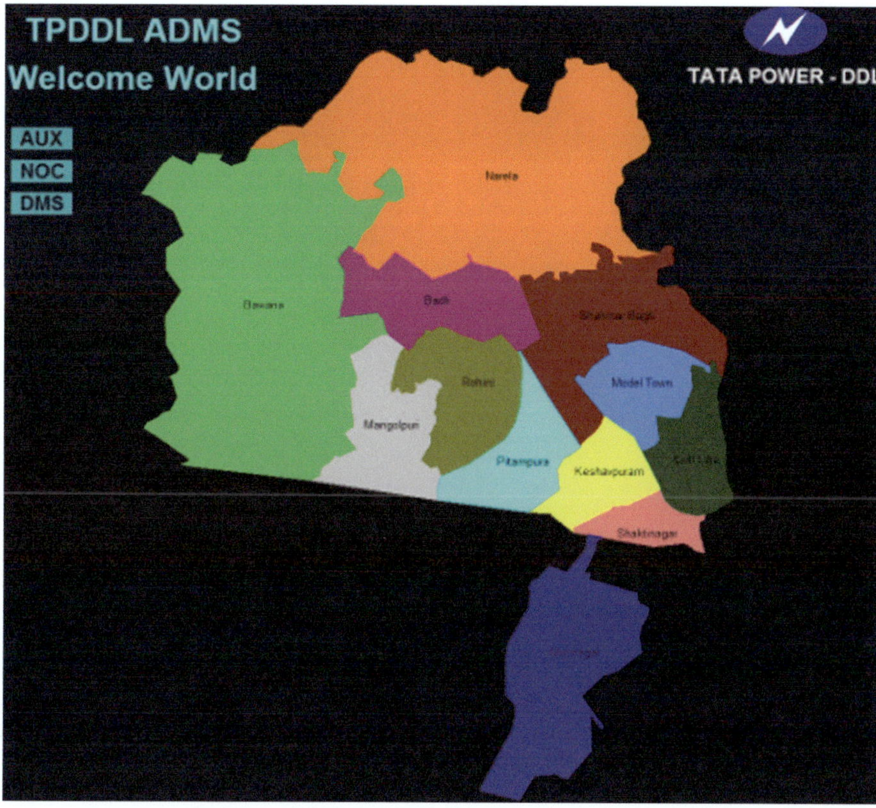

Fig. 2 TATA POWER-DDL world of ADMS

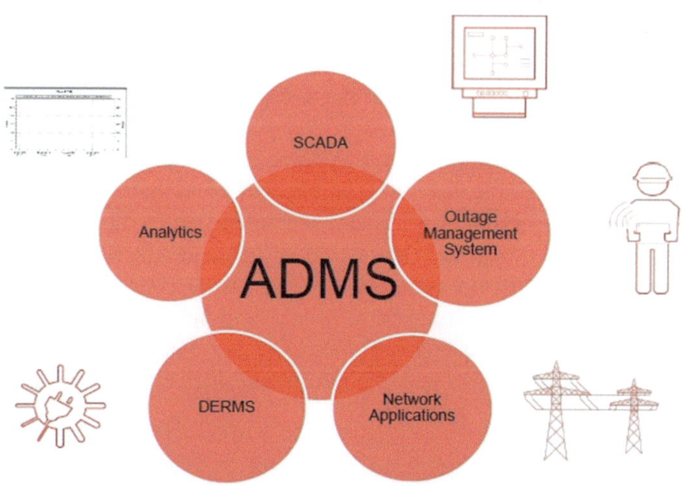

Fig. 3 Unified Platform with one User Interface for SCADA-EMS-DMS-OMS

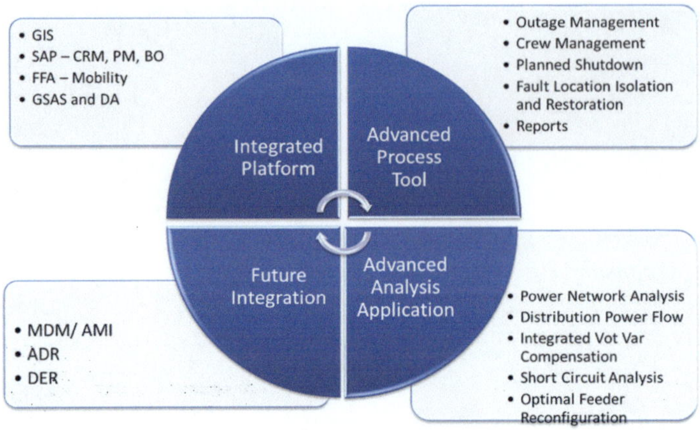

Fig. 4 Overview of ADMS

ADMS has capability to interface with new smart grid elements like Smart Meters technologies, ADR, DER and FFA etc. It is equipped with improved analysis and planning tools which provides better monitoring, effective network management, improved outage planning, proficient outage and crew management. These advanced applications help in maintaining network availability, power quality, reducing outage duration and ensuring better consumer satisfaction (Fig. 4).

4 Effective Network Management

At present, DMS and OMS are different platform properly integrated with each other and maintaining their own network at their end. Due to this reason at some time there is mismatch in both network. Hence network update in DMS, not properly update consumer supply in OMS. But now a single platform, ADMS will serve purpose of DMS and OMS and ADMS is well integrated with GIS. As network is updated in GIS the same will be updated in ADMS. So, network will be updated along with consumers as per actual site condition. In this way supply status of actual consumer will be updated in case of any outage which improve consumer satisfaction. Hence, ADMS provides an efficient visual interactive work environment that integrates all information source on a common real time workspace which helps in accurate estimation of consumer affected and extend of interruption as all work is done on a common GIS based platform and hence maintaining integrity of network.

5 Ensuring Availability

Power and network availability may increase by reducing either number of outages or outage duration through rapid restoration technique. ADMS is decorated with advanced application which will help in fast restoration of distribution network after tripping and assist in creating switching plan of any outage to reduce number of affected consumer and ensuring safe operation. These smart applications provide:

- *Restoring consumer faster*
- *Preventing Equipment Damage*
- *Improving Quality of Power Supply*
- *Effective use of field crews.*

5.1 Automated Power Restoration System (APRS)

APRS is an algorithmic approach for Fault Detection, Isolation and Restoration (FDIR). APRS uses real time data from network to locate fault. It is capable of executing a sequence of switching actions or can commend the same to isolate the fault and restore power to rest of the healthy network and consumers as many as possible. When a fault occurs on the network and an APRS automation program is invoked, the following occurs:

Fault Verification: A downstream trace starts from the tripped device to verify the fault.

Fault Location: The location of the fault is determined based on information provided by telemetered Fault Passage Indicators (FPI).

Fault Switching: Switching actions are created or executed to squeeze the extent of the fault in following way:

- *Isolate upstream then restore upstream by closing the tripped device.*
- *Isolate downstream, then attempt to restore each downstream outage by shifting consumers to neighbouring feeder.*

5.2 Switching Advisor (SA)

SA simplifies the creation of switching plans in either control or advisory mode. It allows you to select from one or more recommended switching actions and subsequently creates a work package with the proposed switching operations. It produces switching steps once a fault has been detected, using either DA and/or manual switches. The switching steps can result in the reduction of the fault area or in the proposal of alternative, manual restoration options to the Control Centre Engineer.

- *Proposing switching steps enabling us to reduce the affected area of network and to restore supply to consumer remaining off by 'squeezing' the fault.*
- *Proposing switching operations that include steps to isolate and earth the area of network to allow a permit to work to be issued.*

Beside this SA can also be used to propose a set of switching operations to maintain supply to customers in case of a planned outages. It creates a work package with proposed switching operations which can give confidence to the Control Centre Engineer that switching will not damage the network or violate any limits.

- *Maintaining supply to as many customers as possible during planned outages.*
- *Proposing switching steps to isolate and earth equipment. It also helps to create restoration switching plan once the required scheduled job has been completed.*

These smart switching functionalities will help us in maintaining power and network availability by reducing consumer interruption and subsiding outage duration. Hence reliability will be improved which ultimately results in improved consumer satisfaction (Fig. 5).

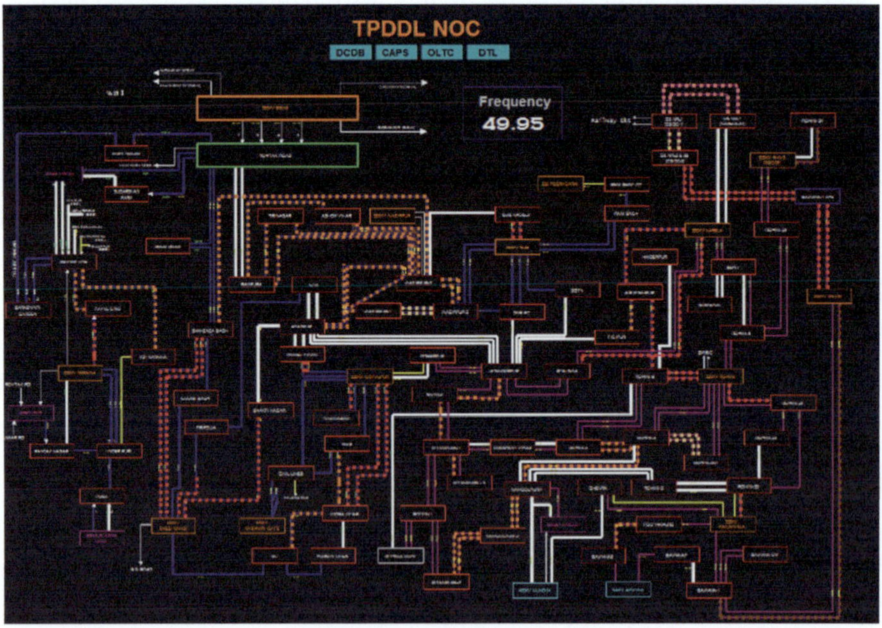

Fig. 5 TATA POWER-DDL NOC world of ADMS for fast switching operations

6 Ensuring Quality

ADMS is equipped with such applications which perform desired functions for maintaining quality of power supply like voltage magnitude, voltage balance, power factor (pf) etc.

Integrated Volt/VAR Control (IVVC): IVVC enable us to find an optimized set of switching recommendations for operation of Capacitor Banks and On Load Tap Changers (OLTC) integrated in our system to achieve greater efficiency and desired operational parameters. During optimization, the application examines the selected section of the network or complete network, taking into account device states, operational constraints and expected load conditions in order to achieve a certain objective. Multiple objective functions can also be added with different weightage so that algorithm tries to satisfy each criterion according to its relative weighting. Following are the different objective functions:

Flatten Voltage: This objective function optimizes voltage and reactive power within normal or emergency voltage limits to minimize the voltage drop in the circuit.

Maximize Power Factor: This objective function optimizes the voltage and reactive power within normal or emergency voltage limits to achieve power factor to set point.

Optimize Reactive Power: This objective function optimizes reactive power to pre-set value. This function is also capable to minimize the reactive power, maximize inductive reactive power and maximize capacitive reactive power.

Optimize Real Power: This objective function optimize real power to pre-set value. This function is also capable to minimize and maximize real power.

6.1 Optimal Feeder Reconfiguration (OFR)

OFR enables to perform analysis of the network, or specific areas of the network, to identify the optimal positioning of Normal Open Points (NOP) in order to achieve voltage regulation or reduce technical losses. Reduction in technical loss may reduce power requirement of utility which transform into reduction in total Power Purchase Cost and hence helpful in Tariff Control. This functionality can be based on combinations of the following—Transformer Utilization, Feeder Load and Consumer per feeders.

Improving Transformer and Feeder utilization means network can be utilized in more effective manner and additional consumers can be served without any additional capital investment in network infrastructure.

Under fault conditions, a balanced network presents more options for reconfiguration, potentially with less switching. Also, NOP positioning can be arranged to balance customer counts so that the Customer Minutes Interruption (CMI) under fault conditions is minimized.

These smart applications will help us to provide power to our consumer within specified limit of power quality parameter like voltage, power factor etc. and also help to improve CMI. Hence these all will collectively improve consumer satisfaction.

Enhancing Satisfaction: Consumers are the end user of electricity and hence last mile in any Distribution Network. That is why it is the prime responsibility of every distribution utility to improve consumer delight. All the initiatives, technology adoption and process improvement are done with keeping it in their mind. So, ADMS also has some advanced and progressive feature to improve consumer satisfaction.

6.2 Outage Prediction

For faster identification of fault, intelligent Outage Prediction logics are set in ADMS. ADMS consist of Call Grouping Engine which processes calls by firstly analyzing the LV network, then, if it cannot locate a tripped device then HV network is analyzed. The prediction of an open device is determined by comparing the number of *no supply call* downstream of the device, against the prediction trigger count of the device, after taking into consideration the timestamps associated with each calls. Protective devices are considered while Escalating and Deescalating Prediction. Besides Customer Calls, Meter Events and SCADA status are also used to accurately predict and update the scope of an outage. Field crew can directly be dispatched to the predicted device. This will result in saving time of finding fault and can help in minimizing repairing and restoration time. Estimated Restoration Time (ETR) are also calculated automatically based on current conditions including type of fault, travel times and time-of-day and updated as crew are allocated and arrive on-site.

ADMS also helps to manage power outages on their networks by providing timely and updated information to consumers. As switching or device operation is confirmed, the system automatically updates power status of the consumer and the information is recorded for analysis and shared with the Call Centre. ADMS is also capable of conveying information of schedule outages to their consumer some day ahead. Hence consumer will be well informed of their power outages, its severity along with expected time of restoration.

7 End to End Integration

7.1 *Adms-AMI*

The AMI Interface allows the ADMS system to obtain meter data by interrogating the AMI system using the MultiSpeak protocol. The user can request readings from any meter and display the data on-demand in the ADMS map. Users can also define schedules for the ADMS system to automatically interrogate the AMI system for specific meter data. The received data is stored in the analog points. This data can be used by other applications such as voltage readings for voltage reduction algorithms.

7.2 *Adms-AVL*

AVL is an interface that enables the location of GPS transponder equipped vehicles to be displayed on a map within ADMS. The location of the vehicle is updated real-time within ADMS so the user can see the actual location of the vehicle on the system map.

7.3 *Adms-MWM*

This interface provides the ability to connect to a third-party Mobile Workforce Management System. With this interface, operators can track the status of jobs and crews related to outage cases. Utilities also have the option of dispatching jobs to an external MWMS; the MWMS would track jobs and crew availability and update the OMS with changes.

7.4 *Adms-GIS*

This interface provides the ability to push the network from Geographical Information System (GIS) to ADMS based on Common Information Model (IEC Standard model to exchange data) (Fig. 6).

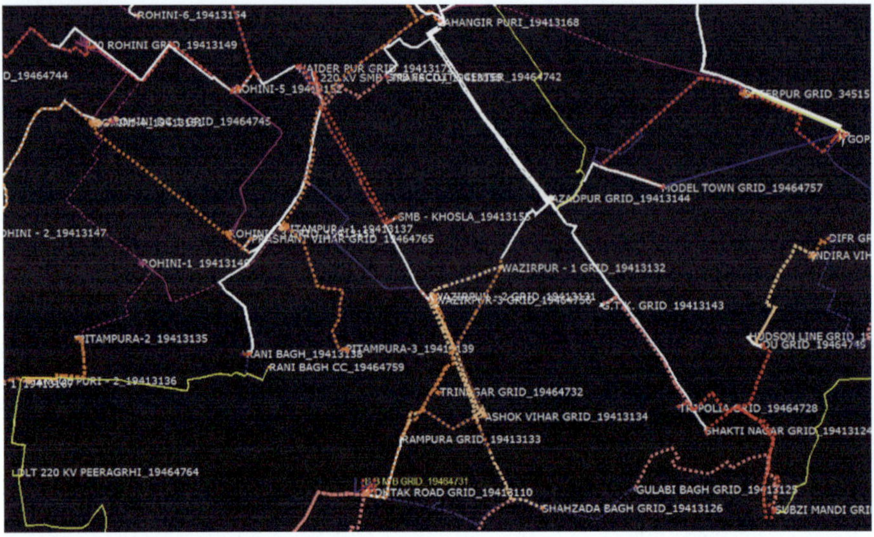

Fig. 6 TATA Power-DDL ADMS CIM World view

7.5 *Adms-DER*

It offer technological solutions to integrate:

- Increasing penetration of distributed energy resources (DER), including distributed generation, distributed storage, demand response, and Micro-grids
- New regulatory regimes
- Heightened physical and cyber security threats
- Aging utility infrastructure and workforce
- Potential and actual structural changes to the distribution business model.

8 Observation

With adaption to Smart Grid Technologies, Tata Power-DDL is moving forward to become a SMARTER DISCOM by adding valuable infrastructure, enhancing consumer services and improve efficiency through implementing ADMS. ADMS is a leading edge technology, packed with progressive application which will help in improving reliability, ensuring power quality and achieving consumer delight.

APRS: helps to reduce outage duration.

Switching Advisor: helps to squeeze the extent of the fault and reduce customer outage.

IVVC: helps to optimize voltage and reactive power to improve quality of power delivered.

OFR: helps to optimize NOP to reduce technical loss, improve voltage profile and enhance equipment utilization.

Outage Prediction: helps to subside fault repair time and improved outage management.

ADMS also helps in provide necessary information of power status to their consumer.

9 Conclusion

Amid stride to achieve vision of Smarter DISCOM, ADMS is a technological milestone which helps in improving network reliability by reducing impact and duration of scheduled and unscheduled outages, increase operational productivity by ensuring operational safety and enhancing consumer satisfaction by updating consumer of its power outage and other necessary information timely. Hence ADMS is an application which will help TATA POWER-DDL to suffice all requirement to face upcoming challenges due to integration of smart grid technology and to meet fore coming expectation of their consumers as well as regulators.

References

1. Power On Advantage—Advanced Distribution Management by GE Grid Solutions
2. Article on Advanced Distribution Management System by Artur R. Avazov, Liubov A. Sobinova, National Research Tomsk Polytechnic University
3. White paper on Advanced Distribution Management by Oracle Utilities
4. IEEE Xplorer. http://ieeexplorer.ieee.org/

Damage and Exposure Prevention of Energy Infrastructure Through Strategies in PEER: Evolving Trends for Efficiency, Reliability, and Resiliency

Ishaq Sulthan

Abstract Power is a basic necessity for every human being, if lost or interrupted causes a huge discomfort. Often these interruptions are caused due to external damages and exposure to the environment. In recent times, cities all over the world are facing big challenges due to climate change causing unprecedented rainfall/hurricane leading to flood. During such catastrophic events, communication networks and power supply is essential to keep oneself protected, safe and reach-out for emergency support. Apart from these extreme weather conditions, interruption in grid network is also happening due to tree contact, animal contact, accident due to vehicles or human interference and fire. All these events can cause a significant human & revenue loss if unprepared. Under this circumstances, it is critical we build resilient & reliable grid systems that help us during natural disaster, human error & threats. The damage that power interruption events caused from a practical and economic standpoint is a key concern for customers. Measuring, reporting, trending, and benchmarking these metrics will assist with identifying poor performance and provide the justification for investment to reduce sustained interruptions and improve reliability. Performance Excellence in Electricity Renewal (PEER) is an effective framework that looks comprehensively on Reliability & Resiliency with specific focus towards damage and exposure prevention. The credit implies on the project to focus on the identification of external threats and measures to mitigate them. It also specifically mention about the hardening of power systems from floods and storms, references to be considered while designing or retrofit, thus reducing the likelihood of equipment failures due to any external & physical threats.

Keywords Exrternal damages · Natural disaster · Power system hardening · Reliability · Safety · Threats

---

I. Sulthan (✉)
GBCI, New Delhi, India
e-mail: isulthan@gbci.org

R. K. Pillai et al. (eds.), *ISGW 2018 Compendium of Technical Papers*,
Lecture Notes in Electrical Engineering 580,
https://doi.org/10.1007/978-981-32-9119-5_2

1 Introduction

Worldwide, cities and communities are utilizing new and innovative technology to address natural and man-made threats. Often these energy infrastructure is tested by blackouts, natural disasters, and increasingly cyber-threats such as the Hurricane Harvey & Irma that damaged the US grid infrastructure in 2017 and in December 2015 attack that took down the Ukraine power grid. While there have been significant strides in some respects, grid inefficiencies and risks are costly, having multiplier effect and likely to increase in future. According to NOAA's National Centers for Environmental Information, the United States has suffered 212 weather and climate disasters since 1980 that have cost more than $1 billion, totaling $1.2 trillion [1]. Power interruptions cost European Union businesses €150 billion each year [2]. In India, at Tamilnadu, during Vardah cyclone in 2016 the state utility Tamilnadu Generation and Distribution Corporation (TANGEDCO) has lost 10,000 electric poles and 800 transformers got damaged causing a loss of more than 1000 crores [3]. Apart from these natural disasters causing damages to the energy infrastructure, there are other threats such as tree falling, animal or vegetation intrusion, human interference causing outages locally. A recent EU study found that €4 billion could be saved annually from using smart grids that includes right from Advanced Metering Infrastructure (AMI) capability to power system hardening techniques.

This paper talks about how by leveraging the PEER (Performance Excellence in Electricity Renewal)—Damage and Exposure Prevention strategies these challenges can be better planned and managed across campuses, communities and cities. PEER is a certification program that measures and improves power system performance and electricity delivery systems.

The comprehensive credit based programme is classified under 4 key categories;

- Reliability & Resiliency
- Energy Efficiency and Environment
- Operations, Management & Safety
- Grid Services.

2 Common Risks and Threats

Common risks and infrequent threats are conditions that occurred in historical data disrupting the quality of electricity delivery such as voltage swings, faulted lines or blown fuse caused due to tree or animal contact or human interference etc. These types of events can realistically be expected to occur periodically in the region where the project grid is located; for example cyclones are historically recorded in

Balasore, Nellore but not in Hyderabad. Hence, by doing this assessment, the grid operator can take meaningful decisions to design or make improvements to modernize the electricity grid.

PEER defines the following as common risks and threats that can affect the reliability performance of a grid infrastructure.

- Supply interruptions—originating from;
  - Tree or branch interference
  - Vines causing short circuits
- Unintentional damage to equipment—including damage from;
  - Animal interference
  - Traffic accidents
  - Unintended public exposure
- Extreme weather—depending on the location of the project grid this may include;
  - Flooding/Rainfall
  - Windstorms, Sandstorms, Ice storms.
  - Earthquakes.

The key objective is to support projects with considerations for common risks and threats, and raise awareness of strategies and measures that can be used to improve reliability by reducing the damages to the grid infrastructure.

3 Damage and Exposure Prevention Strategies

Projects around the world are often looking for advanced technologies and innovative solutions to mitigate common risks and threats through prevention of external damages and exposure to the environment. PEER program proposes the following techniques or strategies to prevent from external damages and threats;

Option 1. External Damage Prevention

A. **Prevention from Tree Contact**: It is observed that relatively high percentage of cable failures were caused by tree fall in bad weather. Tree management near distribution lines is an important adaptation action needed to reduce risks of power distribution system outages. The objective here is to prevention of cables and power system from tree contact.

Tree damage can be organized into five categories:

- Broken branches
- Trunk bending
- Splitting of main or co-dominant stems

- Complete trunk failure
- Tipping or up-rooting.

Following are the two prevention techniques of cables and power system from tree contact;

1. *Up gradation of overhead distribution lines*—Upgrading of overhead distribution equipment, with the aim of making the system more resilient against damage downed trees and limbs. Redesigning wires to provide better protection from falling tree limbs, and to detach more easily when force on the wire is more extreme to reduce the likelihood of damage to poles and other pole-top equipment. Creating greater tree clearances around distribution facilities near substations and critical infrastructure.
2. *Hazard Tree program and vegetation management*—Identify trees that are tall enough to contact the overhead distribution system and are also dead, declining, diseased, or otherwise structurally unsound. Some of the important programs in hazard tree management program include;

    - Designating responsible individuals,
    - Identifying and prioritizing the sites to be examined,
    - Performing and documenting the inspection,
    - Performing the necessary actions to reduce the hazards (Vegetation management),
    - Maintaining the records of inspection & actions taken,
    - Recording tree failures. The program should be compatible with available resources (personnel and funding).

B. **Prevention from Animal/Bird Contact:** Animals such as squirrels and monkeys and birds such as crow, sparrow are probably one of the most causes of distribution faults. Below are the list of problems caused in the grid infrastructure by animals/birds;

- Animals in substations cause a variety of problems, such as faults, which can result in power outages, reduced equipment life, or severely damaged equipment.
- Squirrels, raccoons, snakes and other creatures that climb fences, structures and equipment may eventually come into contact with energized parts of equipment.
- Rodents chewing through power and control cable insulation can also cause outages and damage.
- Bird build nests in substation structures and equipment.
- Some birds have wing spans that can bridge phase-to-phase distance and cause an outage.
- Bird droppings can also be a major problem.

Some of the mitigating methods are;

- **Fence barriers**—To deny animals access at the fence line for an outdoor substation requires consideration of possible access paths over, through, and under the fence [4].
- **Fake predatory animals**—Plastic owls, snakes, falcons or other objects can be placed conspicuously in the substation structure or on top of an enclosure.
- **Disturbing noises**—Various types of disturbing noises can be generated to drive off animals. They consist of ultrasonic, loud, or danger signal noises.
- **Chemical repellents**—Chemical repellents and additives can be used to treat the food supply, or as baits in traps to kill animals if they enter or live within the substation.
- **Perching and climbing deterrents**—Anti-perching features or conductor covers to prevent birds from perching, nesting, and shorting exposed overheard conductors. A common deterrent involves the use of a sticky gel. This gel can be applied as a deterrent, to surfaces where birds perch such as structures, incoming wires, and guy wires.
- **Screened nesting locations**—Deterrence of birds and other animals from the general area can be accomplished if access to the preferred nesting sites within the substation is denied.

C. **Prevention from Human interference**: Substations and transmission lines that are close to the highway or on the adjacent to pathway is often prone to human attacks or vehicle accidents.
 Design requirements should include damage exposure prevention and protection to all potentially exposed major/Critical equipment such as but not limited to

  - Fencing around the structure
  - Pad-mounted equipment along roadways, walkways, and bicycle paths
  - Transformers (Indoor and Outdoor Type)
  - Construction of wall around the structure with secured access.

D. **Prevention from Extreme Weather conditions:** Power outages due to severe weather events (floods, heavy winds, earthquakes, tropical cyclones) has increased over the last decade, many utilities/campus projects sought to reduce the weather related outages by "hardening" the systems to flooding, heavy winds, and earthquakes by making the major electrical equipment less susceptible to damage. Considering the power system hardening strategies during the initial stages of project can significantly contribute in reducing the total operational and damage cost incurred from severe weather events.

PEER propose following design considerations and/or infrastructure to harden power systems against flooding, storms, and other extreme events, to maintain continuous availability of power supply;

1. **Flood plain avoidance** – Implement one of the following strategies to prevent damage to electrical equipment and assets (e.g., substations, diesel gensets, transformers, OH cables) and ancillary equipment (e.g., pumps, compressors), based on a 100-year flood mark or flood map. Protect stored fuel to meet or exceed the requirements set by the authority having jurisdiction.
   - Strategy 1: Build a permanent storm water drainage system to protect critical power assets from inundation.
   - Strategy 2: Install a standalone pump to pump water from low-lying areas around the electrical systems. The pump should be operable in the absence of power supply.
   - Strategy 3: Permanently relocate or increase the height of critical power assets in the flood-prone area as per your local flood zone map.
2. **Storm protection**—Ensure that the outdoor equipment can withstand three-second wind gusts up to 140 mph or equivalent. This may include electric poles, substations, transformers and over-head line.
3. **Seismic protection**—Have in place seismic restraint–certified equipment for critical electrical systems and/or install a seismic restraint structural support for critical electrical systems, based on the seismic zone. The critical electrical system would include substation, Transformers, Poles and other infrastructure depending on the importance of the project.
   Apart from this, NFPA 1600 defines general criteria for organizations to use for developing disaster management programs [5] that focus on the safety measures to be considered during normal and emergency conditions like an extreme events.

E. **Undergrounding**—One of the old strategy of power system hardening is Undergrounding. Placing electrical cables underground, eliminates their susceptibility to heavy winds, ice and lightning damage. In many countries, undergrounding has been proposed as a solution for hardening the transmission and distribution system.

The graph below derived from European report CEER Benchmarking Report 5.1 [6]—CoS demonstrates that undergrounding has a significant positive impact on reducing system outages (Fig. 1).

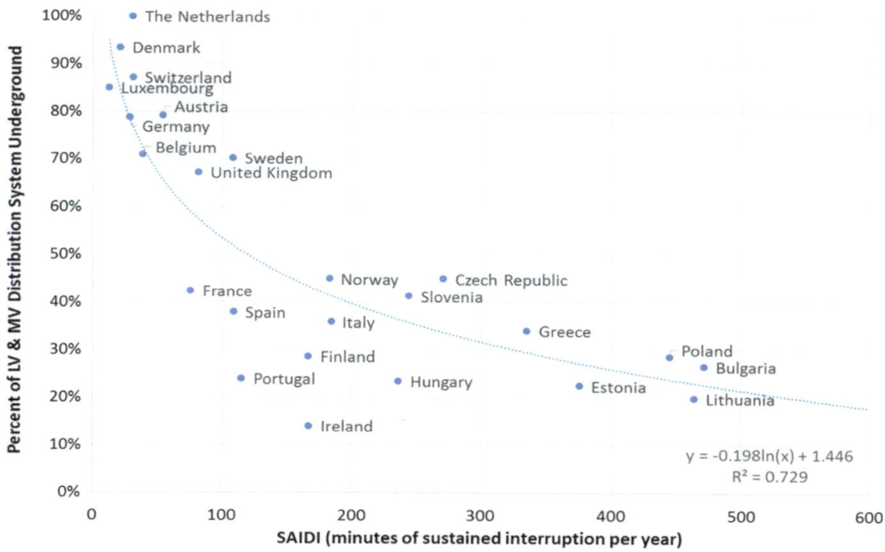

Fig. 1 Showing the reduced SAIDI based on increased underground cabling across Europe

4 Conclusion

Failures of network components could significantly affect the reliability of power systems. Component failures are caused by different factors. Analysis of the failure causes has a practical value since the knowledge of failure causes could help power utilities to take remedial actions and determine appropriate method for failure reduction. Damage and exposure prevention design considerations and prevention strategies reduce the likelihood of equipment failures due to external physical and environmental threats.

Over a period of 2003–2012, weather-related outages are estimated to have cost the U.S. economy an inflation-adjusted annual average of $18 billion to $33 billion [7]. The costs of outages take various forms including lost output and wages, spoiled inventory, delayed production, inconvenience and damage to the electric grid. Continued investment in grid modernization and resilience will mitigate these costs over time—saving the economy billions of dollars and reducing the hardship experienced by millions of Americans when extreme weather strikes. Thus, in a current scenario power system hardening and prevent from external damages has become crucial for both developed and developing countries like India which has seen quite a few weather-related outages in recent years. Modernizing these grid infrastructure to make them resilient will be the need of hour and through PEER strategies and techniques explained in this paper Cities, Communities and buildings can better prevent from external damages & risks.

References

1. Vox.com (www.vox.com/explainers/2017/9/18/16314440/disasters-are-getting-more-expensive-harvey-irma-insurance-climate)
2. GEgridsolutions (www.gegridsolutions.com)
3. Indo-global social service society (www.igsss.org)
4. Electrical power transmission and distribution—Code of practice for animal deterrents for electric power supply substations
5. National Fire Protection Association (2013) NFPA 1600: Standard on Disaster/Emergency Management and Business Continuity Programs. Retrieved from (http://www.nfpa.org/codes-and-standards/document-information-pages?mode=code&code=1600) (http://indiaclimatedialogue.net/2015/02/24/indian-coal-power-plants-worlds-most-polluting/)
6. CEER Benchmarking Report 5.1
7. Economic Benefits Of Increasing Electric Grid Resilience To Weather Outages

Protection Scheme for Smart Distribution Networks with Inverter Interfaced Renewable Power Generating Sources

Anoop V. Eluvathingal and K. Shanti Swarup

Abstract Conventionally in distribution systems, protection is designed for radial networks, which instigates sensitivity and coordination issues with the integration of DERs. Protection switches lockout for faults in the network, which imposes outages to large areas and loss of revenue if DERs are not allowed to remain connected subsequent to a network outage. This demands new protection methodologies for smart distribution networks. The possibilities of active network management schemes, island operation opportunities and new grid codes imposes challenges in protection system design for modern distribution networks. This paper discusses an advanced protection scheme for active distribution network with high large number of DERs connected. The proposed protection scheme is investigated in a distribution feeder with large number of PV sources.

Keywords Distribution network · Distributed energy resources · Distribution network protection · Intelligent electronic device

1 Introduction

The integration of Distributed Energy Resources (DER) to Distribution Networks (DN) are significantly increasing all around the world. Conventionally, many DNs are designed as radial networks with unidirectional power flows. Main feeder substation is the only source of fault current in a radial distribution networks. Thereby the fault current profile will be same all along one feeder for a downstream fault and the magnitude of fault current will be inversely proportional to the distance of fault from the source substation. This radial structure of distribution networks is considered in time graded settings of overcurrent relays. With the introduction of

A. V. Eluvathingal (✉) · K. Shanti Swarup
Department of Electrical Engineering, IIT Madras, Chennai, India
e-mail: ee14d001@ee.iitm.ac.in

K. Shanti Swarup
e-mail: swarup@ee.iitm.ac.in

R. K. Pillai et al. (eds.), *ISGW 2018 Compendium of Technical Papers*,
Lecture Notes in Electrical Engineering 580,
https://doi.org/10.1007/978-981-32-9119-5_3

DER on the load ends, the system is no more radial and fault profile changes. The selectivity and sensitivity of protective devices are affected due to this.

The regulatory bodies of different regions have introduced new grid codes with the emergence of increased integration of distributed renewable sources [1]. According to new standards the DER should maintain connected for wide changes in frequency. With proactive control DERs have to the injection of active power during over-frequencies according to P/f droop characteristics defined by standards. The sources with large capacities are permitted to inject real power when frequency is decreasing. DERs should be able to ride trough voltage sags, which is defined as Fault ride through (FRT) characteristics which is defined as a voltage over time graph. In future smart grids for increasing efficiency and reliability active network management solutions will be used. Active network management solutions aids in utilizing assets more effectively [2]. This active management will affect protection settings of existing relays in distribution networks [3].

In conventional overcurrent protection in distribution networks, fault current magnitudes are sensed for identifying faults. The network protection limits fault currents by opening circuit breaker thereby limiting the fault currents crossing the thermal limits of conductors [4]. In inverter based sources the fault currents are internally regulated by the controller design to safeguard the power semiconductor devices. The overcurrent protection devices fail to detect this small fault currents thus failing the operation. The protection issue with the large scale integration of distributed sources is addressed in various literatures [5]. Zero sequence and negative sequence currents can be used to differentiate which is discussed in [6]. A protection scheme at the interface of microgrid is proposed in present work. Fault is sensed by change in voltage. Voltage sag is estimated for fault identification. The instantaneous positive sequence components of voltage sag is estimated and threshold is checked. The change in current is detected for identifying redundant tripping decision. The fault induced current components give the fault current without having the load affect. The direction is determined for increasing selectivity.

2 Protection Issues in Radial Distribution Networks with DER Integration

Reclosers and fuses are used in distribution system for overcurrent protection due to economic considerations. Fuse saving scheme is traditionally used to avoid fuses interrupting temporary faults. The fault current contribution from distributed PV sources challenges this protection scheme. Various other protection issues are identified in [7].

2.1 *Fuse Saving Scheme in Distribution Networks*

According to statistics 70–80% faults in distribution network are temporary in nature. To avoid expensive rewiring of the fuse for clearing temporary faults the recloser operating characteristic is made lower than fuse minimum curve so that recloser will operate before fuse. Temporary faults will be cleared after recloser's first fast reclosing operation. In times of permanent faults, fault still exists after first reclosing. The second trip characteristics of recloser are above fuse characteristics, thus allowing the fuse to clear the fault. The operational characteristics of a typical fuse saving scheme are given in Fig. 1.

To increase reliability microgrid has to be allowed to operate as an island in the absence of utility grid. In the islanded mode also protection system should work without fail. During islanded operation fault current will be different from grid-connected mode. In islanded mode, the distributed sources being the only contributing sources for fault, the fault current magnitude variation will largely depend on source type, size, available power, grid integration method and limitation of fault currents for self-protection. The variation of fault current magnitudes saw by a protective relay for a fault in its zone, for different penetration levels on both sides of the protective device is summarized in LVRT curve.

2.2 *Fault Current Contribution from Inverter Based Sources*

The Voltage Source Inverters (VSI) are popularly used to integrate PV sources to distribution network. The inverter controller responds very fast during a fault in the host grid and the fault current depends on the inverter controller design. The

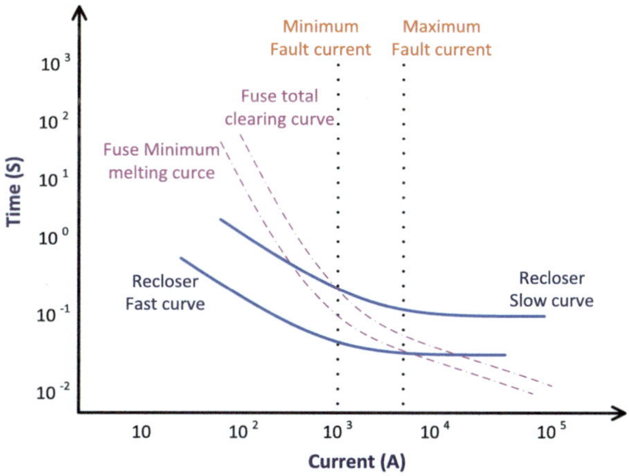

Fig. 1 Fuse saving scheme in distribution networks

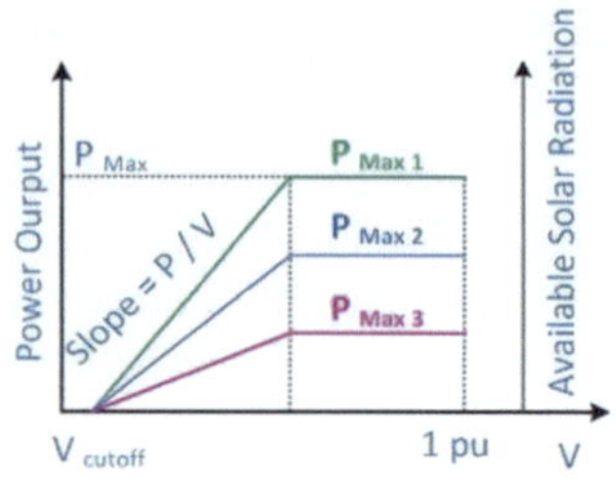

Fig. 2 Fault current limiting from sources

magnitude of fault current contribution also depends on the availability of solar input. The variation of fault current contribution from a typical PV source is represented in Fig. 2.

3 Proposed Protection Scheme for Active Distribution Network

The distribution network will be installed with Intelligent Electronic Devices (IED) which realizes protection for smart distribution networks. The network is divided into protection zones and various protective devices considered are as in figure. Each protection zone can be considered acting as a single controlled unit (microgrid) and is considered by an equivalent source and lumped load. Protective devices are installed at the interface points of microgrid with the utility grid which is termed as interconnection protection. For all fault in that zone interconnection protection will operate and sources will shut down instantaneously. Load feeder protection replaces the conventional low voltage side fuses of distribution transformer. For faults in load feeder zone, interconnection protection relay trips instantaneously. Load feeder protection relay has to selectively trip for faults in its zone and block other load feeder protection relays. Interface protection relay discussed in reference [8] with modifications is used for present work (Figs. 3 and 4).

The implementation of proposed protection philosophy for various fault cases is explained below [9, 10].

In case of fault case F1

- DERC will operate in reverse direction and disconnect the DER microgrid intentionally from the utility network.
- DERC sent a signal to LV microgrid DER units to change their control mode after islanding.
- LVMC will operate in forward direction and also sends simultaneous interlocking signal to Feeder Protection controller and other LVMcs (Fig. 5).
- LVMC send communication-based transfer trip disconnection signals to other DER units directly connected to LV feeder.

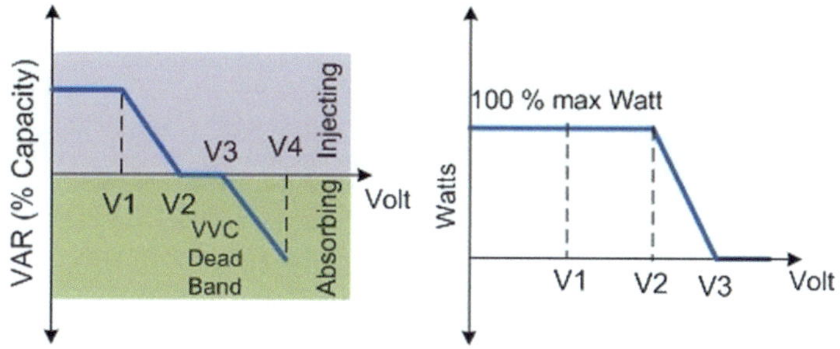

Fig. 3 Volt-Var and Volt-Watt characteristics of DG sources

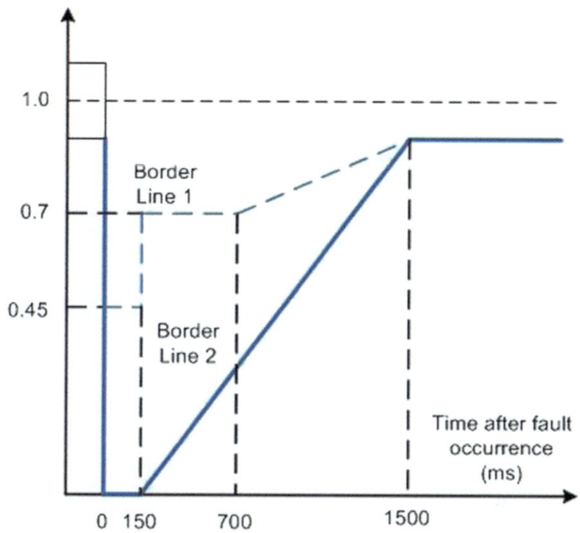

Fig. 4 LVRT requirement of DG sources

- Other DER in LV feeder and adjacent microgrids will remain connected according to LVRT curve.

In case of fault case F2

- LVMC operate with the differential function to isolate the faulty section and send the simultaneous blocking signal to Load feeder protection and feeder protection controller.
- Load feeder protection will operate in the forward direction for backup.
- LVMC operate in reverse direction and disconnect part of the LV feeder for island operation.
- In case of active and reactive power unbalance, LVMC CB2 sent transfer trip signal to DERC and also to DER unit for changing control mode.

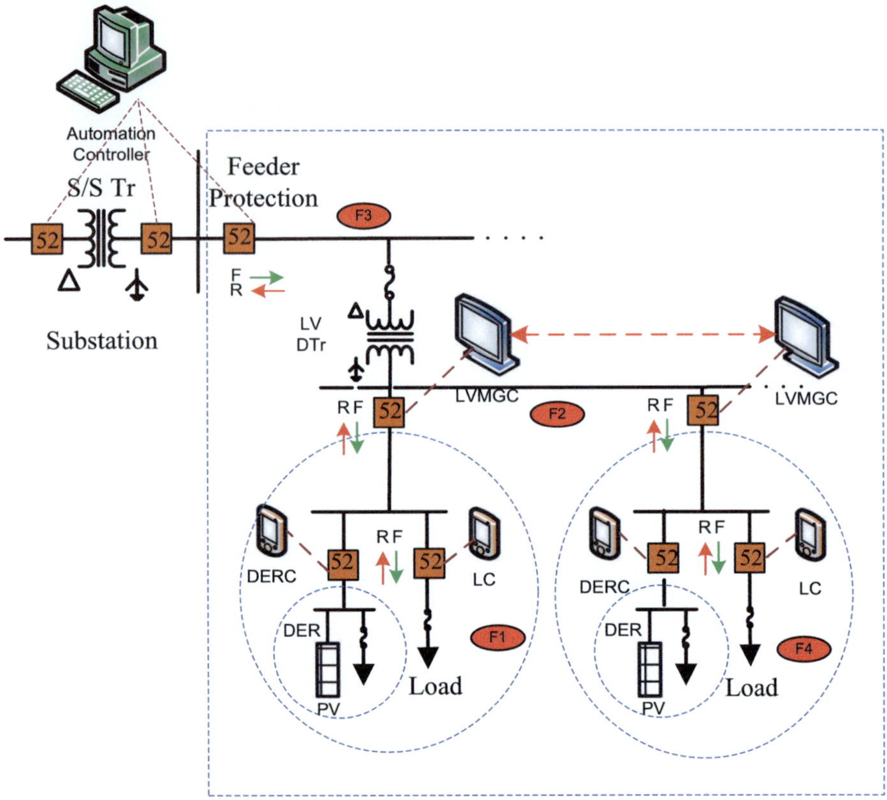

Fig. 5 Protection methodology for smart distribution networks

- For back up, DERC will operate in reverse direction with definite delay to disconnect the LV microgrid intentionally from the utility network for unbalance in active and reactive power unbalance.
- Other DER in MV feeder and adjacent microgrids will remain connected according to LVRT curve.

In case of fault case F3

- Feder protection controller will operate in forward direction.
- Load feeder protection will operate in reverse direction and simultaneous blocking signal to LVMcs.
- For active and reactive power unbalance at load feeder signal is send to LVMC for isolation (Fig. 6).
- LVMC operate in reverse direction, with definite time delay for intentional island operation. For active and reactive power blocking signal is sent to DERC.
- DERC operates in reverse direction, with definite time delay for intentional islanded operation.

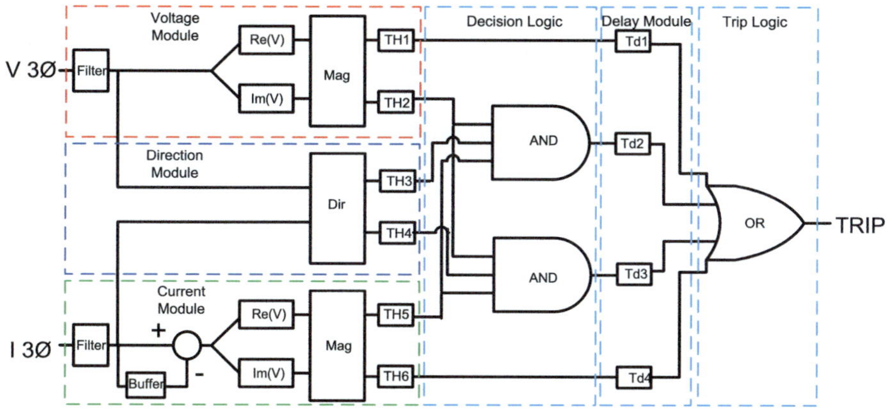

Fig. 6 Interconnection protection relay

In case of fault case F4

- LVMC will operate in forward direction and also sends simultaneous interlocking signal to Feeder Protection controller and other LVMcs.
- Other DER in LV feeder and adjacent microgrids will remain connected according to LVRT curve.
- Other LVMC operate with definite delays for backup protection.
- DER in healthy LV feeder and adjacent microgrids will remain connected according to LVRT curve.

4 Simulation Studies on Case Study Network

The case study system under consideration for simulation studies in present work is a radial distribution feeder discussed in [11] with distributed loads. Distributed solar PV sources are integrated into the existing system and can operate in an islanded mode in times of intentional power outages or in times of contingencies like faults. The size of the microgrid is considered variable for ensuring load balance and increasing reliability. An 11 kV distribution feeder of a 66/11 kV substation is selected for the case study. This system is a typical example of the weak distribution grid. The feeder consists of sixteen distribution transformers with unbalanced distributed loads to which solar PV sources are envisaged to be integrated. Proposed protection schemes in the presence of distributed sources in islanded and grid-connected modes are studied for this system.

4.1 *Fault Currents Variation from Different Branches with IIDG Sources*

To study the variation in fault current contributions from various inverter based sources three-phase faults are simulated at the end nodes of feeders. The variation in

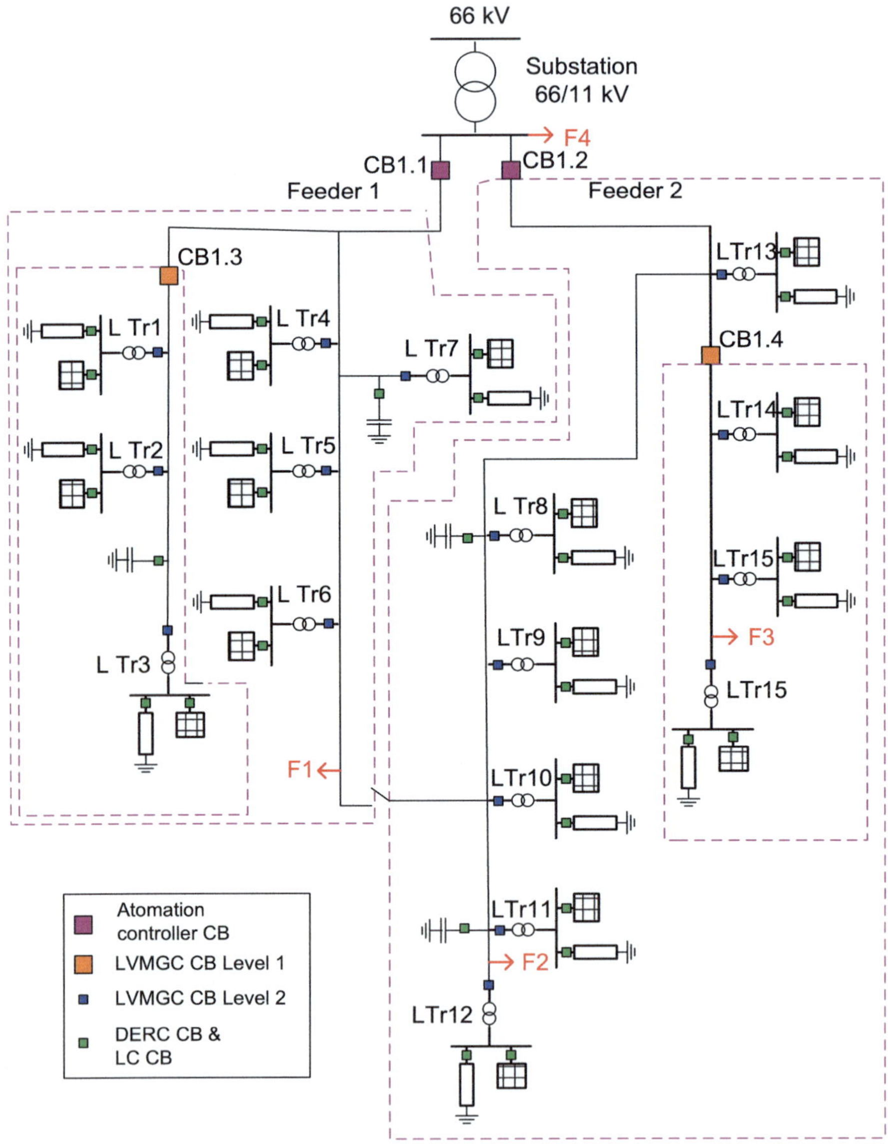

Fig. 7 Case study network for simulation studies

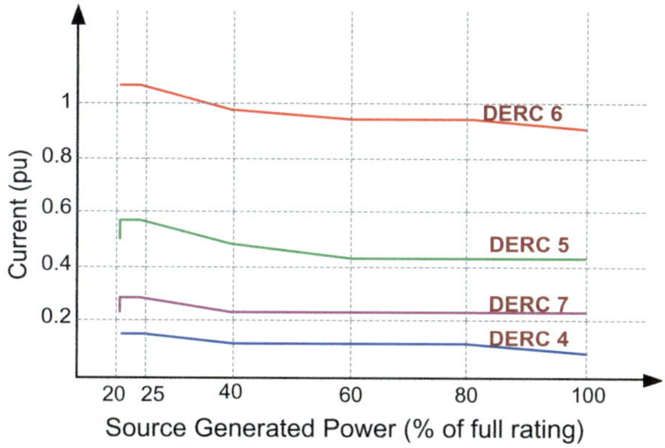

Fig. 8 RMS value of fault currents measured at various DERCs in feeder 1 for three phase fault at F1

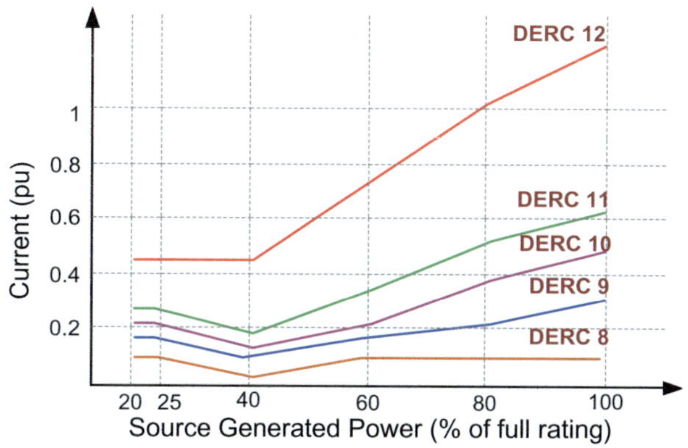

Fig. 9 RMS value of Fault currents measured at various DERCs in Branch 1 of feeder 2 for three phase fault at F2

RMS value fault current measured by various DERCs is plotted in graphs as shown. The difference in fault current contribution is because of load composition and their response to voltage dips due to the fault. Additionally, the self-protection by the controller of IIDG limits the fault currents. From studies, it is shown that the fault current contribution from the various branches cannot be predicted with precision. This demands smart protection system for ensuring flawless protection in the distribution network with IIDG sources (Figs. 7, 8, 9 and 10).

4.2 Simulation Results Studying Proposed Protection Scheme

Various fault cases are simulated and the response time of protection IEDs for different fault cases are studied. The results for two fault cases are tabulated as below. Table 1 gives results of a fault case F3. It is assumed that the generation of

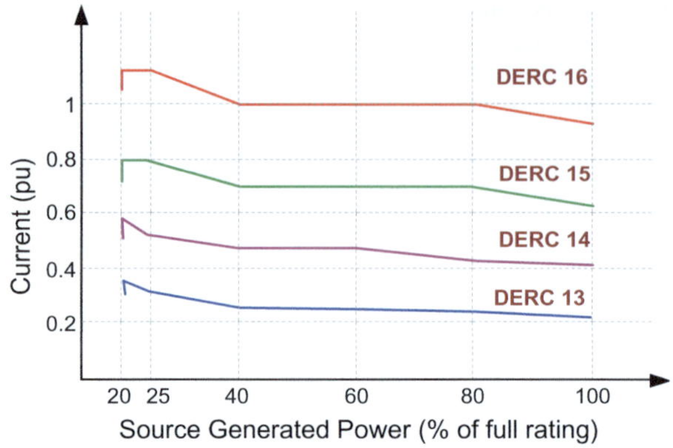

Fig. 10 RMS value of Fault currents measured at various DERCs in Branch 2 of feeder 2 for three phase fault at F3

IIDG is not meeting load demand. With the fault at F3, the protection Low Voltage Microgrid Controller(LVMGC) at CB 4 will operate in forwarding direction. This serves the primary protection. LVMGC of CBS 14–16 operate in reverse mode sensing reverse fault. After isolating the load microgrid, as generation is not meeting loads in each microgrid the Distributed Energy Resource Controller (DERC) for IIDG sources 14–16 will operate with load mismatch. Hence isolating the fault. Synchronization after fault clearance is nor considered in present studies. As a second case, a fault in 11 kV substation bus is considered. The LVMGC of CB 1.2 will trip with voltage protection. Transfer trip is given to LVMGC of CB 1.4. It is considered that generation is sufficient for islanded operation of the isolated area. The DERC of other sources in the feeder will trip because of non-feasibility of islanded operation. The results of studies so obtained for two cases is summarised in Tables 1 and 2 (Fig. 11).

Table 1 Operation time of IEDs for fault at F3

Fault type	Fault at F3			
	Direction	IED NO	Function	Time (S)
L-G	FWD Pri	LVMGC CB1.4	Iz	0.02
	REV Pri	LVMGC CB14-15	Ip	0.03
	FWD Pri	DERC CB 14-15	df/dt	0.05
L-L	FWD Pri	LVMGC CB1.4	Ineg	0.02
	REV Pri	LVMGC CB14-15	Ip	0.03
	FWD Pri	DERC CB 14-15	df/dt	0.05
LLG	FWD Pri	LVMGC CB1.4	Iz	0.02
	REV Pri	LVMGC CB14-15	Ip	0.03
	FWD Pri	DERC CB 14-15	df/dt	0.05
LLLG	FWD Pri	LVMGC CB1.4	Ip	0.01
	REV Pri	LVMGC CB14-15	Ip	0.02
	FWD Pri	DERC CB 14-15	df/dt	0.04

Table 2 Operation time of IEDs for fault at F4

Fault type	Fault at F4			
	Direction	IED NO	Function	Time (S)
L-G	REV Pri	LVMGC CB1.2	dVn or LV	0.018
	REV Pri	LVMGC CB1.4	TransTrip	0.018
	REV Pri	DERC CB 13	Ip	0.03
L-L	REV Pri	LVMGC CB1.2	dVn or LV	0.018
	REV Pri	LVMGC CB1.4	TransTrip	0.018
	REV Pri	DERC CB 13	Ip	0.03
LLG	REV Pri	LVMGC CB1.2	dVn or LV	0.018
	REV Pri	LVMGC CB1.4	TransTrip	0.018
	REV Pri	DERC CB 13	Ip	0.03
LLLG	REV Pri	LVMGC CB1.2	LV	0.01
	REV Pri	LVMGC CB1.4	TransTrip	0.01
	REV Pri	DERC CB 13	Ip	0.03

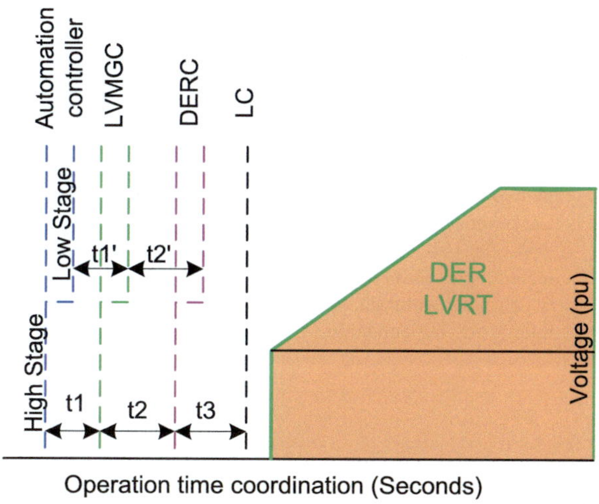

Fig. 11 Operational time coordination for various IEDs

5 Conclusion

The controllability and protection of conventional distribution networks are challenged by the increased integration of DERs. The issues with network protection are considered as serious concern with the increased integration of distributed PV sources in the distribution networks. An advanced scheme for protection in distribution network with high integration of distributed PV sources is investigated in present work. The proposed protection scheme is studied in a case study distribution feeder with high penetration of distributed PV sources.

Acknowledgements The authors acknowledge gratefully, the support provided by Maschinenfabrik Reinhausen (MR) through the IGCS Research Programme on Sustainable Power Engineering.

References

1. Hooshyar A, Iravani R (2017) Microgrid protection. Proc IEEE 105(7):1332–1353
2. Laaksonen H, Suomi F (2013) New functionalities and features of IEDs to realize active control and protection of smart grids
3. Jiang Z, Fan J, Mcdonald J (2012) Advanced distribution automation management for active distribution systems. CIGRE Session 2012:26–30
4. ALSTOM (Firm) (2002) Network protection & automation guide. ALSTOM
5. Mirsaeidi S, Said DM, Mustafa MW, Habibuddin MH, Ghaffari K (2014) Progress and problems in micro-grid protection schemes. Renew Sustain Energy Rev 37:834–839
6. Nikkhajoei H, Lasseter RH (2006) Microgrid fault protection based on symmetrical and differential current components. In: Power system engineering research center, pp. 71–74
7. Walling R, Saint R, Dugan RC, Burke J, Kojovic LA (2008) Summary of distributed resources impact on power delivery systems. IEEE Trans Power Delivery 23(3):1636–1644
8. Zamani MA, Sidhu TS, Yazdani A (2011) A protection strategy and microprocessor-based relay for low-voltage microgrids. IEEE Trans Power Delivery 26(3):1873–1883
9. Zarei SF, Parniani M (2017) A comprehensive digital protection scheme for low-voltage microgrids with inverter-based and conventional distributed generations. IEEE Trans Power Delivery 32(1):441–452
10. Baran ME, El-Markaby I (2005) Fault analysis on distribution feeders with distributed generators. IEEE Trans Power Syst 20(4):1757–1764
11. Hooshyar H, Baran ME (2013) Fault analysis on distribution feeders with high penetration of pv systems. IEEE Trans Power Syst 28(3):2890–2896
12. Laaksonen H (2012) New multi-criteria-based algorithm for islanding detection in smart grids. In: 3rd IEEE PES international conference and exhibition on innovative smart grid technologies (ISGT Europe). IEEE, pp 1–8
13. Li Z, Shahidehpour M, Aminifar F, Alabdulwahab A, Al-Turki Y (2017) Networked microgrids for enhancing the power system resilience. Proc IEEE 105(7):1289–1310

Wide Area Monitoring System (WAMS): The Foundation Block of Smartgrids for Event Analysis

A. R. Kulkarni, S. S. Rajurkar and M. S. Ballal

Abstract Indian Electric Grid working as ONE GRID comprising of high capacity transmission corridors, characterizing rise in penetration level of Renewable Energy (RE) has resulted into complexity in grid Operation, Monitoring and Control. This calls the need to adopt and utilize SMARTGRID technology for Grid management and event analysis. This paper discuses in brief various aspects related to SMARTGRID technology development in the area of Wide Area Monitoring Systems (WAMS) in Indian context. It discusses in detail how WAMS system can be used as a Wide Area Disturbance/Event Analysis System, which is a very important building block while developing SMARTGRIDS. Considering the fact that any occurrences/disturbances/events in power system create oscillations. This paper discuses how Phasor Measurement Unit's (PMU) data as a foundation block of WAMS system can be utilized to understand these events. Paper further presents learnings from MAHATRANSCO SMARTGRID project on WAMS experience in understanding various events in the system along with analysis on type of oscillations seen using PMU data with the help of case studies. This shall assist in improving grid monitoring and operation in a smarter way.

Keywords PMU · Synchrophasors · WAMS · Grid monitoring · SMARTGRID · Oscillation in power system

A. R. Kulkarni (✉)
PAC-Department, Mahatransco, Mumbai, India
e-mail: amk4.ind@gmail.com

S. S. Rajurkar
Transmission (O&M) Department, Mahatransco, Mumbai, India
e-mail: ssrajurkar@gmail.com

M. S. Ballal
VNIT, Nagpur, India

R. K. Pillai et al. (eds.), *ISGW 2018 Compendium of Technical Papers*,
Lecture Notes in Electrical Engineering 580,
https://doi.org/10.1007/978-981-32-9119-5_4

1 Introduction

This section presents mainly brief information on developments in synchrophasor technology [1] along with works done in this area and its future developmental aspects in Indian context. Power system being one of the most complex dynamic system characterises changes in network topology, loads, generation etc., this causes oscillations in the system. If these oscillations are not mitigated, they grow and may further result into catastrophic events like blackouts as observed in case of famous North American blackout of 2003 [2]. The post mortem analysis of this blackout has stressed the importance of angle monitoring for grid management and improving situational awareness, further giving boost to development of synchrophasor technology not only in USA but also in India [3]. Wide Area Measurement System (WAMS) project implementation initially started in India on pilot basis as explained in [3–5, 7]. Northern Region of India's pilot project on WAMS related commissioning experience alongwith basics of synchrophasor technology is explained in [3]. The advantage of synchrophasor technology over SCADA is explained in [4]. This paper gives information on tools used to monitor and archive the PMU data. It also reports events witnessed on Central Transmission Utility (CTU) of India PGCIL network on WAMS system [4]. Comparative analysis of occurrence in the Indian grid with the help of Synchrophasor system, Numerical Relaying system and that of SCADA is explained in [5]. It also explains first of its kind of integration of WAMS and Situational Awareness System in India alongwith details of how situational awareness of the grid can be improved [5]. Low frequency oscillation mode identification in Western and Northern region of India based on PMU data is discussed in [6]. The Operational experience of the first real time Oscillation Monitoring System (OMS) in India in Southern Regional Load Despatch Centre (SRLDC)-Bangaluru is explained in [7]. The implementation of national WAMS project named "Unified Real Time Dynamic State Measurement" (URTDSM) in India is also under progress alongwith its application developments [8].

This paper covers Five sections. Section 2 gives brief description of MAHATRANSCO and its initiatives in WAMS development, Sect. 3 discuses utilisation of WAMS system for analysing the grid events alongwith case studies. Section 4 discuses utilisation of WAMS for understanding oscillations in the system during occurrence. It also covers using Prony analysis for this purpose. At the end, Sect. 5 presents conclusion.

2 Mahatransco Infrastructure and Initiatives in Development of Wide Area Measurement System

Figure 1 shows 400 kV network overview of MAHATRANSCO system. MAHATRANSCO is the largest State Transmission utility in India with 652 EHV substations, 119,212.5 MVA of transformation capacity and 44,715.1498

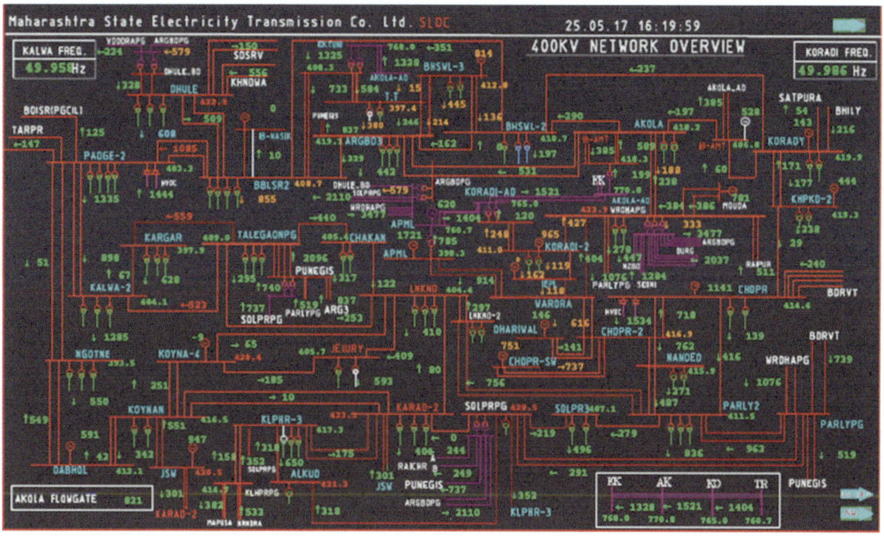

Fig. 1 400 kV Network overview of MAHATRANSCO

Circuit Kilometers of EHV Transmission lines. MAHATRANSCO has transmitted 138,613 MUs. in the year 2016–17.

This paper discuses in brief initiatives taken by MAHATRANSCO in Indian context towards utilization of synchrophasor system which is one of the foundation block of SMARTGRID technologies. It covers utilization of WAMS for grid monitoring and event analysis. Figure 2 shows architecture of WAMS infrastructure in MAHATRANSCO system. Present MSETCL's Wide Area Measurement Systems (WAMS) infrastructure includes Phasor Measurement Units (PMUs), GPS-Clock and antenna installed at various locations at twelve 400 and three 220 kV level substations. Central Phasor Data Concentrator (PDC), Visualization software and Historian software is installed at State Load Despatch Centre (SLDC) of Maharashtra located at Kalwa. The Real time data transfer of parameters like Voltages(V), Currents(I), Active Power(P), Reactive Power(Q), Frequency(*f*), Rate of Change of frequency (*df/dt*) and Delta is done from PMUs located at different important locations like near to Generating Plant, Grid Substations, substations with regional interconnections, Critical substations in the grid is sent to PDC located at State Load Despatch Centre (SLDC)-Kalwa.

3 Utilisation of Synchrophasor System for Grid Event Analysis and for Detecting Oscillations

Synchrophasor technology is considered as an important building block of SMARTGRID development. This section presents two case studies. The first case study explains the manner in which WAMS technology can be utilized to

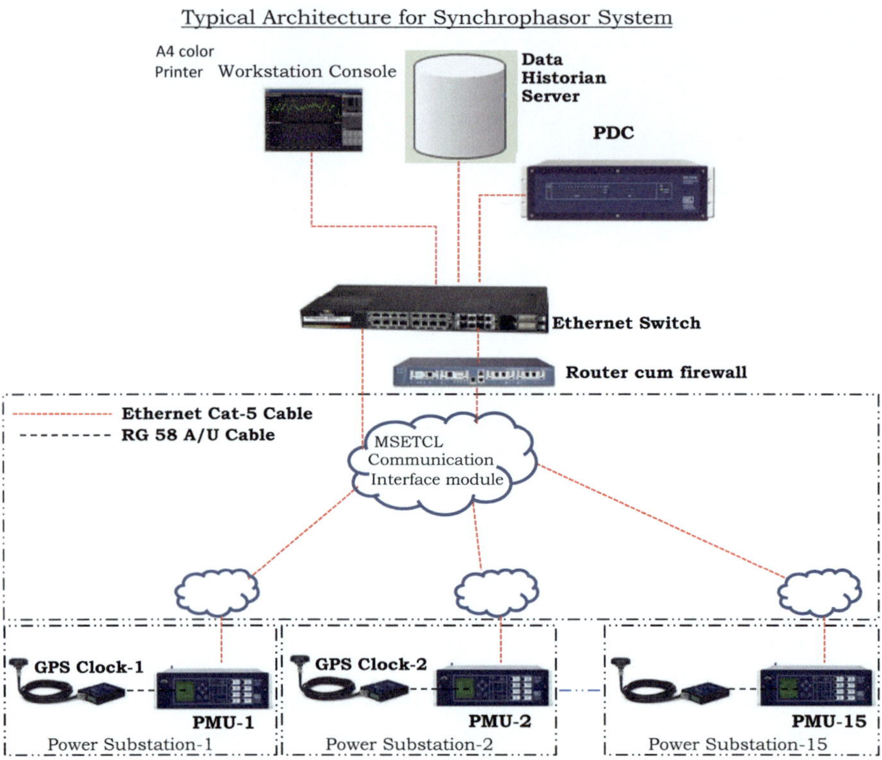

Fig. 2 Architecture of WAMS infrastructure in MAHATRANSCO

understand wide spread parametric variations in the grid using this technology. The second case study describes in detail analysis of the occurrence in MAHATRANSCO grid using synchrophasor technology. It also covers in brief information on kind of oscillations observed during this event.

(A) ***Case Study-1: Identification of Low Voltages in the grid using WAMS***

This section gives information on how Wide Area Monitoring System (WAMS) can be utilized to monitor power system parameters. The effective monitoring of this helps in better grid management in real time as well as for offline studies.

It covers analysis of low voltage scenario observed in MAHATRANSCO system on 12th, April 2017. This phenomena of low voltages in MAHATRANSCO grid was observed at some of the important 400 kV substations in MAHATRANSCO grid. Figure 3 shows dip in frequency observed at 400 kV Girawali substation in MAHATRANSCO system during this incidence, when frequency at other locations was seen to be normal and synchronized. Figure 4 shows voltage dips as observed at 400 kV Kolhapur substation on synchrophasor system. The first dip was observed at 12:53:16:880 h. During this voltage dipped to the extent of 356 kV was

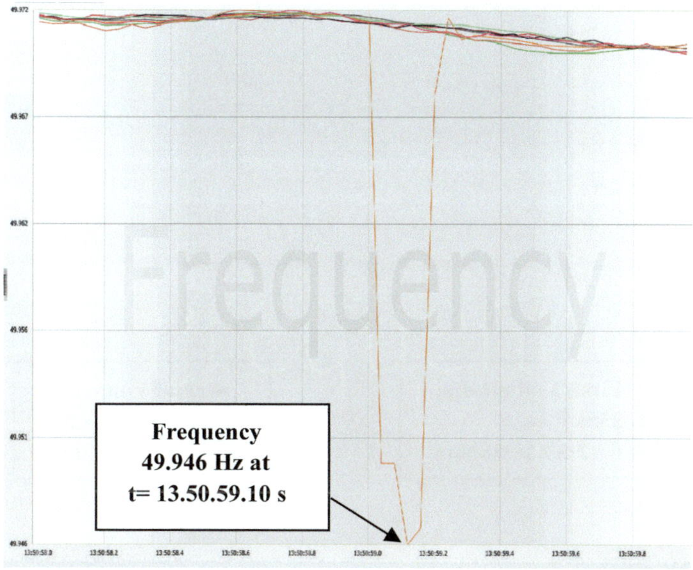

Fig. 3 Frequency dip at 400 kV Girawali substation in MAHATRANSCO system

observed momentarily. Whereas the second dip was observed at 12:53:17:160 h. During this voltage dipped momentarily to the extent of 357.22 kV as observed for A-phase. During this incidence it was observed that, voltage restored to near to normal at 394 kV at 12.53.17.440.p.m. It was also observed that, initial dip started at 12.53.16.800 p.m. and lasted for 640 ms as indicated by WAMS system. A-phase voltage has seen maximum dip as indicated in Fig. 4, followed by C and B -phases respectively. Similarly Fig. 5 shows voltage momentarily going to 376 kV for C-phase at 12:49:59:500 h. Thus WAMS system had indicated system wide parametric variations in the grid.

(B) ***Case Study-2: Occurrence analysis at 400 kV Babhaleshwar substation using WAMS***

On 25th, May 2017 some line trippings and fault in the 400 kV Babhaleshwar substation in MAHATRANSCO system occurred when heavy storm and raining was experienced in the region.

Figure 6 shows two distinct instances of line tripping in MAHATRANSCO system on 25.05.2017 and A-phase voltages as indicated by synchrophasor system. The System was normal around 16:26:06.440 h. During first incidence 400 kV Ektuni-2 line tripped in Z-1 distance protection. Voltages observed in B and C phases [On ABC Nomenclature] at 400 kV Padghe, Kalwa and Lonikand substations as indicated by synchrophasor system before this line tripping event around 16.26.06.440 h and after this line tripping event around 16.26.06.520 h are given in Table 1.

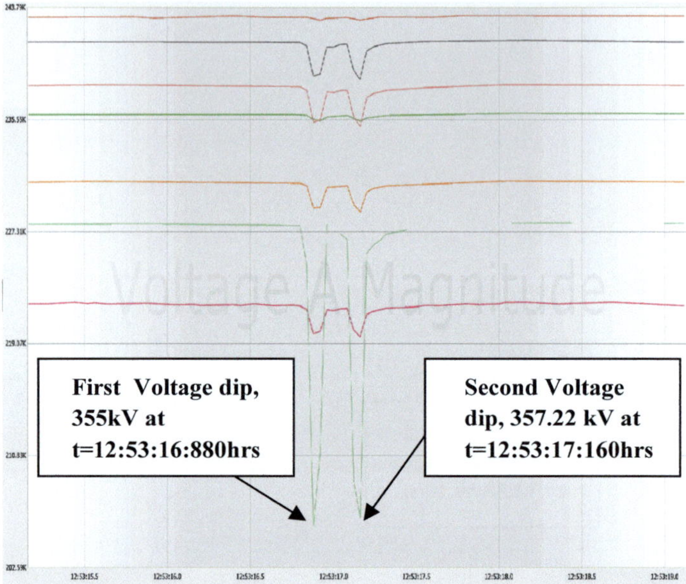

Fig. 4 Voltage dip at 400 kV Kolhapur substation in MAHATRANSCO system

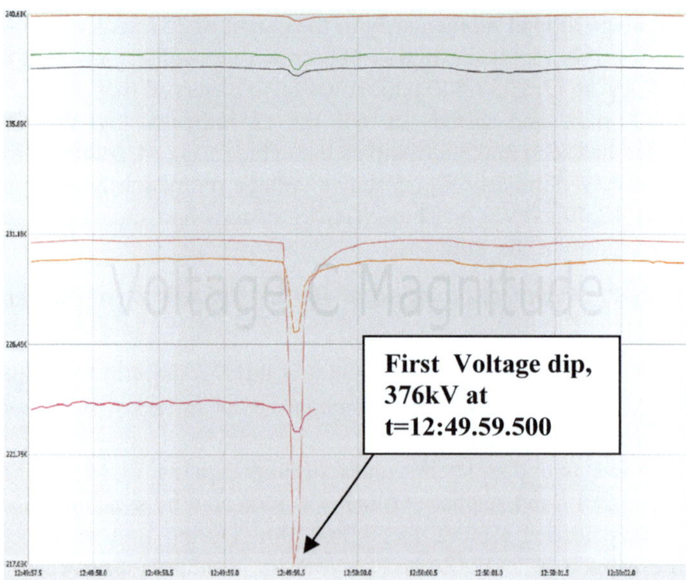

Fig. 5 Voltage dip at 400 kV Lamboti substation in MAHATRANSCO system

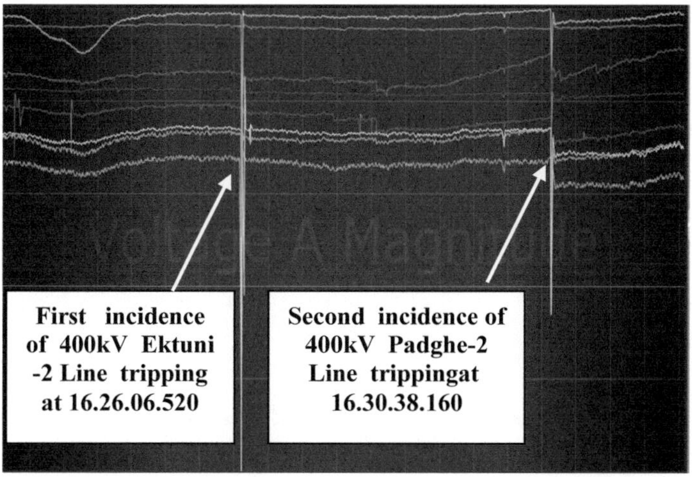

Fig. 6 A-phase voltages as indicated by synchrophasor system in MAHATRANSCO

Table 1 Voltages as indicated by synchrophasor system

Voltages observed	Locations	Time: 16:26:06.440 h (kV)	Time: 16:26:06.520 h (kV)
Vb	400 kV Padghe	414	327.34
	400 kV Kalwa	407	346
	400 kV Lonikand	410	371
Vc	400 kV Padghe	400	362
	400 kV Kalwa	407	372
	400 kV Lonikand	405	385

Figure 7 shows dip in voltages at various locations of PMU placement in MAHATRANSCO system showing 400 kV Padghe voltage dip to the extent of 327 kV momentarily. During this incidence of 400 kV Ektuni-2 line tripping current in A-phase of 400 kV Padghe-Babhaleshwar line increased to 1.89 kA at 16.26.06.560, whereas current in B-phase observed to be 2.19 kA at 16.26.05.680 as shown in Figs. 4 and 5 respectively. Rise in currents in A and B phases during fault indicate A–B phase fault. There was comparatively less current rise in C-phase.

During second incidence of 400 kV Padghe line tripping on 25.05.2017, A-phase voltage at Padghe was observed to be normal i.e. 401 kV around 16.30.36.000 h, but it reduced to 367 kV momentarily at 16.30.38.160 h as shown in Fig. 8.

C-phase voltage at Padghe was observed to be normal i.e. 398 kV around 16.30.36.320 h, but it reduced to 334.27 kV momentarily at 16.30.38.240 h as

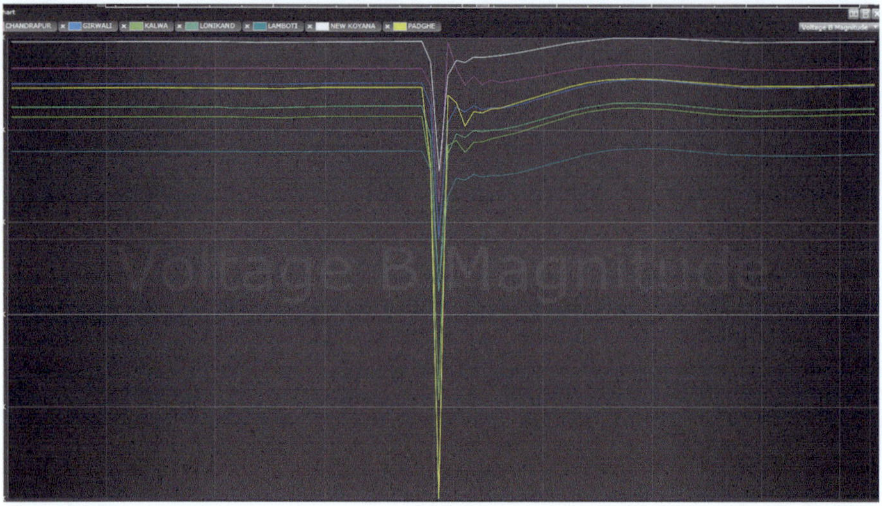

Fig. 7 B-phase voltage dip as indicated by synchrophasor system

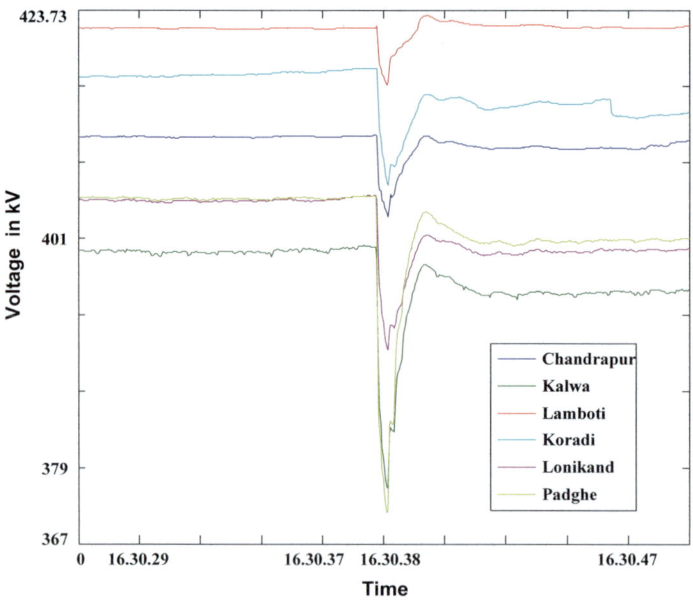

Fig. 8 A-phase voltage during second fault

shown in Fig. 9. During second fault A-phase current 'Ia' was increased to 1.023 kA at Padghe as shown in Fig. 10. Rise in C-phase current 'Ic' which has increased to 2.05 kA at 16.30.37.960 h is shown in Fig. 11.

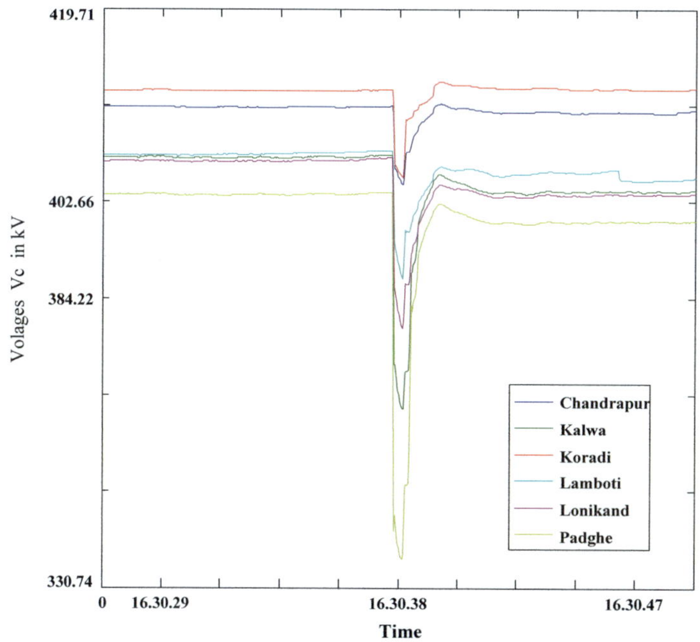

Fig. 9 C-phase voltage during second fault

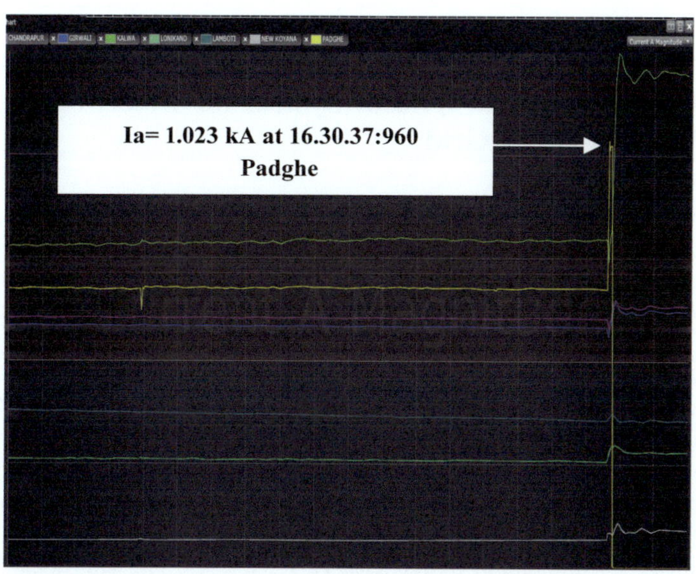

Fig. 10 A-phase current rise during second fault at 400 kV Padghe substation

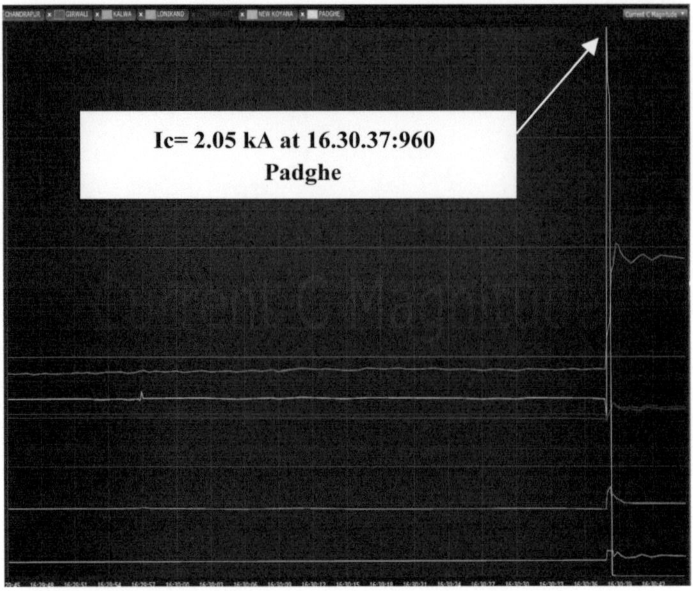

Fig. 11 C-phase current rise during second fault at 400 kV Padghe substation

Thus this case study explained the manner in which WAMS system can be utilized for occurrence/event analysis in the grid.

4 Utilisation of Synchrophasor System for Detecting Oscillations in the Grid

This section explains utilization of PMU data for analysis of oscillations in the system during occurrences. The occurrence of 400 kV Babhaleshwar explained in the case study-2 is considered for understanding oscillations seen during this event. This is done by using Prony analysis technique. The 20 s duration data for frequency from PMUs at 220 kV Eklahre, 220 kV Boisar within MAHATRANSCO grid is used for this purpose. Figure 12 shows frequency response as indicated by PMU data of 220 kV Eklahre substation during this occurrence. Figure 13 shows specific data range of signal of 220 kV Eklahre considered for Prony analysis.

Figure 14 shows plot for Prony approximate and PMU measured frequency response for 220 kV Eklahre substation. Table 2 tabulates important results based on Prony analysis. It comprises of amplitude, damping and frequency in Hz. The results indicate oscillations of frequencies 1.31.4 and 1.6 Hz with negative damping. Similar analysis is also done for 220 kV Boisar substation and other substations within MAHATRANSCO grid. This study indicated oscillation around the frequencies as indicated in Table 2.

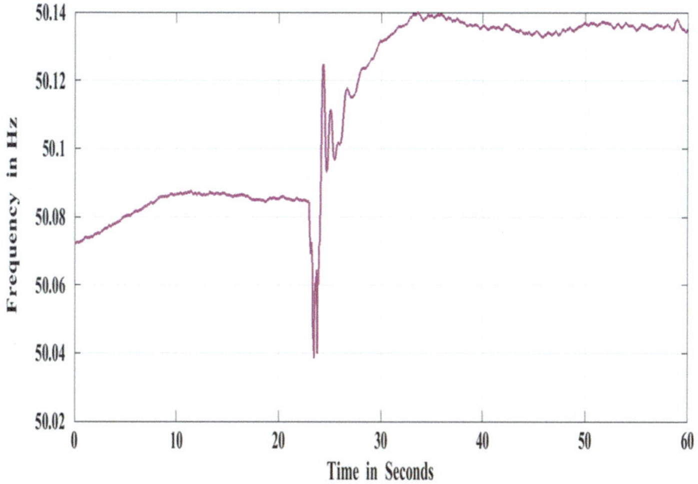

Fig. 12 Frequency response as indicated by PMU at 220 kV Eklahre substation

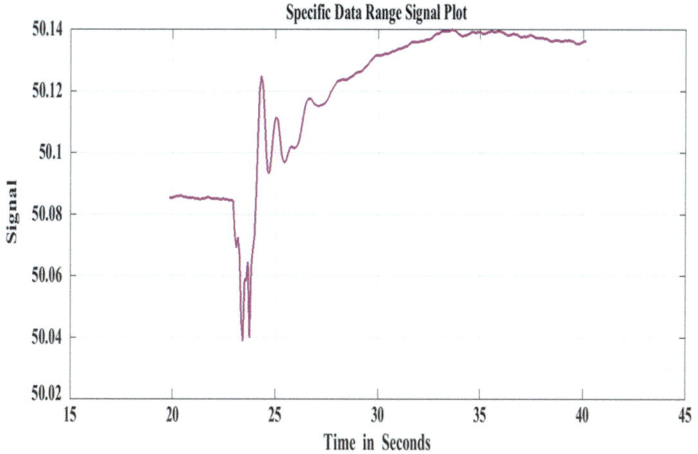

Fig. 13 Specific data range signal for analysis

Figure 15 shows frequency response as indicated by PMU data of 220 kV Boisar substation during this occurrence.

Figure 16 shows plot for Prony approximate and PMU measured frequency response for 220 kV Boisar substation in MAHATRANSCO grid. These results are also verified with Fast Fourier Transform (FFT) analysis. These studies indicate plant mode of oscillations in the range of 1.0–2.0 Hz with negative damping. This can be confirmed with the field observations that during this occurrence system around Nashik and Pune area was affected and had not affected far away system.

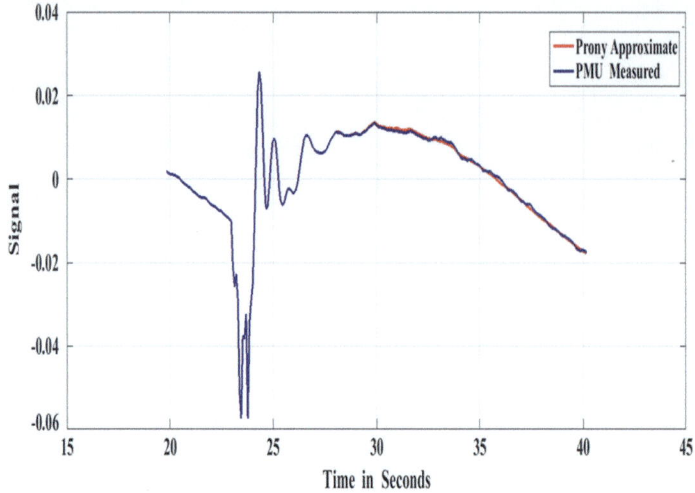

Fig. 14 Plot for Prony approximate and PMU Measured frequency response for 220 kV Eklahre substation

Table 2 Important results for prony analysis based on PMU data for 220 kV Eklahre substation

Sr. No.	Amplitude	Damping	Frequency in Hz
1.	1.6	−2.4	1.3
2.	0.19	−0.76	1.4
3.	0.055	−0.73	1.6

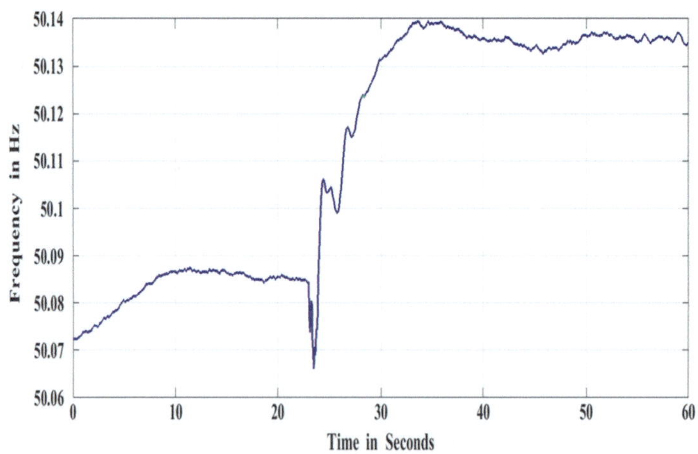

Fig. 15 Frequency response as indicated by PMU at 220 kV Boisar substation

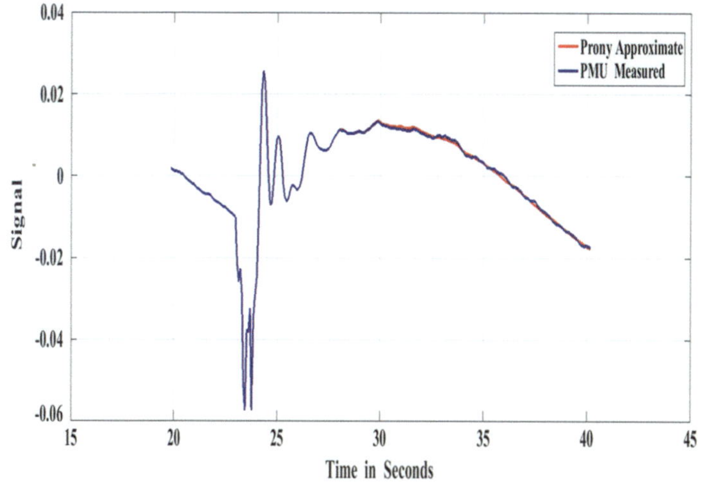

Fig. 16 Plot for Prony approximate and PMU Measured frequency response for 220 kV Eklahre substation

5 Conclusion

Wide Area Monitoring/Measurement (WAMS) system is the foundation block of SMARTGRID developments. This paper describes the manner in which this technology can be effectively used for detecting small changes in system parameters as explained in case study-1. It discuses how synchrophasor technology can be used for occurrence analysis in case study-2. It also covers oscillation detection using synchrophasor system and Prony analysis.

References

1. Phadke AG (1993) Synchronised phasor measurement in power system. Comput Appl Power, IEEE 6(2):10–15
2. Andersson G, Donalek P, Farmer R, Hatziargyriou N, Kamwa I, Kundur I, Martins N, Paserba J, Pourbeik P, Sanchez-Gasca J, Schulz R, Stankovic A, Taylor C, Vittal V (2005) Causes of the 2003 major grid blackouts in North America and Europe and recommended means to improve system dynamic performance. IEEE Trans Power Syst 20(4):1922–1928
3. Agrawal VK, Agarwal PK, Kumar R (2010) Experience of commissioning of PMUs Pilot Project in the Northern Region of India. National Power System Conference-2010, India
4. Manyam V, Harikrishna A, Singh V (2013) Synchrophasors redefining SCADA system. In: GRIDTECH-2013 international conference and exhibition, New Delhi, India
5. Kulkarni AR, Ballal MS, Rajurkar SS (2016) Indian experience of utilisation of synchrophasor system and its integration with situational awareness system. In: 6th, IEEE international conference on power systems, 2016 (ICPS-2016), IIT-Delhi, New Delhi-India

6. Rampurkar V, Mangalvedhekar HA, Kazi F, Pentayya P (2015) Synchrophasor based oscillation Detection: a case study for indian power grid. Int J Eng Res Technol 4(04)
7. Cherian S, Biju AS, Jose N, Raman R Operational experience of real time oscillation monitoring system in India. In: CIGRE colloquium on SMARTGRID under study committee D2 for information systems and telecommunication, Mysore, Karnataka, India

A Decentralised Approach for Load Management in an Islanded Distribution Network During Contingency

P. T. Seenath Beevi and R. Harikumar

Abstract Electricity plays a vital role in both day to day life of people and the economic development of a nation. To ensure sustainable supply of electricity and energy security, the efficient management of the electricity distribution network through the use of intelligent devices, which can monitor and control power flows in real time, has become necessary. The major grid features sustainability, energy security, reliability etc. became the challenges of the 21th century grid management because of the grid interactive distributed renewable generation (DGs) and the increasing number of Electrical Vehicles. In order to manage the distribution network during a contingency condition, various options are explored for the maintenance of critical loads by fixing priority based on the criticality of loads. The restoration strategy formulated by integrating the DGs and Electric Vehicles (EVs) in a decentralized operation. Multi Agent System (MAS) have capabilities like standard interfaces, standard agent communication language, reactivity etc. that can be used for solving the problem associated with supply restoration when DGs are present. The energy restoration process is formulated as a multi objective optimization problem. This is formulated by maximizing the critical loads and minimizing the number of switching operation subject to the constraints voltage, current, Power, frequency and state of charge of EVs battery within the permissible limit. Algorithm is formulated to achieve the objectives. The power quality parameters System Average Interruption Duration Index (SAIDI) and System Average Interruption Frequency Index (SAIFI) values are compared with the values of existing utility outage management system (OMS). The impact of V2G concept for supply restoration was assessed along with the benefits of deploying V2G facility of the EVs for restoration of power were analysed.

P. T. Seenath Beevi (✉)
IT Unit, Kerala State Electricity Board Ltd, Trivandrum, India
e-mail: jdmistvm@ksebnet.com; ee.ritutvm@kseb.in

R. Harikumar
Department of Electrical Engineering, College of Engineering Trivandrum, Trivandrum, Kerala, India

R. K. Pillai et al. (eds.), *ISGW 2018 Compendium of Technical Papers*,
Lecture Notes in Electrical Engineering 580,
https://doi.org/10.1007/978-981-32-9119-5_5

Keywords DG-Distributed generator · EVs-Electric vehicles · G2V-Grid-to-Vehicle · MAS-Multi agent system · SoC-State-of-charge · V2G-Vehicle-to-Grid

1 Introduction

The traditional electrical network is radial and the power flow is unidirectional. In modern power system the grid management is important due to the presence of large number of intermittent energy sources such as grid interactive DGs. In most of the critical power applications, there is a need to maintain minimal operating capability under fault conditions.

The traditional electric grid will need to build additional layers of automation, communication and IT system to transform it into a smarter grid. Supervisory Control and Data Acquisition Systems (SCADA) with Energy Management Systems (EMC) and Distribution Management Systems (DMS), Enterprise IT network covering all substations and all field offices with reliable communication systems, Enterprise Resource Planning (ERP)/Asset Management Systems, Geographic Information System (GIS),—mapping of Electric network assets and consumers on GIS maps, modernization of substations with modern switchgear and numerical relays, Automated Metering Infrastructure (AMI) with Meter Data Management System (MDMS), Billing and Customer Care system, Distribution Automation (DA) and Substation Automation, etc. are application perspective.

The greatest Environmental pollution in future will be due to the communication infrastructures, microwave towers and Wi- fi hotspots, etc. All the IT systems and SCADA systems require the need for a reliable and secure communication network. National Optic Fibre Network (NOFN) project envisages fibre backbone network up to 33/11 kV substations and National Information Infrastructure (NII) up to distribution network. Also as part of digital India vision, a lot of communication programs need to be evolved for transmission and distribution sectors and also for implementing the fibre to home concept. Especially in Kerala in addition to all the centrally declared projects Kerala Fibre Optic Network (K-FON) state wide ariel communication network is being planned. The K-FON project is a mega infrastructure project of Government of Kerala which aims to build a high speed, reliable and scalable fibre network across the state to provide connectivity to Government institutions, households, by averaging the transmission and distribution infrastructure of Kerala State Electricity Board Ltd. (KSEBL).

The importance & evolution of the new concept of Smart Grid (SG) is to provide 24×7 quality and reliable power supply to all the customers.

To restore the load during fault conditions, a series of actions need to be carried out such as detection of the fault, restoration strategy to be evolved depending on the type of fault, DG loads, EVs in the system, etc. These activities can be performed by adding additional layers of automation, communication and IT systems that are imperative for the operation. The SG development in distribution grid is a challenge because of the non standard construction. As part of Restructured Accelerated Power Development and Reforms Programme (R-APDRP), Part A (IT) and SCADA, MoP, GoI have launched lot of IT infrastructure and software applications thereby prepared the building blocks for SG.

The IEEE Standard 1547.2-2008 proposed the implementation of intentional islands powered by DGs in case of power outages in the system [1]. The IEEE Standard 1547.4-2011 provided guidelines for the design, operation, and integration, validating the use of DGs to help sustain power flow in selected essential loads in islands.

This paper, focused on the agent based method for power restoration with the help of DGs (PV& EV). The Foundation for Intelligent Physical Agencies (FIFA) is an IEEE standard organization that promotes the agent based technologies. The major contributions of this paper are:-

1. Prioritized the loads (node points) using an algorithm based on sensitivity.
2. Demonstration of the decentralized MAS approach for the restoration of critical loads, as a multi objective optimization problem, in distribution network during contingency condition based on real time scenario.

2 Objectives

- Prioritization of loads (node points) with provision for setting priority, based on individual sensitivity and category.
- To develop an algorithm for the decentralized MAS to strategically restore critical loads during fault situations in a distribution network, thereby making the distribution network smarter through decentralized load management.
- Conceptualize the V2G integration facility in the algorithm.

3 Proposed MAS Framework for Supply Restoration

Agents are software or hardware entity situated in some environment that is able to autonomously react to changes in an environment. Autonomy means that the agents in an environment function on its own. Multi Agent approach is utilized in many power system applications such as system restoration, congestion management, energy trading etc. Multi Agent System (MAS) have capabilities like standard interfaces, standard agent communication language, reactivity etc. that can be used for solving the issues associated with SG, micro grid solutions and protection when DGs are present. Agents are having decision making capability.

3.1 *Distributed Artificial Intelligence*

Distributed Artificial Intelligence (DAI) considers the provision of small subsystems working together to achieve a common goal. Sometimes a group of computational agents like computers, sensors, robots, AI vehicle along with people need to work together to find solution to a common problem. Mainly the areas of research in DAI are classified into two, namely Distributed Problem Solving (DPS) and MAS.

3.1.1 Distributed Problem Solving

DPS considers solving a problem by splitting among a number of modules. The modules must be such that they cooperate among themselves in dividing tasks and sharing information. Cooperation must be there in evolving into solutions as well. All interaction between modules are considered as a part of the system. The available knowledge among modules need to be shared.

3.2 *Multi Agent System*

MAS concerns with collection of already existing agents aiming at finding a solution for a particular problem. MAS can be considered as network of agents that work together to solve a problem that is unable to be solved by them individually. Agents are autonomous and heterogeneous. MAS have several advantages over a single central control like fast solution by using parallelism, communication gets decreased as raw data is not send to a central location, flexibility increases as there are agents with different capabilities to solve a problem, more reliable and agents take responsibility for failure. In the decentralized MAS architecture, four types of agents are proposed to represent the four entities which participate in the restoration process [2].

4 MAS Development

At present, a number of platforms are available for the development of MAS like agent builder, FIFA-OS, ZEUS, JADE, etc. The proposed MAS is developed in JAVA agent development environment which is a FIFA compliant open source software for interoperable multi-agent system. The task to be executed by an agent is represented by its behaviour. These behaviour can be added at any time during the development phase or in the process stage.

5 Agents and Their Characteris-Tics

5.1 Load Agent

Each LA is pre-set by its agent ID, its local name, the node it is attached to, the load at the node, the priority of the load, ID of the Agregator Agent (AA) at the node, ID of the nearest DG agent, and ID of the neighbouring LAs. During normal system operation, the LA (at each node) continuously monitors the voltage and current profile at its node. Load agent initiates the process of restoration by sending message to the AA and DG agent to meet the load requirement [2].

5.2 Aggregator Agent

The AA represents the charging location (parking lot) at each node where the EVs are charging/discharging. Each AA is pre-set by the information such as its agent ID, its local name, and the number of EVs connected to the node for charging/discharging. Aggregator agent receives message from the LA and aggregates the power provided by the EVs through V2G facility and transfer it as content [2].

5.3 DG Agent

The DG agent is initialized with the agent ID, local name, generation capacity, and ID of the other DG agents. The DG agent constantly monitors the power generated by the DG it represents. DG agent Receives message from the load agent, decide the island range by optimizing the node to be restored by interacting with the neighbouring DG agents and accept or decline the message to the load agent and send message to the SA [2].

5.4 Switch Agent

The SA is initialized with agent ID, local name, and the node numbers to which it is connected. The SA is placed at each node and controls the switching of all the line switches connecting the node to its nodes. The switch agent receives operation command, identify the switches to be operated by the node number of the connecting node [2].

6 Modelling of the Distribution Network

A typical network considered for the simulation has 25 number of nodes and the load in the feeder is 160 A evening peak and 140 amperes day peak. The network is modelled using GIS/AutoCAD/PSS or any network builder.

The distribution SCADA system is introduced as a decentralised approach, over and above the SCADA remote control operation and monitoring. The behaviour of each agent reflects an individual step of the restoration strategy and this is demonstrated with the help of case studies as per GUI in Fig. 1. It includes 65 nodes, 25 LAs, 3 AAs 4 DG agents.

To formulate the network, study is conducted to find out the positions where the feeder terminated, the existing network equipments and the parameters and analysis of the load pattern for selected feeders and substations. The priority concerns are finalised based on sensitivity of loads. The nodes are prioritised using an algorithm.

The electric power restoration strategy was formulated based on the concept of islanding and implemented by using decentralised MAS. The Multi objective optimisation problem is framed as maximisation of priority load and minimisation of switching operation and maintain the voltage and current within the permissible limits. The database created using a static network in open source platform and the coding were done in JAVA. An interface between MAS and power system simulation software by FIPA compliant language (JAVA) is developed. The process of supply restoration is based on the prioritised nodes, by using various agents and algorithm. The impact of integration of V2G for service restoration was also considered [2].

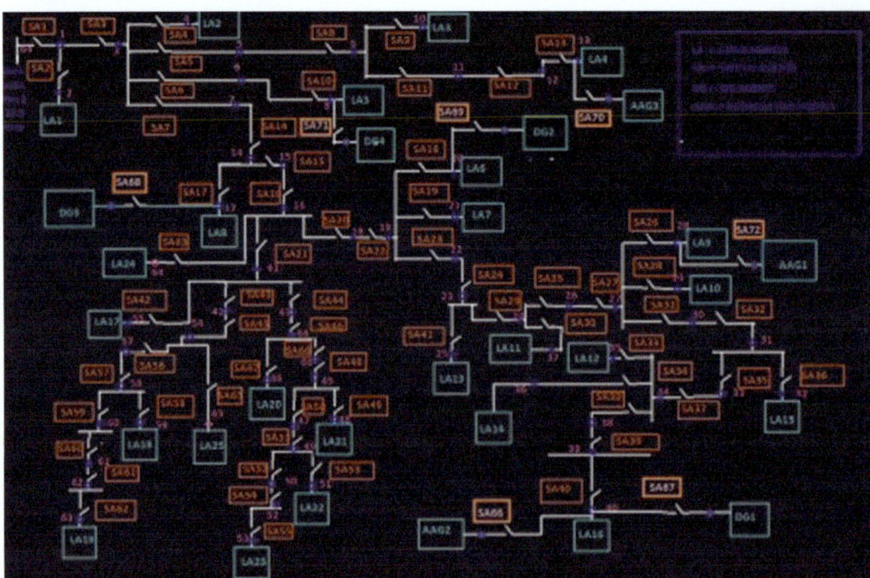

Fig. 1 Graphical user interface

7 V2G Power from the EVs

Vehicle to Grid (V2G) describes the system in which the stored energy in the EVs (Batteries of EVs) supplies energy to the grid, whenever requested by the grid operators. The EVs parked in the parking station would provide a considerable amount of energy storage, like a distributed Energy recourse. The V2G technique can reduce the stress on overloaded distribution transformers during peak hours as well as in contingency conditions such as system faults namely transformer failure, line faults etc.

Through a communication link, transaction utility grid operator (LA) communicate with the plugged vehicle operator (AA) and the transaction can be initiated to manage the Critical load, the aggregator act as a commercial operator. The aggregator can be formed in between the fleet of vehicles and load agents, and will take real time decisions of charging or discharging of EVs under it, depending on the prevailing grid conditions. A control technique evolved for the charging and discharging of EVs, the V2G controller placed on a particular node and is a part of distribution network. An optimal aggregator to be placed for the integration of the V2G concept (Fig. 2).

In the study it is assumed that three numbers of charging stations with 3–5 number of vehicles parked in each station, each vehicle have the capacity of 25 KW, the State of Charge (SoC) of the vehicle is hundred percent and minimal level of discharge is 15%. In addition the EVs available at each parking lot are assumed to have lithium-ion batteries. The V2G power supply, available for restoration, is calculated using in a step by step procedure [3].

8 Result

The proposed method for the power restoration problem was solved by the algorithm developed in the JAVA based platform. Simulation is initiated by triggering a fault at the beginning so that full load can be used for simulation and this was helped to demonstrate the algorithm in full functionality and has more visibility to the application.

At the time of fault, all affected LA will broadcast restoration request message to its associated AA and DG with the load to be restored in a predefined protocol. Upon on successful reception of this request, respective AAs will process the request and intimate to respective DG regarding the load that can be fed from its sources based on its SoC. DG will initiate an iterative algorithm for the selection of suitable candidates for restoration of loads among the requested agents. The iterative algorithm at each DG center aims at maximizing the loads to be restored under its association. The loads to be restored are iterated based on assigned priority of each load center such that the switching operations required for such restoration is minimized.

Agent Id	Local Name	Node attached	Load	Priority	Nearest Aggregator ID	DG Id	Neighboring Loadagent Id	LA Voltage	LA Current
1	DENTAL	2	100.0	5	1	1	2	11000.0	100.0
2	TAGORE GARDEN	4	100.0	1	1	1	1	11000.0	100.0
3	THAMARABUGAM	10	100.0	5	1	1	4	11000.0	100.0
4	ADRACK HOTEL	13	155.0	4	1	2	3,5	11000.0	100.0
5	ATHIRA APARTMENTS	8	100.0	1	1	1	2	11000.0	100.0
6	NIKUNJAM	20	100.0	1	1	1	7	11000.0	100.0
7	KUMARAPURAM UPS	21	100.0	1	1	1	6	11000.0	100.0
8	DEVI SCANS	17	100.0	1	1	1	17	11000.0	100.0
9	COSMO	28	100.0	1	1	1	10	11000.0	100.0
10	MURIJAPALAM	29	100.0	1	1	1	9	11000.0	100.0
11	INDRUM	37	100.0	1	1	1	13	11000.0	100.0
12	KESAVADASAPURAM OFFICE	35	100.0	1	1	1	14	11000.0	100.0
13	BIVASATH	25	100.0	1	2	2	11	11000.0	100.0
14	AC SCH	36	100.0	1	1	1	12	11000.0	100.0
15	POTTAKUZHY	32	100.0	1	1	1	12	11000.0	100.0
16	VYDYUTHI BHAVANAM	40	100.0	1	2	1	14	11000.0	100.0
17	SIMI AJ HALL	55	100.0	1	1	1	18	11000.0	100.0
18	ORIENT	59	100.0	1	2	1	17	11000.0	100.0
19	NIKUNJAM ORIENT	63	100.0	1	1	1	18	11000.0	100.0
20	POTHUJANAM	46	100.0	1	1	1	21	11000.0	100.0
21	CHETTI	48	100.0	1	1	1	20	11000.0	100.0
22	KARIKKA	51	100.0	1	1	1	21	11000.0	100.0
23	VALLYVI	53	100.0	1	1	1	22	11000.0	100.0
24	KUMARAPURAM	64	100.0	1	1	1	17	11000.0	100.0
25	SUHA RESIDENCY	65	100.0	1	1	1	18	11000.0	100.0

Aggregator Agents

Agent ID	local Name	Node Connected To	EV Load	Total EVs
1	ELECTRIC 1	28	50.0	1
2	ELECTRIC 2	40	50.0	1
3	ELECTRIC 3	13	50.0	2

DG Agents

Agent ID	Local Name	Capacity	DG POwer	Other DG Ids	Node Connected
1	DG1	200.0	200.0	2,3,4	40
2	DG2	200.0	200.0	1,3,4	20
3	DG3	200.0	200.0	1,2,4	17
4	DG4	200.0	200.0	1,2,3	8

Switch Agents

Agent ID	Nodes Conneceted to	Local Name	Switch Status
1	64,1	SWITCH1	Open
2	1,2	SWITCH2	Open
3	1,3	SWITCH3	Open
4	3,4	SWITCH4	Open
5	3,5	SWITCH5	Open
6	3,6	SWITCH6	Open
7	3,7	SWITCH7	Open

Fig. 2 Database table

The total power consumption of the restored load should not be more than the power provided by the distributed energy resources i.e., DGs and EVs.

$$\Sigma P_i \leq P_{Dtt} + P_{EV} \tag{1}$$

The voltage is within the range of +6 and −9% and the frequency is within the range of 49.2 and 50.3 Hz is considered for case studies. These are the allowable ranges as per grid code 2010.

9 Conclusion

In this paper an open source database is created and developed to demonstrate the decentralised MAS approach used to solve the supply restoration problem as a multi objective optimisation technique to maximize the priority load and minimise the switching operation based on the concept of islanding.

The algorithm first check the priority order to solve the optimisation technique and the impact of V2G concept for supply restoration and the benefits of deploying V2G facility of the EVs for service restoration is explained. The advantage of the solution is the reduction of outage time and frequency of interruption, leading to the reliability improvement and thereby enhancing customer satisfaction and revenue to the utility.

10 Future Scope

The developed system is scalable and have capability to integrate to the GIS application. Piloting of the said technology is required in a real time scenario by deploying proper hardware. In addition an iterative algorithm to find the optimal placement of DGs for managing more load in an efficient manner and reducing the duration of restoration time can be developed. Financial benefits on account of the supply restored by the critical load can be assessed.

Acknowledgements The authors would like to thank Sri. Aravind. L, Asst. Engineer, Regional IT Unit, Tvm for sharing his expertise in the software development.

References

1. IEEE Recommended Practice for Interconnecting Distributed Resources with Electric power Systems Distribution Secondary Networks. IEEE Standards Coordinating Committee 21 IEEE Standard 1547 June 2011
2. Sharma A, Srinivasan D (2015) A decentralised multi agent system Approach for service restoration using DG islanding. IEEE Trans Smart Grid 6(6)
3. Kopka H, Daly PW (2013) A multi agent solution to distribution system management by considering distributed generators. IEEE Trans Power Syst 28(2)
4. Li H, Sun H, Wen J (2012) A fully decentralised multi-agent system for intelligent restoration of power distribution network incorporating distributed generators. IEEE Comput Intell Mag
5. Logenthiran T, Sreenivasan D, Khambadkon AM (2012) Multiagent system for real time operation of a microgrid in real time digital simulator. IEEE Trans Smart Grid 3(2)
6. Ren F, Zhang M, Sutanto D (2013) A multi-agent solution to distribution system management by considering distributed generators. IEEE Trans Power Syst 28(2)

Seenath Beevi Received B.Tech Degree in Electrical Engineering from Kerala University, India during 1987, M.B.A Finance from Indira Gandhi Open University, India from 2007, M.Tech Power system from Kerala University, India during 2018. Currently working in Kerala State Electricity Board Limited as Exe. Engineer. She is joined in KSEBL during 1990 as Asst. Engineer. Published a paper in International conference on Smart Grid during 2018.

Roadmap for Digital Control System (DCS) Implementation in Utility Substations

Narayanan Rajagopal, Ajit Khandekar, Amrit Mishra and Sriram Shankara Narayanan

Abstract Traditional protection systems in substations were predominantly based on standalone electromechanical relays. During the 1970s, with advancement in technology and increasing requirements from grid operations, static relays and RTU based Substation Automation Systems (SAS) evolved. Further, the 90s saw the evolution of Intelligent Electronic Device (IED) s with much higher processing capability and advanced protection functionality based on digital processing techniques. IEDs also performed additional tasks such as fault recording, reporting, remote control and metering. The need for communicating with the IEDs became a necessity which resulted in development of proprietary protocols creating an issue of interoperability among the various products and applications. The 2000s saw remarkable progress in the field of substation automation with advent of IEC 61850 Edition 1, which brought standardized configurability, object oriented data models, simple architecture, the promise of interoperability and adaptability with evolving future smart grid developments making it a popular option for substation automation. To leverage the benefits of new standards, electrical utilities implemented IEC 61850 based SAS. Progressively IEC 61850 evolved overcoming the shortcomings and technical issues associated with Edition 1 and new functionalities emerged in Edition 2 to cater to the requirements of grid operations. Utilities which implemented SAS based on IEC 61850 Edition 1 face challenges pertaining to standardization of SAS design/implementation, in defining maintenance strategies and most importantly in defining future strategy to leverage the benefits of latest

N. Rajagopal (✉) · A. Khandekar · A. Mishra · S. S. Narayanan
Tata Consultancy Services Ltd, Nirmal Building, 9th Floor, Nariman Point, Mumbai 400021, Maharashtra, India
e-mail: narayanan.rajagopal@tcs.com
URL: www.tcs.com

A. Khandekar
e-mail: ajit.khandekar@tcs.com

A. Mishra
e-mail: mishra.amrit@tcs.com

S. S. Narayanan
e-mail: sriram.narayanan@tcs.com

R. K. Pillai et al. (eds.), *ISGW 2018 Compendium of Technical Papers*,
Lecture Notes in Electrical Engineering 580,
https://doi.org/10.1007/978-981-32-9119-5_6

technological development in standards and bringing SAS implementation to a common/standard design. An important step in defining the roadmap for digitally enabling SAS is to assess the current landscape and identify technologies and standards to be implemented. This paper presents the assessment of Digital Control System (DCS)/SAS generally implemented across power Utilities, based on the IEC 61850 Edition 1, bringing the view of global SAS implementations and industry best practices. The assessment is with respect to electrical functionalities, system architecture, time synchronization and cyber security. The paper further defines a strategy for future implementations to leverage the benefits of latest technological developments in IEC 61850, to stay at par with industry trends/standards and most importantly to bring SAS implementations to a standard design. A maintenance strategy for handling multivendor SAS products, operational cost optimization, dealing with ageing infrastructure, extending functionality required to meet power system infra-structure enhancements with minimal effort and cost, maintaining high availability of equipment is also discussed.

Keywords SAS · DCS · IEC 61850 · Process Bus · PRP · Time synchronization · Cyber security

1 Introduction

Substation Automation System (SAS) are which designed to acquire data from generating stations and substations, for control and monitoring purpose with minimum human intervention. In early 1970s, data acquisition were done through hardwired transducers and contacts wired to Remote Terminal Units (RTU). Data refresh rate in these systems were in the range of several seconds to minutes.

Further, the 90s saw the evolution of Intelligent Electronic Device (IED) s with much higher processing capability and advanced protection functionality based on digital processing techniques. IEDs also performed additional tasks such as fault recording, reporting, remote control and metering. The network of IEDs reduced requirement of additional transducers, input and output contacts and even RTUs. The need for communicating with the IEDs became a necessity which resulted in development of proprietary protocols which restricts selection of SAS products to a particular OEM. Several advancements came along with the introduction of open communication protocols.

The beginning of 21st century saw remarkable progress in the field of substation automation with advent of IEC 61850 Edition 1, which brought standardized system engineering, object oriented data models, simple architecture, the promise of interoperability and adaptability with evolving future smart grid developments making it a popular option for substation automation. It is a robust communication protocol which contains a collection of multiple protocols, concepts and component standards. To leverage the benefits of new standards, power utilities started implementing IEC 61850 based SAS.

The paper discuss about implementation of IEC 61850 in power Utilities, maintenance strategy for handling multivendor SA products and the technologies to be leveraged to keep it future ready.

2 Assessment and Future Strategy for SAS

This section elaborates functionalities of SAS implemented in Utilities and technologies which can be implemented as a part of future strategy.

SAS assessment is based on following broad categories

A. Electrical Functionalities
B. System Architecture
C. Time Synchronization
D. Cyber Security.

2.1 Electrical Functionalities

Electrical functionalities such as Measurement, Interlock, Protection systems are analyzed.

1. **Measurement**: Parameters like voltage and current from Voltage Transformer (VT) and Current Transformer (CT) respectively, need to be sent to IEDs for protection and measurement. For sending of these parameters, three types of approaches are followed, which are:

   - Conventional approach
   - Sampled Value approach (Process Bus).

At present, most power Utilities follow **conventional hardwired** approach for sending measured parameters like voltage and current from VTs and CTs to IEDs. This approach is not very cost effective and causes difficulty in isolation procedure of components such as IEDs. The risk of mal-operation during maintenance, in terms of human error is very high because of huge number of cables and connections.

It is suggested that Utilities can implement IEC 61850 based Sampled Value approach through Process Bus, using non-conventional instrument transformers and merging units for their new substations.

The various components of a process bus enabled substation is shown in Figure 1.

Implementation method 1: Implementation of merging unit with conventional CT and VT.

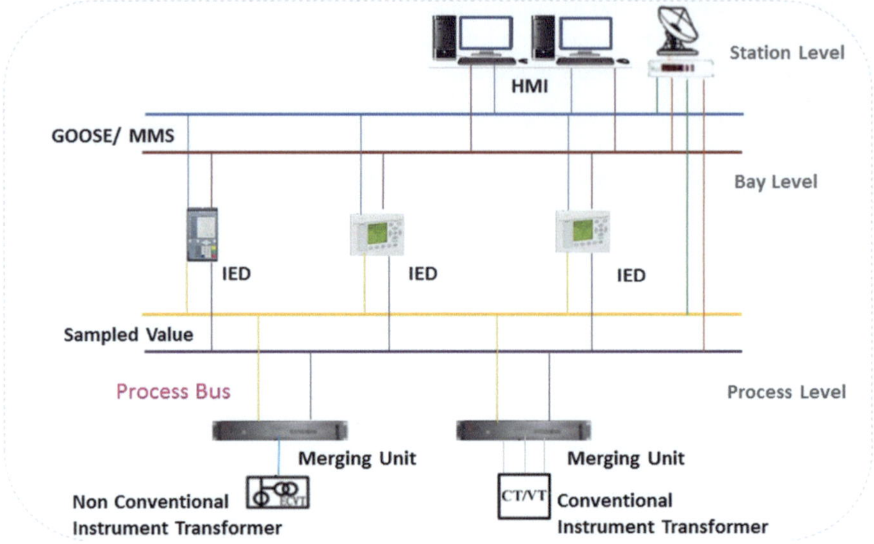

Fig. 1 Process bus overview

If utility decides to retrofit, at initial stage only SAS without the replacement of primary equipment such as CT and VT, then process bus can be implemented using merging units compatible with conventional CT, VT and Process Bus compatible IEDs. The approach is followed in substations where life of secondary equipment has come to an end but primary equipment is still in healthy condition, considering life span of primary equipment is more than that of the SAS components in general.

Implementation method 2: Implementation of Non-conventional CT and VT.

While retrofitting of primary equipment, conventional CT and VT can be replaced with the non-conventional CT and VT which will be connected with the merging units.

Benefits: The main benefits a utility can realize with the implementation of Process Bus are:

- Handling of secondary CT and VT circuits at control panel level will be eliminated, as there will be only Fiber Optic (FO) cables at control panel.
- Isolation procedures between IEDs and secondary circuit of instrument transformers will be simplified. As all connections at secondary level are based on FO, replacement of IEDs can be easily done without any hazardous conditions such as open CT secondary.
- Cabling cost shall be reduced by approximately 40% in Process Bus implementation as compared to conventional CTs/VTs hardwired approach [1].
- CT burden will be independent from connected devices.

2. **Interlock**: Interlocking in switchgear ensures safe operation of equipment using the correct procedures, with almost no chance of error. Utilities can implement

fast Interlocking schemes using Generic Object Oriented Substation Event (GOOSE) functionality of IEC 61850, eliminating the need of hard-wired signals. As GOOSE works on Publisher-Subscriber Model, each message will be repeated several times to ensure reliable delivery of messages in Ethernet network.

3. **Protection systems**: Nowadays Utilities has several digital substations where IEC 61850 has been implemented up to a great extent. Yet some of the substations still have electromechanical/static relays installed for protections typically for line differential protection.

Electromechanical technology has certain disadvantages like lack of self-supervision, drifting functional characteristics due to component ageing, need for periodic testing at frequent intervals, lack of communication features resulting in limited automation functions and above all product being discontinued due to obsolescence of the technology leading to challenges in life cycle management.

It is recommended for these Utilities to have numerical IED's which offers feature such as its improved operating characteristics ensuring secure operations, possibility to include other local measurements for decision making in case of communication failure, flexibility and self-supervision and so on. For example, numerical line differential protections are available which are based on fiber optic connections, unlike electromechanical relays which work based on pilot wire between stations.

2.2 System Architecture

In general, IEC 61850 based SAS architecture can be logically divided into four levels as described below.

- Control Centre level: It provides overview of several substations acquiring data from gateways of individual substations using protocols like IEC 60870-5-101, IEC 60870-5-104 and so on.
- Station level: Station level mainly consists of the station servers, Operator and Engineering Workstations, GPS receiver and gateway interfaces for remote communication.
- Bay level: This level consists of all devices responsible for protection (IED), monitoring and control of substation equipment (energy meter, disturbance recorder).
- Process level: The purpose of this level is to acquire data for protection and control. The equipment found at this level includes all the switchgear, current and voltage transformers and merging units.

So IEC 61850 based substations has two levels of Ethernet communication networks, as described below.

- Station Bus: Communication network responsible for the connection between station level devices such as station computer, gateway and bay level devices such as protection and control IEDs
- Process Bus: Communication network responsible for the connection between process level devices such as CTs and VTs and bay level devices such as protection and control IEDs.

Many power Utilities have implemented redundancy in their Station Bus communication network through ring architecture, where the IEDs are connected in a ring. It offers greater advantages such as self-healing systems and increased availability. Implementation of a typical ring architecture is shown in Figure 2. Dis-advantages with such type of implementation is mentioned below:

- Only supports single point of failure.
- Failure of more than one IED in a ring would affect communication with other IEDs connected between the failed IEDs.

Some Utilities have used combination of star and ring topology where all Bay level devices such as IEDs are connected with Ethernet switches in a star architecture and the Ethernet switches of the substation are connected in ring architecture with the implementation of Rapid Spanning Tree Protocol (RSTP) for managing redundancy.

For substation with significantly high number of IEDs, combination of star and ring topology can be implemented. In this architecture, fault in one IED will not disrupt the communication with other IEDs. Also maintenance is easy, since the faulty IED, can be isolated from the rest of network for maintenance, without affecting the functioning of rest of the network. However this topology also possesses some limitations as mentioned below:

- Failure of one switch will disrupt the communication of all IEDs connected to that switch.
- As an IED is connected to the Ethernet network via a single port, failure of the communication port or cable will disrupt the communication of the IED.

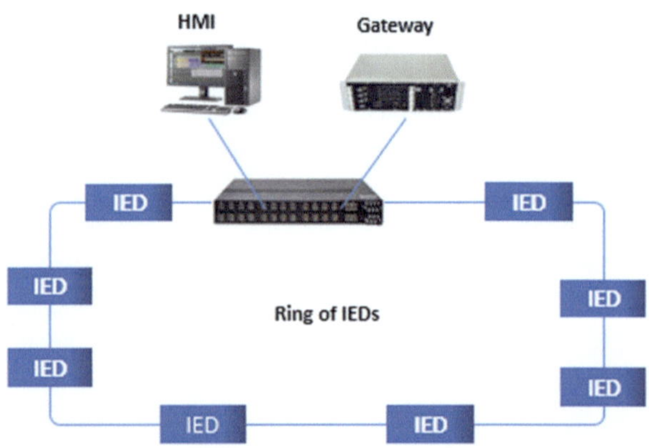

Fig. 2 Ring architecture

Topology must be selected depending on the network recovery time requirements. Time for which the substation tolerates an outage of the automation system is called the "grace" time. The network recovery time should be lower than the grace time.

As per IEC 61850-5:2013 standard recovery time requirements are mentioned in Table 1 [2].

To meet these requirements, IEC 62439-3:2016 [3] lists Parallel Redundancy Protocol (PRP) and High level Seamless Redundancy (HSR) which increases the availability of automation network through redundancy and offering zero recovery time. The same is also referred in IEC 61850-90-4:2013. These protocols are applicable to substations of any size and topology for the station bus as well as for the Process Bus. Typical example of PRP network is shown in Figure 3.

Table 1 Recovery time requirements

S. No.	Communication	Service	Bus type	Recovery time (ms)
1	SCADA to IED e.g. command information	IEC 61850-8-1 MMS	Station bus	100
2	IED to IED interlocking	IEC 61850-8-1 GOOSE	Station bus	4
3	IED to IED reverse blocking	IEC 61850-8-1 GOOSE	Station bus	4
4	Bus bar protection	IEC 61850-9-2 Station bus	Station bus	0
5	Sampled values	IEC 61850-9-2 Sampled values	Process bus	0

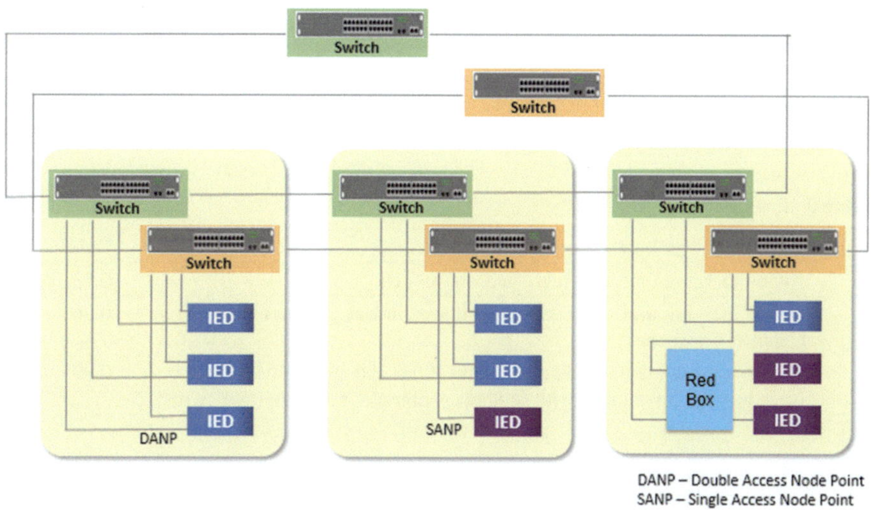

Fig. 3 PRP network architecture

The features of PRP and HSR are listed in Table 2 [4].

In conventional hardwired substations at the process level, the voltages and currents are scaled down using voltage and current transformers and are sent to protection relays through hardwired cables.

In substations having Sample Value based Process Bus, the analogue current and voltage signals are replaced by digital signals known as SV as per IEC 61850-9-2:2011. Measured values from CTs, VTs and switchgears are continuously transmitted over the process bus using devices such as Merging Units, which are used by all connected devices for protection, measurements, metering or monitoring.

There are various ways in which redundancy can be applied for the devices connected in the Process Bus. Reliability for redundancy implementation of devices involved with Process, for 3 types of architecture are given Table 3 [5].

Some of the challenges in implementing process bus are:

- High bandwidth requirements ranging from 4.2 to 5.8 Mb/s. [6].
- Time synchronization accuracy of 1 μs as per IEC 61850-9-2 which can be supported by IEEE 1588.

It is recommend to implement Process Bus which has greater reliability, safety when compared to conventional wiring.

Table 2 PRP & HSR comparison

S. No.	Feature	PRP protocol	HSR protocol
1	Topology	Double star	Ring
2	Flexibility of topology	High	Less compared to PRP
3	Redundancy	Best	Best
4	Cost	High	Low
5	Bandwidth utilization	Low	High
6	Process bus	Possible	Possible

Table 3 Process bus reliability

S. No.	Network topology	Reliability
1	One merging unit, one relay and one switch	0.931
2	One merging unit and one relay with redundant Ethernet ports, and two switches	0.965
3	Two merging units (redundant acquisition) and two relays (redundant protection) with redundant Ethernet ports and two switches	0.999

2.3 Time Synchronization

Utilities can implement three different types of time synchronization methods,

- Inter-range instrumentation group time codes B (IRIG-B)
- Simple Network Time Protocol (SNTP)
- Precision Time Protocol (PTP)—IEEE 1588/IEC 61588:2009.

The typical implementations of IRIG-B timing protocol provides accuracy in the range of 1 ms. IRIG-B installations require dedicated twisted pair cable to transport the timing signals and a single output can drive a limited number of devices, depending on device load & cable length. The SNTP can synchronize computers over a local area network with time synchronization accuracy of 1–10 ms. But the accuracy of IRIG-B and SNTP will not be sufficient for applications such as process bus or synchrophasors. The process bus implementation requires 1μs time synchronization accuracy. The IEEE 1588 standard defines Precision Time Protocol (PTP) that provide time synchronization accuracy in the range of 1 μs.

Typical SNTP method of implementation architecture is shown in Figure 4.

Typical IEEE 1588 method of implementation architecture is shown in Figure 5.

The comparison of different time synchronization methods available for substations are shown in following Table 4 [7].

It is recommended to implement IEEE 1588 time synchronization standard for substations where Process Bus implementation is planned.

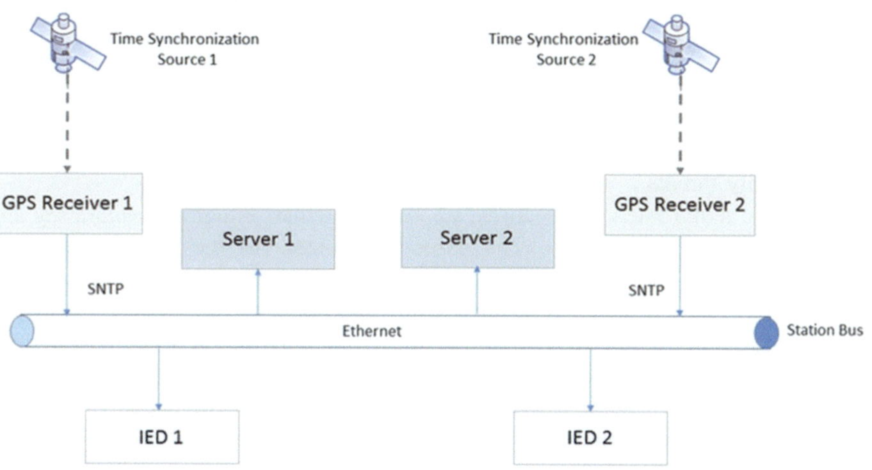

Fig. 4 SNTP architecture

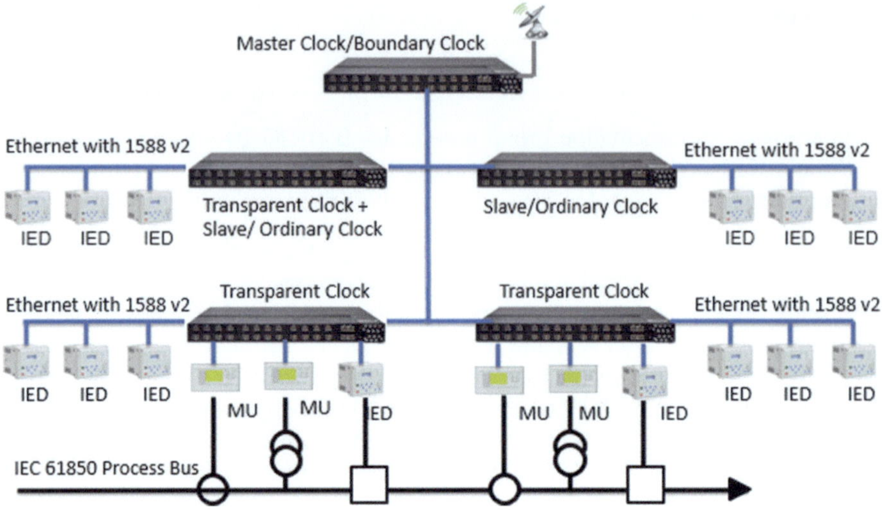

Fig. 5 IEEE 1588 architecture

Table 4 Time synchronization accuracy

S. No.	Time synchronization method	Typical accuracy
1	IRIG–B	1 ms
2	Network time protocol/SNTP	1–10 ms
3	IEEE 1588	1 μs

2.4 *Cyber Security*

IEC 61850 implementation provides advanced features for control & monitoring of substations increasing availability and flexibility. At the same time it introduces cyber security threats.

The basic requirements of security are confidentiality, integrity, availability and non-repudiation of SAS data. The associated threats to the above mentioned requirements and possible attacks for corresponding threats are shown in Figure 6.

The standards and methods developed for traditional IT may not be applicable for SAS, for example the installation of anti-virus software on server can significantly affect the performance of a SAS. The technical specification of IEC 62351 aims to secure IEC 61850 based communications [8, 9]. The scope of the IEC 62351 series is information security for power system control operations with a primary objective of undertaking the development of standards for security of the communication protocols defined by the IEC 61850 series.

General methods of cyber security controls for SAS, to avoid/minimize cyber security attacks are mentioned below along with Figure 7.

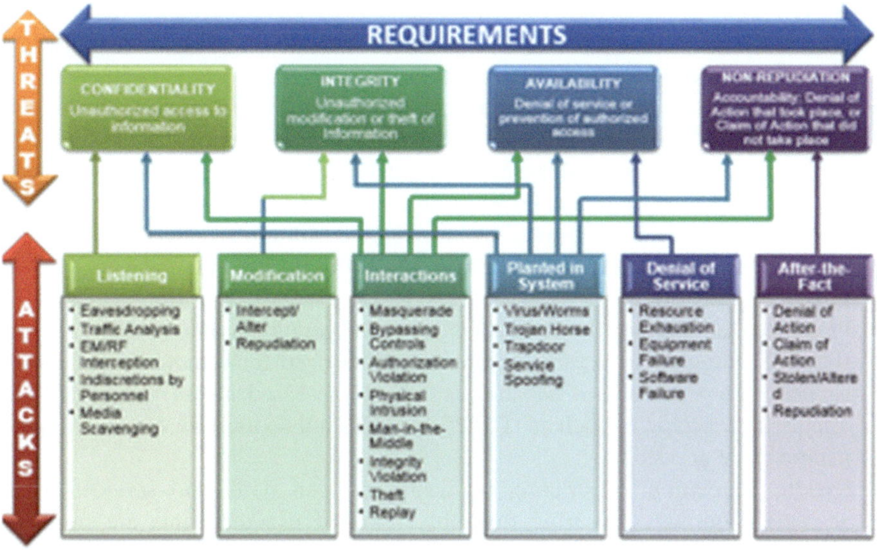

Fig. 6 IEC 62351 cyber security requirements

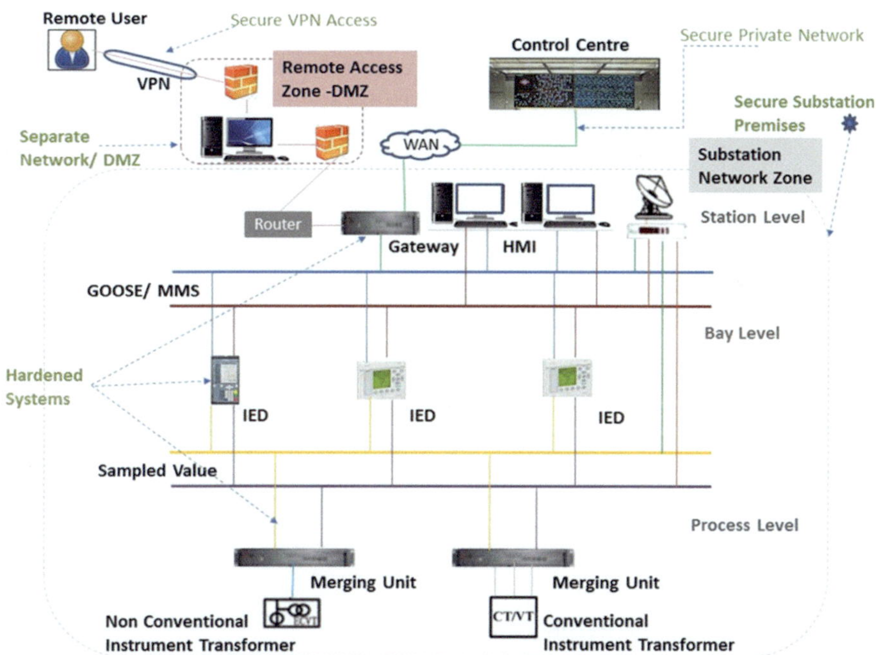

Fig. 7 Cyber security implementation

- Access Control: It is based on the principle of least privileges for users or process. It involves centralized authentication and logging, comprehensive password protection at different levels.
- Secured Remote Access form External Network: The substation automation network should be clearly separated from external network to restrict data access. In order to authenticate accessing entities firewall with a Virtual Private Network (VPN) [10] gateway can be implemented. A more secure multi-layer architecture can be implemented by creating De-Militarized Zone (DMZ) which separates two networks and serves as a proxy between external network and substation network.
- Protected Communications: Details of configuration/maintenance of IEDs should be logged for security/future use. HMI used for remote access should also have proper anti-virus installed as well as role-based access control enabled. The security protocols such as HTTPS or SSH shall be used for engineering and maintenance access.
- System Hardening: Restricting applications, open ports and services to an absolute minimum. Ensuring that only required user accounts are installed. Enforcement of strong authentication with regular password management, privilege control etc.
- Backup and Recovery: Regular backups of critical data, to restore business operation with minimum effort, in case of cyber security attack. Maintaining multiple copies of backup in multiple locations. Business continuity plan identifying risks and response scenarios.
- Encryption: Data exchanged on a serial link between the IEDs and the gateway can be secured by encryption devices. To secure data exchanges with control centres and remote users, use of VPN form of encryption in substation gateways, which enables IP traffic to travel securely over a public TCP/IP network by encrypting all traffic from one network to another.

3 Maintenance Strategy

The objective of SAS maintenance strategy is to ensure system availability, reliability, safety and reduced maintenance effort. To achieve the objective, appropriate approach should be formulated which involves implementation of different maintenance methods and advanced technologies.

The main contributing methods/technologies followed across industry for SAS maintenance are shown in Figure 8.

3.1 Configuration Management

The configuration management is a methodology for maintaining backup of necessary configuration files, application software and as-built documents to increase

Fig. 8 Maintenance strategies

traceability. Configuration management will track the changes/modifications made to the SAS over the time.

3.2 Knowledge Repository and Training

A centralized database stores data from different substations and helps in analysis of faults, changes occurred to the SAS and decision making. The central knowledge repository maintains files/records such as Configuration backup files, Log files, Manuals, training documents, Architecture diagrams, SLDs, cable schedules, logic diagrams/documents, as-built documents and so on.

3.3 Reliability Centered Maintenance (RCM)

RCM utilizes predictive/preventive maintenance techniques with root cause failure analysis to detect and pinpoint the precise problems to avoid or eliminate re-occurrence of problem. In RCM methodology, all equipment in a substation are not considered as of equal importance as the failure rates, criticality and cost of each equipment types are different. It also considers component health condition assessed using component life cycle management.

3.4 Testing practices

The appropriate testing practice needs to be defined for SAS components to verify the functionality of equipment. For hardware/software modification particular part of Site Acceptance Test (SAT) procedure can be followed for verification of maintenance.

3.5 Spares Management

The objective of spares management is to achieve high availability of component and services pertaining to SAS. Utility should define spares for SAS taking into account following strategies:

- Mandatory Spares
- FSN Method
- Asset Health Assessment.

3.6 Advanced Maintenance

Utilities can implement Computerized Maintenance Management System (CMMS) using ERP tools for material management, plant maintenance to perform and track maintenance activities. Implementation of Virtual Private Network (VPN) based remote maintenance can also be enabled to quickly troubleshoot problems and perform maintenance activities.

3.7 Benchmarking/Metrics

Effectiveness of maintenance strategy implementation needs to be evaluated at predefined time interval mostly annually. Metrics such as equipment availability, corrective maintenance percentage, maintenance overtime, maintenance expenditure and so on should be monitored for controlling.

4 Summary

Standards and technology evolution are important aspect to for defining roadmap to digitize SAS in Utilities. This paper has focused on an assessment framework used to identify interoperability requirements, standards and guidelines for selection of underlying technology components. This will serve as practical reference for arriving at a customized approach to define SAS roadmap. Current implementation of DCS/SAS based on IEC 61850 standards along with their limitations are discussed in the paper. Also the technological advancements/trends in SAS, which can be adopted by the Utilities for future implementation are suggested along with the benefits of such implementation.

There is scope for further work in defining the SAS architecture, considering integration/extension of IEC 61850 based substation network with various Smart grid applications such as Distributed Energy Resources (DER), electrical storage systems, synchrophasors and so on.

Acknowledgements The authors would like to show their gratitude to Dr. Krishna Prasad, Mr. Manish Singh and Mr. Mangesh Pawar for their valuable support and guidance in completing the work.

References

1. https://library.e.abb.com/public/c1256d32004634bac1256e19006fd99c/PAPER_2002_02_en_The_Cost_Benefit_of_Modern_Substation_Automation_in_Electrical_HV_Installations.pdf
2. https://webstore.iec.ch/publication/6012
3. https://webstore.iec.ch/publication/24447
4. http://www.industry.siemens.com/datapool/industry/automation/Tech-Art/2013/FAV-336-2013-IA-SC/FAV-336-2013-IA-SC-V01_EN.pdf
5. Dutra CA, Zimath SL, Rolim JG (2014) Process bus reliability analysis. Published in: 12th IET international conference on developments in power system protection (DPSP 2014)
6. Ingram DME, Senior Member, IEEE, Campbell DA, Member Test and evaluation system for multi-protocol sampled value protection schemes, IEEE, Pascal Schaub, Member, IEC TC57 WG10, and Gerard Ledwich, Senior Member, IEEE
7. http://w3.siemens.com/mcms/industrial-communication/en/rugged-communication/technology-highlights/ieee-1588-precision-time-synchronization-solution-for-electric-utilities/pages/ieee-1588-precision-time-synchronization-solution-for-electric-utilities.aspx
8. IEC 62351- Part 1: Communication network and system security –Introduction to security issues
9. https://www.smartgrid.gov/files/IEC_62351_Parts_18_Information_Security_for_Power_System_Con_200907.pdf http://www.cisco.com/c/en/us/products/collateral/security/asa-5500-series-next-generation-firewalls/prod_white_paper0900aecd804fb79a.html

Utility Power Transformers' Performance Enhancement in Smart Grids

Ajit Khandekar, Amrit Mishra, Hardik Shah and Bhargav Dave

Abstract Power Transformers (PT) are one of the most critical components of a Transmission & Distribution (T & D) Utility in terms of its impact on network availability. Apart from criticality, it is one of the costliest equipment in a substation comprising up to 60% of total investment. Multiple operational and environmental factors such as overloading, moisture, heat, progressive ageing, and so on lead to deterioration of insulation (winding and bushing), oil quality and so on result in decreasing PT operational efficiency and lifespan. It is crucial to assess the health condition of such PTs continuously to realize improved efficiency, early fault detection & diagnostics, prediction of maintenance requirements, and reinvestment strategy. The solution is developed to monitor and predict the performance of PT. Data Historian/Cloud Platform acts as a data lake and model repository to fetch online parameters and offline test results integrating with multiple data sources such as SCADA, online Dissolved Gas Analyzer device, Bushing monitor and the Enterprise Resource Planning (ERP) system. Asset Health Index (AHI) for PT is calculated using a multi-criteria analysis approach to derive a quantifiable figure which represents the condition based health of the PT. AHI calculation not only considers online parameters such as DGA, Load, Bushing tan delta capacitance, but also the offline test results such as power factor, Degree of Polymerization, Polarization Index, Winding Resistance and so on. Duval Triangle and IEC Ratio techniques employed in the solution for detecting the incipient fault and diagnosing nature (thermal/electrical) and location (paper/oil) of fault when it occurs.

A. Khandekar (✉) · A. Mishra · H. Shah · B. Dave
Tata Consultancy Services Ltd., Nirmal Building, 9th Floor, Nariman Point, Mumbai 400021, Maharashtra, India
e-mail: ajit.khandekar@tcs.com
URL: https://www.tcs.com/

A. Mishra
e-mail: mishra.amrit@tcs.com

H. Shah
e-mail: hs.shah1@tcs.com

B. Dave
e-mail: bhargav.d@tcs.com

R. K. Pillai et al. (eds.), *ISGW 2018 Compendium of Technical Papers*, Lecture Notes in Electrical Engineering 580,
https://doi.org/10.1007/978-981-32-9119-5_7

The deployed solution can help the Utilities in: (1) Adopting condition based maintenance strategy moving away from schedule based maintenance; (2) preventing unwanted breakdowns; and (3) Prioritize planning to achieve optimal balance between maintenance and replacement strategy. This paper discusses about the challenges faced by the Utilities in managing the PTs, which has crossed significant life span and the solution to resolve the challenges, taking into consideration maintenance practice followed in Utilities, availability of measurement/test data in Utilities and recommendations of international standards.

Keywords Power transformer · Asset performance management · Asset health index · Duval triangle · IEC ratio · DGA

1 Introduction

Power Transformers (PTs) are one of the most critical components of a Transmission & Distribution (T & D) power grid. Failure or breakdown of it has great impact on power network, since it may lead to power outage. Apart from its criticality, it is one of the costliest equipment in substation comprising up to 60% of total investment.

The average lifespan of transformer is about 25–40 years [1, 2], however, it depends on characteristics of operation and maintenance. Multiple operational and environmental factors such as overloading, moisture, heat, progressive ageing, and so on lead to deterioration of insulation (winding and bushing) and oil quality results in decreasing operational efficiency and lifespan.

This paper describes the challenges faced by the Utilities for operating and maintaining the PTs and the solution developed to improve asset health by early fault detection & diagnosis. Asset Health Index (AHI) is calculated for assessing health condition of PTs, considering only the parameters, which are measurable/quantifiable.

2 Challenges Faced by the Utilities

PT operate at high voltage level and supply power to a large/critical customer base within a Utility. The challenges faced in maintaining PT are:

- Ageing PT require frequent maintenance.
- Lack of visibility of actual asset health results in poor asset planning, in terms of defining continuation/replacement strategy.
- Unpredictable failure/breakdown leads to high loss of revenue.
- Schedule based maintenance causes unnecessary shutdown/unavailability of asset.

3 Solution

To address these challenges, the solution is developed for Health Monitoring and Analysis of PT, after studying availability of data and the maintenance practice of Utilities in general.

High level architecture of the solution is shown in Fig. 1. Data Historian/Cloud Platform acts as Data Lake and model repository, having integration with various sensors/devices/systems for fetching online parameters and offline test results as mentioned below:

- Dissolved Gas Analyzer: Real time concentration of dissolved gases in transformer oil.
- SCADA: Real time voltage, current, loading, oil and winding temperature.
- Bushing Monitor: Real time capacitance, tan delta, power factor.
- ERP System: Results for various tests performed offline such as Polarization Index, Turns Ratio, Winding Resistance and so on.

The solution is scalable, based on open standards and platform agnostic. It can be extended to any Data Historian and/or Cloud (IOT platform) integrating with most of the sensors/devices/applications.

The analytics developed in the solution are as follows:

- Transformer overloading analysis.
- Transformer On-Load Tap Changer (OLTC) analytics
- Transformer winding Hot Spot analysis
- Fault Identification and diagnosis using Dissolved Gas Analysis (DGA)
- PT Asset Health Index
- Asset Planning and Management.

The analytics which are of prime importance are described in the subsequent sections.

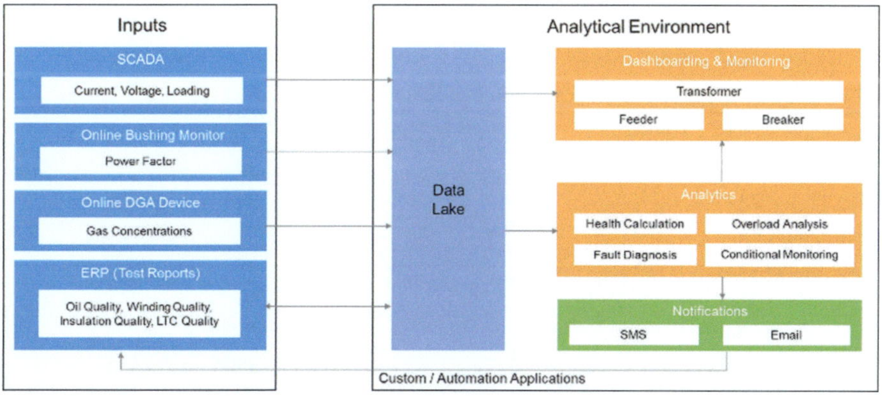

Fig. 1 Solution architecture

4 Dissolved Gas Analysis

The DGA method is used to analyze the concentration of gases in transformer oil, formed by decomposition of oil and paper insulation [3, 4].

DGA technique is a reliable method and widely used to detect incipient faults which may occur in transformers such as partial discharge, thermal fault and electrical fault [2, 5, 6].

To derive insight regarding nature or type of probable fault from concentration of gases, two widely accepted methods: Duval Triangle and IEC Ratio techniques are implemented in the solution. Out of the recommendations given by two methods, the Utility should consider the recommendation which is critical in nature, being in the conservative side.

4.1 DGA by Duval Triangle Method

Duval Triangle is one of the accurate and dependable method to analyze DGA [2, 5]. The method provides visual representation (different Triangles) of developing fault in transformer insulation and recommends nature, severity and location of the fault.

Each side of Triangle represents one particular gas. Each Triangle is divided into predefined regions based on concentration of the gases. Each region is associated with certain type of fault and severity. Based on the concentration of gases measured from the DGA device, the resultant point will be plotted in specific region of the Triangle, signifying the nature of fault. Before applying the method, at least one of the individual gases must satisfy some preconditions in terms of gas concentration (L1) and gas generation rate (G2) beyond safe limit as mentioned in Table 1 [2, 5].

Duval Triangle 1 uses 3 gases CH_4, C_2H_4 and C_2H_2 to predict the nature of fault. As a limitation, Triangle 1 can not specify the fault location. Triangle 4 & 5 are used to further diagnose the fault location, whether in oil or paper [7]. The fault in paper will always have higher severity as compared to fault in oil. Duval Triangle 4

Table 1 Gas concentration and generation rate limits

Gas	Limit L1 (ppm)	Limit G1 (ppm/month)	Limit G2 (ppm/month)
H_2	100	10	50
CH_4	75	8	38
C_2H_2	3	3	3
C_2H_4	75	8	38
C_2H_6	75	8	38
CO	700	70	350
CO_2	7000	700	3500

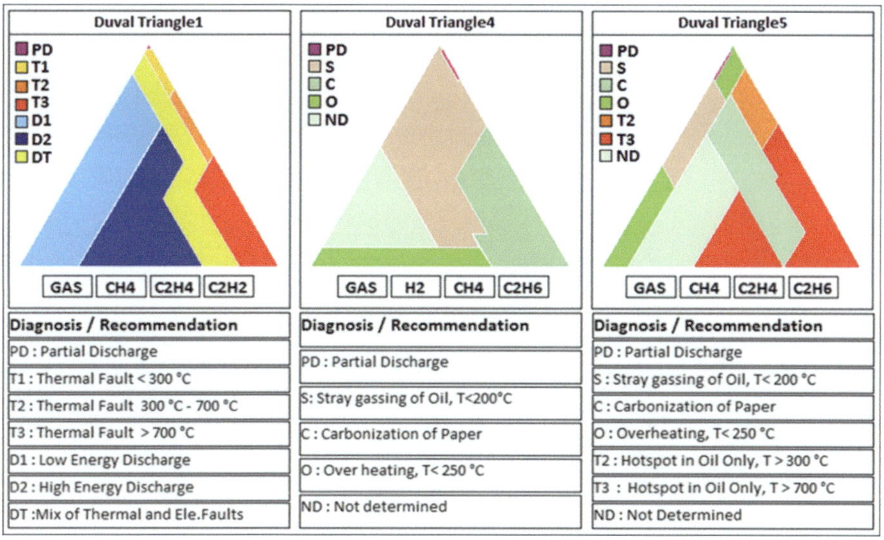

Fig. 2 Fault diagnosis using Duval triangle

uses H_2, CH_4 and C_2H_6 gas that are formed predominantly by faults of low energy or temperature. Triangle 5 uses CH_4, C_2H_4 and C_2H_6 gas that are formed predominantly by faults of high energy or temperature.

Figure 2 shows the Duval Triangle 1, 4 and 5 with different regions and corresponding fault type.

4.2 DGA by IEC Ratio Method

IEC Ratio is another method for fault diagnosis of PT based on DGA. The method compares quantities of different gases by dividing one into the other [8].

At certain temperatures, generation of one gas will be more than another gas. Based on this relationship it can be determined that if a certain ratio exists, then a specific temperature has been reached and accordingly type of fault can be predicted.

IEC Ratio uses three gas ratios, which are: C_2H_2/C_2H_4, CH_4/H_2 and C_2H_4/C_2H_6, where each ratio is quantized to classification code 0, 1 or 2. Out of the total combination possible based on these codes, there are only some combinations for which fault diagnosis is possible and rest of the combinations are non-decisive. The IEC Ratio codes and probable fault types for different combination of codes are given in Tables 2 and 3 respectively.

Table 2 IEC ratio codes

Rages of the gas ratio	Codes of different gas ratios		
	C_2H_2/C_2H_4	CH_4/H_2	C_2H_4/C_2H_6
<0.1	0	1	0
0.1–1	1	0	0
1–3	1	2	1
>3	2	2	2

Table 3 Fault types according to the IEC gas ratio codes

Fault type number	Fault type	C_2H_2/C_2H_4	CH_4/H_2	C_2H_4/C_2H_6
1	No fault	0	0	0
2	<150 °C Thermal fault	0	0	1
3	150–300 °C thermal fault	0	2	0
4	300–700 °C thermal fault	0	2	1
5	>700 °C thermal fault	0	2	2
6	Low energy PD	0	1	0
7	High energy PD	1	1	0
8	Low energy discharge	1 or 2	0	1 or 2
9	High energy discharge	1	0	2

5 Transformer Health Index

Manufacturers often define the anticipated life of PT to be 25–40 years. However, few transformers continue to operate much beyond without any trouble and few transformers breakdown before the anticipated life. So it is very much essential to know the health of the asset for maintenance and replacement planning and strategy.

Several analysis has been done for condition assessment and life management of PT. However, most of them attempt to model the insulation life, mainly the paper insulation only considering temperature and DGA, without taking into account routine test results and maintenance data.

The Health Indexing represents a practical tool that combines the results of operating parameters, field inspections, site and laboratory testing into a quantitative index, providing overall health of asset. It signifies long term degradation or remaining asset life. In turn help in identifying if the asset are at or near to the end of-life and are at high risk of failure. As a result of which, health indexing has become an increasingly popular method for performing asset health assessment.

5.1 Parameters Considered for Health Index Formulation

Some work has already been carried out to derive AHI based on Kinectrics model and recommendations from IEEE, IEC, CIGRE Working Group [1, 9], however trying to be comprehensive these method takes into account many parameters to compute AHI. In practical scenario it may not be feasible for Utilities to have all required data. Hence in the solution developed, the AHI model is tuned as per the data generally available for the transformer, in Utilities.

In the newly developed health model, Degree of Polymerization (DP) is considered in place of Furan. All the parameters related to visual inspection e.g. Infrared Thermography are not considered, as result of these totally dependent on the perception of the person conducting test. Similarly, all the parameters based on count of past corrective maintenance work orders for transformer auxiliary component such as main tank corrosion, cooling equipment, Oil tank corrosion, foundation, grounding, gaskets & seals, connectors, oil leak, oil level are not considered. Reason being, the Utility do not always have separate work orders for auxiliary component and are addressed during major maintenance work/annual maintenance. For the sake of simplicity, the overall condition is considered which takes into account the total count of maintenance work order for the transformer and its increasing trend. The new AHI model takes into consideration only the parameters, which are measurable/quantifiable.

The complete list of parameters considered for calculation of AHI is mentioned in Table 8. Details of few parameters are described below.

5.1.1 Dissolved Gas Analysis (DGA)

DGA technique is a reliable method to access condition of transformer. Temperature rise in the transformer depends upon the nature of fault developed. Formation of gas start at a specific temperature. So concentration of gases inside transformer oil depicts different fault conditions within the transformer.

Recommended alarm level for hydrocarbon gases has been given in different standards such as IEEE, IEC, Dornenburg, and Bureau of Reclamation. Considering the different recommendations, Table 5 introduces a ranking method developed using the DGA data [1, 9].

DGA factor can be calculated by Equation:

$$DGAF = \frac{\sum_{i=1}^{7} S_i \times W_i}{\sum_{i=1}^{7} W_i} \tag{1}$$

where S_i is the score of each gas based on Table 4 and W_i is the weightage factor.

Table 4 Scoring and weight factors for gas levels (ppm)

Gas	Score (S_i)						W_i
	1	2	3	4	5	6	
H_2	≤ 100	100–200	200–300	300–500	500–700	>700	2
CH_4	≤ 75	75–125	125–200	200–400	400–600	>600	3
C_2H_6	≤ 65	65–80	80–100	100–120	120–150	>150	3
C_2H_4	≤ 50	50–80	80–100	100–150	150–200	>200	3
C_2H_2	≤ 3	3–7	7–35	35–50	50–80	>80	5
CO	≤ 350	350–700	700–900	900–1100	1100–1400	>1400	1
CO_2	≤ 2500	2500–3000	3000–4000	4000–5000	5000–7000	>7000	1

Table 5 Transformer rating based on DGA factor

Rating code	Condition	Description
4	Good	DGAF < 1.2
3	Acceptable	$1.2 \leq$ DGAF < 1.5
2	Need caution	$1.5 \leq$ DGAF < 2
1	Poor	$2 \leq$ DGAF < 3
0	Very poor	DGAF ≥ 3

5.1.2 Load History

It is advisable to load the PT near its rated load value. Overloading leads to significant deterioration in insulation health.

In other models, monthly peak load is considered for computation of load history [1, 9], based on the assumption that change in loading of a transformer is not very frequent. To increase accuracy in computation of load history, the method takes into account hourly peak load.

The load history is categorized according to the five groups listed below:

N0: Count of S_i/S_B that are lower than 0.6
N1: Count of S_i/S_B that are between 0.6 and 1
N2: Count of S_i/S_B that are between 1 and 1.3
N3: Count of S_i/S_B that are between 1.3 and 1.5
N4: Count of S_i/S_B that are bigger than 1.5.

Where S_i is the hourly peak load and S_B is the rated loading of the transformer.

Based on values of N0 to N4, Load Factor (LF) is calculated using following equation:

$$LF = \frac{\sum_{i=0}^{4}(4-i) \times Ni}{\sum_{i=0}^{4} Ni} \quad (2)$$

Rating for condition of PT based on load factor is mentioned in Table 6.

Table 6 Load factor rating codes

Rating code	Description
4	LF $\geq$ 3.5
3	2.5 $\leq$ LF < 3.5
2	1.5 $\leq$ LF < 2.5
1	0.5 $\leq$ LF < 1.5
0	LF < 0.5

5.1.3 Degree of Polymerization

Degree of Polymerization (DP) is one of the most dependable means of determining paper deterioration and remaining life of insulation. The cellulose molecule in paper insulation is made up of a long chain of glucose rings which form the mechanical strength of the molecule and the paper insulation. DP is the average number of these rings in the molecule. With ageing and deterioration because of heat and moisture, bonds between the rings tend to break, as a result the number of these rings decreases. Thus value of the DP of the paper indicate its mechanical strength.

Condition rating of PT on the basis of DP value, as mentioned in Table 7, was developed based on the recommendations given in the literature [2, 10] and maintenance practice followed in the Utilities.

5.2 Health Index Calculation

Weightage is assigned for each parameter based on its relative importance. The concerned PT is rated against a set of criteria for each parameter to produce Health Index Factor (HIF), which has range from 4 to 0. Total score is calculated by summing the multiplied value of weightage with corresponding HIF for each parameter. Total score is further divided by the maximum possible score and then multiplied by 100 to compute the AHI. This way the AHI is normalized within 100 to 0, where 100 signifies the perfect condition and 0 signifies the worst degraded condition. The equation for computation of AHI is as follows [1, 9]:

Table 7 Transformer rating based on degree of polymerization

Rating code	Degree of polymerization(DP)	Inference
4	$\geq$ 1000	Paper practically unaged
3	601–999	Paper slightly aged
2	401–600	Paper medium aged
1	201–400	Paper strongly aged
0	$\leq$ 200	Paper extremely aged

$$HI = 60\% \frac{\sum_{j=1}^{11} W_j HIF_j}{\sum_{j=1}^{11} 4W_j} + 40\% \frac{\sum_{j=12}^{13} W_j HIF_j}{\sum_{j=12}^{13} 4W_j} \quad (3)$$

The AHI model give 60% weightage to PT main tank and 40% weightage to OLTC, based on the survey conducted by CIGRE. However, CIGRE suggests that, if the Utility has different statistics regarding failure rate caused by the OLTC, then the weightage should be applied accordingly. The AHI parameters and their weightage are given in Table 8.

Based on AHI, PT condition can be analyzed as shown in Table 9.

5.3 *Result of AHI Model*

The model was applied to transformers which were judiciously selected with wide range of age group ranging from 2 years to almost 35 years to verify the effectiveness of the model.

The AHI result generated by the model are shown in Fig. 3, in the form of a trend with respect to age. Based on the generated data points, a linear regression trend has been plotted.

The "goodness of fit" of the model is represented by the coefficient of determination (R^2). The R^2 for generated set of data is calculated to be 66.2%.

It was observed that the transformers with age between 0 and 10 years, which are in very good condition and well maintained, has AHI around 90.

Table 8 Condition criteria weights for scoring

No.	Parameters of health index	Weightage (W)	Health index factor (HIF) 4 to 0
1	DGA	10	4, 3, 2, 1, 0
2	Load history	10	4, 3, 2, 1, 0
3	Power factor	10	4, 3, 2, 1, 0
4	Oil quality	6	4, 3, 2, 1, 0
5	Overall condition	8	4, 3, 2, 1, 0
6	Degree of polymerization	5	4, 3, 2, 1, 0
7	Transformer ratio	5	4, 3, 2, 1, 0
8	Leakage reactance	8	4, 3, 2, 1, 0
9	Winding resistance	6	4, 3, 2, 1, 0
10	Core to ground	2	4, 3, 2, 1, 0
11	Bushing condition	5	4, 3, 2, 1, 0
12	OLTC DGA	6	4, 3, 2, 1, 0
13	OLTC oil quality	3	4, 3, 2, 1, 0
Health Index (HI)			

Table 9 Condition of transformer estimate from asset health index

Health index	Condition	Requirement
85–100	Very good	Normal maintenance
70–85	Good	Normal maintenance
50–70	Fair	Increase diagnostic testing, possible remedial work or replacement needed depending on criticality
30–50	Poor	Start planning process to replace or rebuild considering risk and consequences of failure
0–30	Very poor	Immediately assess risk; replace or rebuild based on assessment

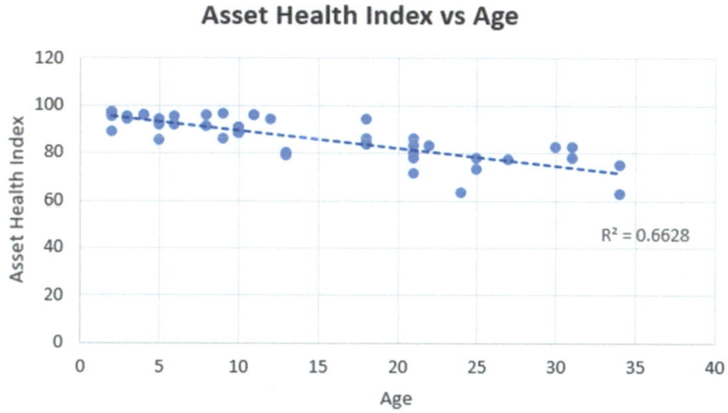

Fig. 3 Trend of asset health index versus age

Another observation shows considerable differences in AHI between assets of a similar age group. For example, a transformer with age 24 years has AHI of 63%. This transformer has low AHI when compared to other transformers of its age group. This was again verified with maintenance experience, past maintenance records and results for the concerned transformer.

So it can be concluded that the AHI is indicative of the accurate condition of a transformer.

6 Asset Planning and Management

It is a task of paramount importance for the Utilities to properly plan and prioritize the PT in hand in terms of maintenance, refurbishment and replacement.

Following the calculation of AHI to get the severity of health condition, it is necessary to find out the criticality of PTs. For computing criticality, factors such as numbers of customers affected, type of customer, cost of asset, availability of replacement/spare, availability of redundant PT/alternate network path and so on are considered. By combining the severity and criticality, risk factor is computed for PT. The risk factor gives a quantitative figure for prioritization of PT for the maintenance strategy/investment planning.

7 Conclusion

The AHI calculation presented here is a very useful tool for representing the overall health of a complex asset such as a PT. The Duval Triangle and IEC Ratio Method helps in detection of incipient fault developing in PT. The solution helps the Utilities in moving towards condition based maintenance and defining the asset maintenance and replacement strategy.

Loss of Life (LoL) of PT can be calculated as a part of future scope of transformer asset analytics.

Acknowledgements The authors would like to show their gratitude to Dr. Krishna Prasad, Mr. Manish Singh and Mr. Mangesh Pawar for their valuable support and guidance in completing the work.

References

1. Jahromi AN, Piercy R, Cress S, Service J, Fan W. An approach to power transformer asset management using health index. Kinectrics Inc., Transmission and Distribution Technologies
2. Facilities instructions, standards, and techniques volume 3-31 transformer diagnostics. United States Department of the Interior Bureau Of Reclamation
3. IEEE Std C57.104-1991 (1991) IEEE guide for the interpretation of gases generated in oil-immersed transformers. Transformer Committee of PES
4. IEC standard for Mineral oil-impregnated electrical equipment in service—guide to the interpretation of dissolved and free gas analysis (1999) IEC 60599 standard
5. Alghamdi AS, Muhamad NA, Suleiman AA. DGA interpretation of oil filled transformer condition diagnosis
6. Ciulavu CA. Techniques for incipient fault diagnosis
7. Duval M. New frontiers of DGA interpretation for power transformers and their accessories
8. Shrivastava K, Choubey A. A novel association rule mining with IEC ratio based dissolved gas analysis for fault diagnosis of power transformers. Int J Adv Comput Res
9. Naderian A, Cress S, Piercy R, Fan W, Service J. An approach to determine the health index of power transformers. Kinectrics Inc., Transmission and Distribution Technologies
10. Nejedly J, Newesely G. Evaluation Of The Extent Of Ageing Of Paper In Oil-Immersed Power Transformers. CIGRE

Roll Out and Learnings of Smart Grids (The French Experience)

M. Laurent Karsenti, M. Christophe Feuillard and M. Yann Toravel

Abstract The development of smart grid solutions offers new opportunities for DSOs to ensure continuity and quality of supply in a cost- efficient way. It will help them face the various challenges related to energy transition and answer the expectations from public authorities, such as the development of renewable energy sources and electric vehicles. This paper details first the approach adopted by Enedis to elaborate and implement its roadmap of Smart Grid technical solutions, from R&D concept to their industrial roll-out. It then describes Enedis roadmap and its two main objectives, which are to modernise network management processes and infrastructures and to provide assistance to stakeholders of the electric power system and regional territories in the French energy transition. Finally, the first industrialised solutions, aiming at bringing tangible benefits for all users of the distribution system, are detailed with their large-scale deployment strategy.

Keywords Smart grids · Smart meter · Roadmap · Roll out · Non technical losses

1 Introduction

India's electricity demand is the third largest globally, exceeding 1100 Twh. Strong economic growth, urbanization, rural electrification together with government's drive to ensure uninterrupted power for all will be key drivers.

In this context, the government of India has launched several programs to support, among others, smart cities in India. Energy projects in those smart cities

M. Laurent Karsenti (✉) · M. Christophe Feuillard
EDF International Networks, Paris, France
e-mail: laurent.karsenti@edf.fr

M. Christophe Feuillard
e-mail: christophe.feuillard@edf.fr

M. Yann Toravel
Enedis/NEDI Digital Program, Paris, France
e-mail: yann.toravel@enedis.fr

R. K. Pillai et al. (eds.), *ISGW 2018 Compendium of Technical Papers*,
Lecture Notes in Electrical Engineering 580,
https://doi.org/10.1007/978-981-32-9119-5_8

programs include smart grid approaches and AMI/Smart Meters. This creates opportunities to address some of the key challenges of Discoms, reduce AT&C Losses and prepare the grid to better integrate Renewables generation.

The French Experience in Smart Grids implementation can help understanding and dealing with those challenges in India.

EDF IN's expertise comes from Enedis, EDF's network operator and data manager. Moving towards digital grid, Enedis, a member of Think Smart Grid association, has successfully rolled-out 8 million Linky smart meters in France, with a target of 35 million by 2021.

DSOs are today facing several challenges which impact the way they manage their networks. Enedis, the French DSO that is operating 95% of the distribution grid in France, is particularly concerned by these challenges, which include among others the growing penetration of renewable energy sources onto the grid or the development of electric vehicles. These latter evolutions are changing the nature of power flows. Distribution grid operations are therefore becoming more complex and will have to be more active and dynamic than in the past.

In France, 90% of Decentralized Generation are connected to MV and LV networks in France. End of 2016, 9.9 GW of wind farms and 5.6 GW of PV plants were connected to the networks operated by Enedis and 110,000 charging points for electric vehicles were available, corresponding to nearly 670 MW. These volumes will dramatically increase in the years to come as a result of recent legislative and regulatory evolutions. The Energy Transition for Green Growth act, which has been adopted in 2015, has indeed set ambitious objectives:

- increase the share of renewable energy sources to 32% of the final energy consumption in 2030 and 40% of the electricity production,
- offer 7 million recharging points for electric vehicles by 2030.

In the context stated above, Enedis has decided to develop Smart Grid solutions [1] to face these challenges, while continuing improving continuity and quality of supply, and fulfilling their contractual and legal obligations in a cost-efficient way. These solutions under development by Enedis also represent opportunities to answer the strong expectations from public authorities, such as the French Energy Regulator (CRE), the Ministry of Ecology, sustainable Development and Energy, the Ministry of Industry and the European Commission.

In order to prepare the future of distribution networks, Enedis is involved in a series of Smart Grid R&D projects and demonstrators in France and Europe, with more than 100 academic and industrial partners. All these projects contribute to the development and experiment of innovative tools, materials and processes designed to improve the reliability of the grid. Enedis is currently participating in 19 Smart Grid demonstrators, including 5 European projects. As results of these experiments are becoming available, the DSO is progressively shifting from the experimental phase to the roll-out process.

2 A Structured Approach Based on Innovation

Enedis has adopted a 3-steps approach to develop a technical base of smart grid solutions (see Fig. 1):

- Develop and scope new solutions using R&D studies and start-up innovations,
- Experiment and evaluate new solutions on the field, in particular through the programme of demonstrators in which Enedis is participating,
- Prepare the roll-out of solutions which are technically and economically relevant, based on evaluations and lessons learned from the experiments.

Step 1: R&D and Open Innovation
The first step aims at exploring a large scale of technical possibilities, as well as accelerating experiments using Proofs of Concept. In addition to its R&D programme, Enedis has organised since 2015 contests with start-ups and innovative small-and-medium-enterprises, which may propose innovations to improve network management on various themes (big data & data analytics, new "3.0" connected tools for field technicians, observability and commandibility of the grid…). The solutions are then evaluated by a jury and the most promising ones receive a prize from Enedis.

Step 2: Experiments and Demonstrators
The purpose of the second step is to assess the technical and local economic value of smart grid solutions, through experiments on the field. The technical evaluation consists in analysing their performances, security aspects, or degree of integration/maintainability for instance.

Step 3: Industrialisation and Roll-Out
The results of these tests and analyses are studied in the third phase, for the promising solutions. Then, Enedis coordinates the necessary actions for the roll-out these solutions on the field, which include:

Fig. 1 Enedis approach for Smart Grid roll-out

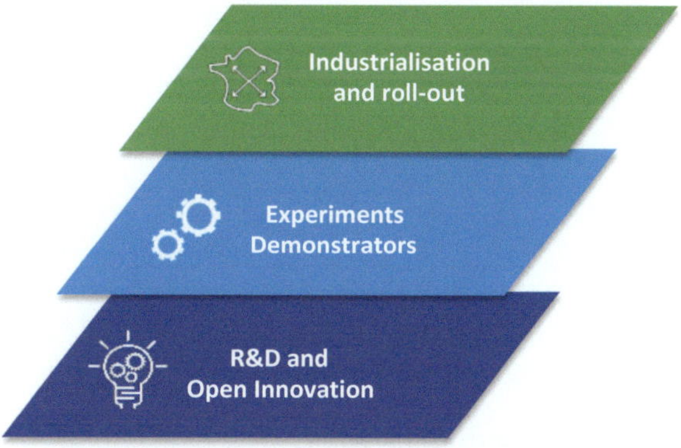

- a cost-benefits analysis on the basis of the benefits that would be generated by the roll-out of the solution at a national scale—on different types of networks (rural or urban areas for instance)—and the target cost of the industrial solution;
- the deployment strategy of the solution and the estimated annual volumes according to the cost-benefits analysis,
- the compliance of the solution deployment with the legal and regulatory framework,
- an analysis of the industrial policy (including sourcing, intellectual property…).

3 Enedis Smart Grid Roadmap

Using this approach, Enedis has elaborated a roadmap of technical Smart Grid solutions, to be implemented by 2020 [2, 3] (see Fig. 2). The DSO has mapped the most promising solutions according to their level of maturity and their position in the previous three-stage process (R&D, experiment, industrialisation), and defined their implementation trajectory. Made by Enedis technical experts and top management, the roadmap has been validated by the executive committee.

The roadmap covers all business domains (network operations, maintenance, assets and infrastructures) and is built around two objectives:

- **Modernise network management processes and infrastructures**, with predictive maintenance solutions (big data & analytics, monitoring of primary

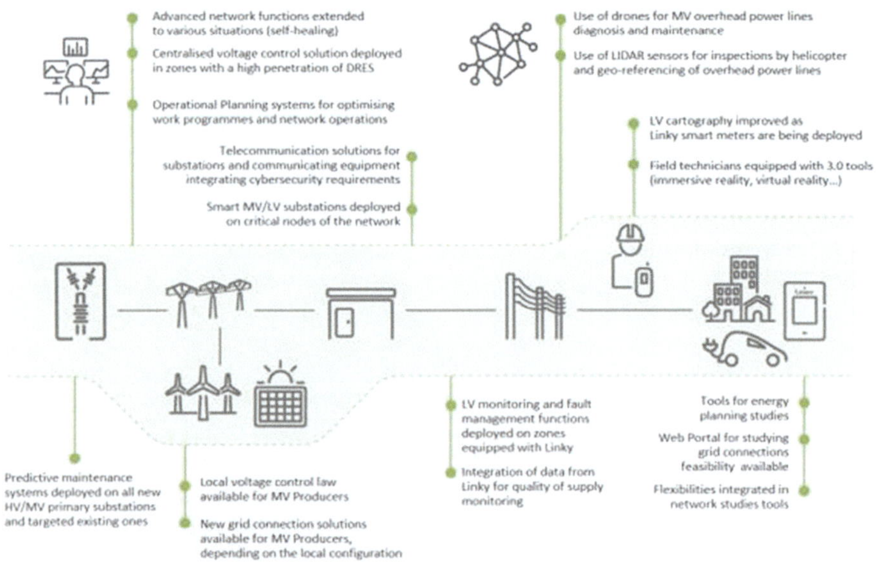

Fig. 2 Enedis smart grid roadmap

substations…), the improvement of its quality of service using new sensors, data, and automations, and the modernisation of its network assets,
- **Provide assistance to stakeholders of the electric power system or of regional territories in the French energy transition**, with innovative grid connection solutions and processes/automations for facilitating the integration of DRES [4].

To achieve these two objectives, Enedis has chosen an organisation based on a portfolio of projects, with their own governance structure.

The different technical solutions included in the roadmap are listed on the graph above.

Linky system, fully developed by Enedis, includes all the information systems: bidirectional data transmission, real time communication, data storage, information system for rollout and maintenance, and secure connection with all energy stakeholders. Investments are paid back by operational workforce adapted to remote operations. Moreover, one of the Enedis' key know-how consists of G3-PLC technology system, supported by an Alliance with more than 80 members worldwide.

These solutions will be deployed in two large-scale industrialisation pilots adopted within the French Electric smart grid programme "REI 6", for a New industrial France and decided by the Ministry of Industry. The projects are SMILE (Smart Ideas to Link Energy) in the West of France and Flexgrid in the South East of the country and their roll-out has already started. They will rely on a basis of technical solutions for the network, as well as the collection and use of data, various projects with regions, and local optimisation of their electrical resources.

The large-scale roll-out aims at preparing and testing the integration of solutions onto Enedis networks, as well as the staff ability to use them, thanks to relevant training programmes. Lessons learned from REI6 will help consolidate a first level of Smart Grid at a national scale by 2020.

Objective 1: Modernise Network Management Processes and Infrastructures
By 2020, Enedis will have integrated technological advances in sensors, telecommunications, and digital processing, in order to improve its industrial performance.

Maintenance will become more predictive and adapted to each kind of equipment, based on their health index. Sensors in primary substations, with data monitoring combined to big data models will help optimise maintenance operations. Drones equipped with optical sensors will facilitate the maintenance of overhead power lines. Thanks to enhanced reality and mobile solutions, Enedis staff will also have access to real time information on the field to operate networks more effectively, with higher safety standards.

New sensors, data, and automations will help improve quality of supply for all customers. New sensors and connected objects will be installed in various places on the grid. Network operations will become more dynamic, thanks to the data collected and to an increased number of information regarding the state of the network and assets. New solutions based on the Linky infrastructure will facilitate fault management on MV and LV networks. Advanced network functions will be extended to various situations, to manage fault in presence of important shares of DRES for instance.

Network assets will be modernised and will integrate technological advances (connected objects, LIDAR, data from Linky Smart meters…), in order to be more efficient and to take into account new requirements such as cybersecurity and data protection.

Objective 2: Provide Assistance to Stakeholders of the Electric Power System or of Regional Territories in the French Energy Transition
By 2020, Enedis will have developed tools and technical solutions to support producers, consumers, prosumers, and regions to meet the energy transition goals.

Innovative grid connection solutions will be available and adapted to the needs of each user: producer, storage facility, self-consumer, and electric vehicle supply equipment. MV Producers will be able to benefit from alternative grid connection solutions, with lower connection costs and delays in return for the capacity to modulate their power injection in case of network constraints.

Different solutions will be available to integrate production from distributed renewable energy sources. They will enable to manage dynamically network constraints and thus avoid systematic grid reinforcements, thanks to:

- an increased observability of production and consumption power flows which helps identifying potential constraints in operational planning, from annual works programme to real-time operations,
- new means of managing active/reactive power and voltage in order to solve constraints on MV networks, with local solutions or system-coordinated ones,
- reinforced coordination with other stakeholders of the electric power system.

Enedis will support regions in their projects and energy transition. They will be able to rely on new network development tools for implementing urban planning policies. Enedis will help regional authorities to make the most of their energy resources, by optimising the development and the real-time management of the network at a local level.

In conclusion, by 2020, Enedis will have industrialised technical smart grid solutions to achieve the two objectives described above.

4 Non Technical Losses and Smart Meter

A business case is studied in this paper concerning Non-Technical Losses (NTL) using Linky system. This use case has been studied in a benchmark Cired report [5].

The reduction of Non-Technical Losses (NTL) is one of the operational gains targeted from the roll-out of 35 Million ‘Linky’ smart meters.

This gain is especially linked with:

- The massive replacement of pretty old and electro-mechanical meters, which are likely to be responsible of previous NTL
- The roll-out of new smart-meters with remote index reading and remote operating, which allow the implementation of "smart processes" enhancing the new data of these meters for a long-term NTL reduction.

As such, Enedis has engaged two complementary innovative approaches on NTL reduction with smart-meters, in order to reach a concrete operational implementation from an experiment trajectory:

1. Analytics and other evolved data management methods associated to smart-meters, aiming at:

- An end-to-end control of smart processes to limit the anomaly cases in energy measurement (e.g., meter reading and billing) or in contract management (e.g., new contract, move-in move-out). These methods are progressively implemented with smart-meter roll-out.
- The development of datamining methods on smart-meters data, for favoring complex anomalies detection (e.g., frauds or meter malfunctions) and optimizing NTL management. These methods will gradually be tested and validated, in order to constitute the new process of "smart NTL detection".

2. A specific instrumentation of local LV network allowing a cross-treatment of both client and network data, aiming to:

- The search of residual NTL located before the meter: either with direct connection, or wit meter by-pass
- The identification of other IT or business anomalies, as connections errors in GIS database or certain types of power cut which are not related to network problems (e.g., failing meter).

These two approaches, either bottom-up from clients meter data or top-down from a global LV network energy measurement, will gradually complete and reinforce each other, for a better knowledge of LV networks.

Therefore, NTL reduction with smart-meters has to be carried out as a natural part of a smart-grid global strategy, with (i) the search of operational and techno-economical synergies in network instrumentation; (ii) the optimized management of all accessible big data (from smart-meters, network sensors, process indicators,…); (iii) the coordination of all processes (meter to bill, new contract, GIS,…) involved in business and IT anomaly treatment.

5 Conclusion

Enedis has started the industrialisation of Smart Grid solutions, by capitalising on lessons learned from implementation of all demonstrators in which Enedis has participated. The DSO is already implementing its Smart Grid roadmap and has rolled-out first industrialised solutions, using a three steps structured approach (R&D, experiments, roll-out preparation).

This approach is dynamic since the duration of each step may change from one solution to another, according to its simplicity/complexity of implementation.

Moreover, new solutions may constantly emerge and become part of the industrialisation process. Enedis Smart Grid roadmap is continuously enriched with new use cases and new solutions. Among new solutions, the use of flexibility levers to solve network constraints is currently studied and will be tested in local experiments. Another solution briefly presented in this paper is the use of the smart meter in order to deal with Non Technical Losses.

In addition, the industrialised technical solutions constitute pillars for smart grid offers to customers (producers, consumers, prosumers, customers with storage).

Our expertise developed through past and ongoing AMI and Smart Grid projects can help other stakeholders to gain efficiency on technical implementation, deployment schedule and cost optimisation for their own development plans.

References

1. Karsenti L, Daguzan P, Colin D, Carre O (2015) A step beyond French demonstrators: first approach in terms of Smart Grid solutions industrial development. The example of voltage regulation solutions tested in VENTEEA demonstrator. CIRED 2015, paper 1328
2. Karsenti L (2016) Enedis roadmap to industrialize new solutions to operate the distribution grid. European Utility Week 2016, Session 2 Technology & innovation thinking ahead, Operating the grid of the future
3. Karsenti L, Daguzan P (2017) Enedis approach for the roll-out of technical smart grid industrial solutions. CIRED 2017, paper 040
4. Beauné F, Carre O, Levaufre S, Karsenti L (2016) New solutions for a better DG integration to MV and LV networks. CIRED 2016, paper 0369
5. Working Group on Losses Reduction, CIRED WG CC- 2015-2, Reduction of technical and non-technical losses in distribution networks, 20/11/2017

Evaluation of Smart Metering Scenario and Demonstration of Smart LED Streetlighting Using PLC

Shobhit Jain, A. Paventhan, R. Mohan Kumar, Balmukund Prasad, Aman Jha and Manoj Kumar

Abstract Smart Metering and Smart lighting application are two major use cases utilizing Internet of Things (IoT) based heterogeneous network infrastructure supporting the two-way communication. In smart meter scenario the metering data is transferred through gateway and Data Concentrator Unit (DCU) at regular intervals and to central servers. The internet connected smart meter network makes it possible for the utility company to get customer's power consumption pattern in real time scenarios for applications like demand response, outage management system (OMS). IoT network for smart meter can be realized using heterogeneous last mile technologies such as power line communication (PLC), 802.15.4 g—2.4Ghz/sub-1Ghz, LoRa, NB-IoT etc. In this paper, evaluation of Device Language Message Specification (DLMS) meter data transport over the different scenario of heterogeneous communication technology in terms of response time, message latency, data throughput, and bandwidth has been analyzed. Further, smart streetlight deployment example is presented using the G3-PLC testbed deployment at STMicroelectronics campus. The testbed is working reliably last four months and exchanging data via G3-PLC.

Keywords Smart meter · LoRA · PLC · GSM/GPRS

S. Jain (✉)
India Smart Grid Forum, Bangalore, India
e-mail: shobhit@indiasmartgrid.org

A. Paventhan
ERNET India, Bangalore, India
e-mail: paventhan@eis.ernet.in

R. Mohan Kumar
Wearables Technologies, Bangalore, India
e-mail: mohan.k.rajagopal@ieee.org

B. Prasad · A. Jha · M. Kumar
System Lab, STMicroelectronics Pvt. Ltd., Greater Noida, Delhi, UP, India
e-mail: aman-kumar.jha@st.com

R. K. Pillai et al. (eds.), *ISGW 2018 Compendium of Technical Papers*,
Lecture Notes in Electrical Engineering 580,
https://doi.org/10.1007/978-981-32-9119-5_9

1 Introduction

Two of the major smart grid applications enabled through IoT networking are Advanced Metering Infrastructure (AMI) and smart street lighting. In India, utilities have already started deploying AMI on a pilot basis funded by Government of India and Ministry of Power. There are various last mile connectivity technologies for smart metering as recommended in BIS IS16444 standard including IEEE P1901 Power Line Communication (PLC) standard, and various low power wireless connectivity options such as IEEE 802.15.4 Low-Power Wireless Personal Area Networks (6LoWPAN), LoRa and NB-IoT cellular technologies. IEC62056 Device Language Message Specification and Companion Specification for Energy Metering (DLMS/COSEM) define the data model and application layer protocol for the metering to transport data by means of various application packets frames [1]. IEEE 802.15.4 g is one of the most popular data link and network layer wireless communication technology to access and communicate with smart meters. PLC offers wired communication option simplifying networking; however, its adoption in countries like India is not widespread.

IEEE has recently initiated PLC initiative [2] to study the PLC opportunities for India and to create PLC testbeds focusing on specific use cases. Powerline Intelligent Metering Evolution (PRIME) is one of the well-known PLC technologies adopted in Europe and PRIME deployments were found cost effective and seamlessly support DLMS/COSEM. The typical AMI architecture is shown in Fig. 1. The bi-directional communication between smart meters and utilities is enabled by Internet of Things (IoT) connectivity using various last mile heterogeneous technologies. The data concentrator is responsible for collecting and transferring all the data from the smart meters to utility Head End System

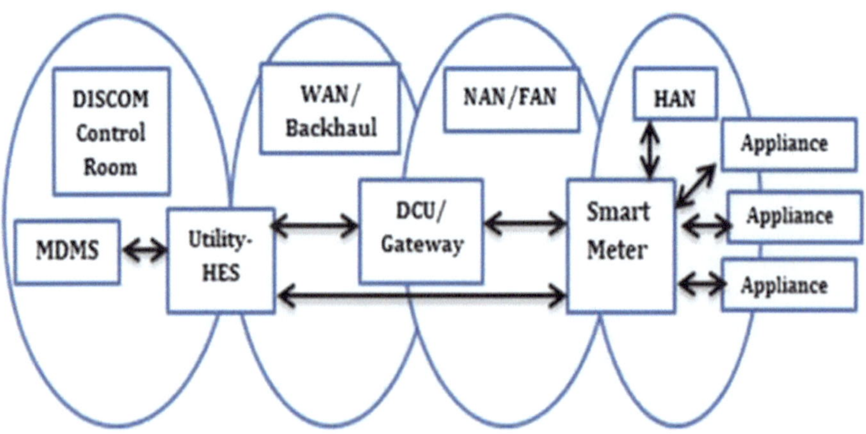

Fig. 1 AMI fundamental architecture block diagram

(HES) and Meter Data Management System (MDMS) using the Neighborhood Area Network (NAN) and Wide Area Network (WAN) [3].

The last mile communication technology either wired or wireless is chosen based on various factors such as technical feasibility, geographical area of deployment, cost etc. The communication module enabled in smart meter may be of PLC or RF for the NAN and cellular technologies or Fiber optics for the WAN [4]. In this paper, an analysis of DLMS data transport independent of various last mile technologies is presented in terms of computation and comparison of DLMS message for various packet size with respect to the MAC layer framing with different baud rates. Further, the deployment example of 20 smart LED streetlight is presented based on the G3-PLC network at STMicroelectronics campus. The power converter used for streetlight is developed in house and meets all power quality parameters with high efficiency and current regulation. The smart LED streetlight is working reliably since last four months as a pilot project.

2 Background and Related Work

Advance Metering Infrastructure (AMI) system is the combination of various technologies that measure, collect, communicate, aggregate and analyze energy usage data from metering devices. AMI system is able to collect data remotely from smart meter also it is used for interval metering for energy audit, theft and outage detection, remote connect/disconnect, web portal for residential and industrial consumers, prepaid metering, meter-to-home for In-home display. Smart meter data is transferred by data concentrator unit (DCU) [5]. The data is then delivered by various methods to the utility head end systems and MDMS for processing at a central location. The utility billing, outage management, and other application systems use the meter data for operational purposes. In a wireless mesh scenario, the smart meters talk to each other (multihop) to form a mesh network connecting to internet through a gateway. The gateway further helps transfer the meter data using various WAN methods to the utility servers. The mesh network for AMI uses 6LoWPAN, an IPv6 adaptation layer that is open standard, light weight, scalable, secured and interoperable [6]. The effort at Internet Engineering Task Force (IETF) working groups [7] in pushing IPv6 towards the idea of supporting Internet Protocol (IP) even in smallest devices has resulted in adoption of IPv6 to various physical layer technologies—Bluetooth Low Energy, DECT Ultra Low Energy, Power Line Communications, LoRaWAN, IEEE 802.15.4 g/Wi-SUN, NB-IoT, SIGFOX. IETF emerging drafts in supporting IPv6 over PLC leverages the inherent advantage in utilizing the existing electricity infrastructure by expanding PLC reach to more devices raising demand for IPv6 in future applications. Each IoT use case deployment will require understanding and evaluation of IPv6 over various link layer technologies considering multiple parameters such as subnet topology (star, mesh), mobility requirement, security, latency/QoS, data rate. For example, one communication technology cannot serve all the AMI deployment scenario.

AMI Deployments by Utilities in India

Some of the successful AMI pilot deployments by utilities in India are discussed below:

Chamundeshwari Electricity Supply Corporation (CESC) Mysore executed a pilot project that is funded by Government of India with industry solution providers supporting the system integration and consultancy. The functionality to be covered under this pilot project include advance metering, distributed automation, outage management, peak load management, condition based monitoring, mobile work force management, integration of renewable and distributed generation [8]. The overall benefits envisaged of this pilot study are loss reduction with key performance indicators (KPI) improvement like reduction in interruption durations, better energy utilization, aggregate technical and commercial losses. The AMI deployments are based on smart meters capable of two way communication with GSM/GPRS capabilities for to support last mile communication features.

The Tata Power along with industry consortium members has executed a pilot project for the deployment of AMI. The 6LoWPAN wireless technology is used for meter data acquisition along with head end server. The meter data management system handles the billing and fault management. The first phase of this project has a deployment of about 5000 smart meters.

Reliance Infrastructure has executed AMI pilot deployment of about 2500 m in Mumbai and Delhi with 99.5% success rate. These are first successful demonstration of smart metering pilots in India based on Low power RF with 6LoWPAN being used for data collection, transaction execution and connectivity. Both 3G cellular and fiber connectivity was used for WAN connectivity to back-end servers in AMI network. This project has leveraged the benefit of IPv6/6LoWPAN in offering open, standards compliant, light weight, scalable and interoperable Internet connectivity to smart metering applications. The internet facing Head-End System is deployed in cloud that can handle millions of transaction from AMI network on a real time basis with minimal latency and high reliability.

3 Analysis of DLMS Data Transport-Methodology & Scenarios

The smart meter communication infrastructure is mostly based on RF mesh network, PLC, cellular network or a combination of these. The communication network is expected to use suitable standards from ITU/IEC/IEEE/CEN/CENELEC/ETSI for NAN and WAN network. Communication network needs to provide reliable medium for two-way communication between various smart meter nodes and HES. RF based network can also be based on emerging sub-1 GHz 865–867 MHz license exempt frequency spectrum in India to support low power long range connectivity. The expected performance level by utilities largely depends on the quality of the network layer services that can be guaranteed. In order to help with an objective interpretation of smart meter data transport independent of various last mile network technologies, the following analysis of various smart meter data transfer scenarios are presented. The estimated time for smart metering data transfer using different profile scenarios are considered.

The metering parameters are identified and classified into the following profiles as per the DLMS specification:

a. Instantaneous parameter
b. Block profile
c. Daily load profile
d. Billing profile.

The measurement and computation of each of these parameters and events are based on standard methods and algorithms as per utilities demand. For each of the identified parameters, the Object Identification System (OBIS) code, interface class and the attributes are followed by each profile. The metering parameters are chosen based on classified estimated time scenario for smart metering using different profile as mentioned in the sections below. The following analysis is based on typical data rate for RF mesh and PLC scenario. However, the study can also help in extending to other communication technologies like LoRa, NB-IoT, LTE-M etc.

3.1 Billing Profile-(IC_3 Parameter = 13)

Billing load profile is an array of load survey data captured as a profile. Data is captured using some of the objects of this block load profile and interface class. The minimum and maximum data size are calculated for this particular profile and data transfer time is estimated assuming 9600 baud rate as shown in Table 1.

3.2 Daily Load Profile-(IC_3, Parameter = 2)

Daily load profile is an array of load survey data captured as a profile. This profile data is generated at the end of every 24 h. In this data capture, parameters and identified objects are based on daily load interface class three. The minimum and

Table 1 Time estimation with different data size

Baud rate	Data size (bytes)		Time (seconds)		Total Time (seconds) (for 100 m node)		15 min interval (for 100 m nodes)	
9600	Min	Max	Min	Max	Min	Max	Min	Max
	116	132	0.09	0.11	9.66	11	2.41	2.75

Table 2 Time estimation with different data size

Baud rate	Data size (bytes)		Time (seconds)		Total Time (seconds) (for 100 m node)		15 min interval (for 100 m nodes)	
9600	Min	Max	Min	Max	Min	Max	Min	Max
	28	42	0.02	0.03	2.33	3.5	0.58	0.87

maximum data size are calculated for this particular profile and data transfer time is estimated assuming 9600 baud rate as shown in Table 2.

3.3 *Block Load Profile-(IC_3, Parameter = 13)*

Block load profile is an array of load survey data captured as a profile. The OBIS code is 1.0.99.1.0.255 based on interface class as 7 but in this assumption the interface class 3 for all profiles. So at the end we have captured some of the objects of this block load profile and also captured few parameters based on interface class three. The minimum and maximum data size are calculated for this particular profile and data transfer time is estimated assuming 9600 baud rate as shown Table 3.

3.4 *Estimated Time of Daily Load Profile Scenarios*

Scenario 1:Single Smart Meter (100 Node) Data sending to Data Concentrator Unit

The data concentrator unit (DCU) is a gateway for communication of data between the Smart Meters. The DCU receives information from the smart Meter on scheduled times as required and stores data. The DCU provides the central link between smart meters by enabling continuous and periodic meter reading and control. DCU is capable of exchanging data from smart meters using RF, PLC and other communication technologies. The minimum and maximum data size for 100 smart meters is computed and by varying the data rate (9.6–512 kbps), the minimum and maximum data transfer time is estimated as shown in Table 4.

Table 3 Time estimation with different data size

Baud rate	Data size (bytes)		Time (seconds)		Total Time (seconds) (for 100 m node)		15 min interval (for 100 m nodes)	
9600	Min	Max	Min	Max	Min	Max	Min	Max
	36	72	0.03	0.06	3	6	0.75	1.5

Table 4 Time estimation with different data size

Data rate	Data size (100 m nodes)		Time required for single DCU	
(kbps)	Min	Max	Min (ms)	Max (ms)
9.6	5700	7900	4750	6583.33
64	5700	7900	712.50	987.50
256	5700	7900	178.12	246.87
512	5700	7900	89.06	123.43

Scenario 2: Data Concentrator Unit to Head End System

The main objective of HES is to aggregate meter data through DCU automatically avoiding any human intervention for onward transfer to MDMS. HES will ensure data integrity checks, for example, checksum, time check, pulse, overflow, etc. on all the meter data. HES is generally developed based on open platform following distributed architecture for scalability without degradation of the performance using additional hardware. HES generally support storage of raw meter data, alarms and alerts for minimum 3 days. The database approach to storage and security features are to be ensured by HES. In our analysis, the minimum and maximum data size for 100 smart meters are computed and by varying the data rate (1024 kbps to 4 Mbps), the minimum and maximum data transfer time is estimated as shown in Table 5. The connectivity between DCU and HES are typically based on cellular or terrestrial links, accordingly the data rate assumption are made.

Scenario 3: Estimation of Total Volume Data for Smart Meter

Smart meters are well known for their ability to provide meter read data at smaller intervals, such as every 15, 30 or 60 min, as well as bidirectional communication and remote operating capabilities. In addition to these features, smart meters also generate hundreds of meter events based on the profile. An event is information that originates from the meters endpoints and can have several attributes, including source and proxy information, severity level and event category. The source is normally the device that generates the event, while the proxy is the device responsible for detecting and communicating the event.

In Table 6, the minimum and maximum data size are computed based on the estimated total volume of the data required at all metering profile for monthly, yearly for both single meter and 100 m nodes.

Table 5 Time estimation with different data size

Data rate	Data size (100 m nodes)		Time required (HES)	
(kbps)	Min	Max	Min (ms)	Max (ms)
1024	5700	7900	44.53	61.71
2048	5700	7900	22.26	30.85
4096	5700	7900	11.13	15.42
8192	5700	7900	5.56	7.71

Table 6 Time estimation with different data size

(a) Billing profile							
Data size (bytes)		Month (bytes)		Yearly (bytes)		Yearly data (bytes) 100 m nodes	
Min	Max	Min	Max	Min	Max	Min	Max
168	244	5040	7320	60,480	87,840	6,048,000	8,784,000
(b) Daily load profile							
Data size (bytes)		Month (bytes)		Yearly (bytes)		Yearly data(bytes) 100 m nodes	
Min	Max	Min	Max	Min	Max	Min	Max
57	39	1710	2370	20,520	28,440	2,052,000	2,844,000
(c) Block load profile							
Data size (bytes)		Monthly (bytes)		Yearly (bytes)		Yearly data(bytes) 100 m nodes	
Min	Max	Min	Max	Min	Max	Min	Max
65	109	187,200	313,920	2,246,400	3,767,040	222,640,000	376,704,000

4 Analysis of Deployment Scenario Over G3-PLC Communication Technology

4.1 G3-PLC Controlled Smart LED Streetlight Installation

This pilot project is based on G3-PLC controlled smart light emitting diode (LED) streetlight solution that work on constant current mode using quasi-resonant (QR) mode converter as a primary power converter. The input to power converter is 90–230 V AC which is supplied by utility power supply. The commercial LED driver, HVLED001A is used for driving the LED [9–12]. The prototype of this LED driver is experimentally verified, manufactured and deployed at STMicroelectronics campus using 20 numbers of existing poles in an area of 200×50 m^2. The conversion efficiency is more than 90% individually with good power quality and running reliably. The streetlights are connected to each other by a G3-PLC network implemented using commercially available STCOMET integrating power line communication (PLC) modem, Metrology and cortex-M4 core from STMicroelectronics [13–15]. The LED driver details are more explained in [8, 9]. The scope of this deployment is to show PLC application on smart lighting as an example, which further enhances the capability of IoT in smart city domain.

4.2 G3-PLC Smart Lighting Control System

The G3-PLC modem architecture is able to support the EN50065, Federal Communications Commission (FCC), Association of Radio Industries and

Businesses (ARIB) compliant PLC applications. Together with the application core, it enables the STCOMET to support the ITU-T G.9001, ITU-T G.9903 (G3-PLC®), ITU-T G.9904 (PRIME), IEEE 1901.2, IEC 61334-5-1 (G1), CLC/TS 50568 (METERS AND MORE®) and other narrow band PLC protocol specifications.

The metrology sub-system is suitable for EN 50470-1, EN 50470-3, IEC 62053-21, IEC 62053-22 and IEC 62053-23 compliant class1, class0.5 and class0.2 AC metering applications.

The G3-PLC uses orthogonal frequency-division multiplexing (OFDM) with adaptive modulation, this creates multiple channel across the spectrum assigned. The sampling frequency of OFDM is 400 kHz. Each channel occupies 25 kHz. The number of one sided carrier in G3-PLC is 36. It is a narrowband PLC technology due to which it can attain better error correction and longer range.

Operating band range of G3-PLC is as follows:

a. European Committee for Electrotechnical Standardization (CENELEC) A Band (35 kHz to 91 kHz): In Europe
b. CENELEC B Band (98–122 kHz): In Europe
c. ARIB Band (155–403 kHz): For Japan
d. FCC Band (155–487 kHz): For the US and others G3-PLC network is based on IPv6 with the G3-PLC link layer frame carrying IPv6 packet. The G3-PLC based smart street light installed in STMicroelectronics campus operates in CENELAC A Band. There are 10 nodes installed in the ST campus of 20 streetlight connected through G3-PLC. Each node consists of PLC modem and LED driver of 80 W. There is one master node with PLC modem that controls all the slave nodes. All the communication is based on G3-PLC.

The STCOMET is used as PLC Modem in both master and save. The STCOMET embeds an advanced hardware Advanced Encryption Standard (AES) peripheral which implements an advanced standard cryptographic algorithm according to the National Institute of Standards and Technology Federal Information Processing Standards (NIST FIPS) 197. The data encrypted at physical layer adds the security level.

It is having ON/OFF/DIMMING control of individual node. All the streetlight is controlled collectively and individually with required brightness as per need.

4.3 G3-PLC Smart LED Street Lighting Deployment

Figure 2 shows the actual views of 20 sets of deployed smart LED streetlight at STMicroelectronics campus. It shows the actual night vision at boundary spread in 200 m length in STMicroelectronics campus. Figure 3 shows the DCU which

Fig. 2 Actual views of deployed G3-PLC based LED streetlight system

Fig. 3 DCU of G3-PLC based LED streetlight

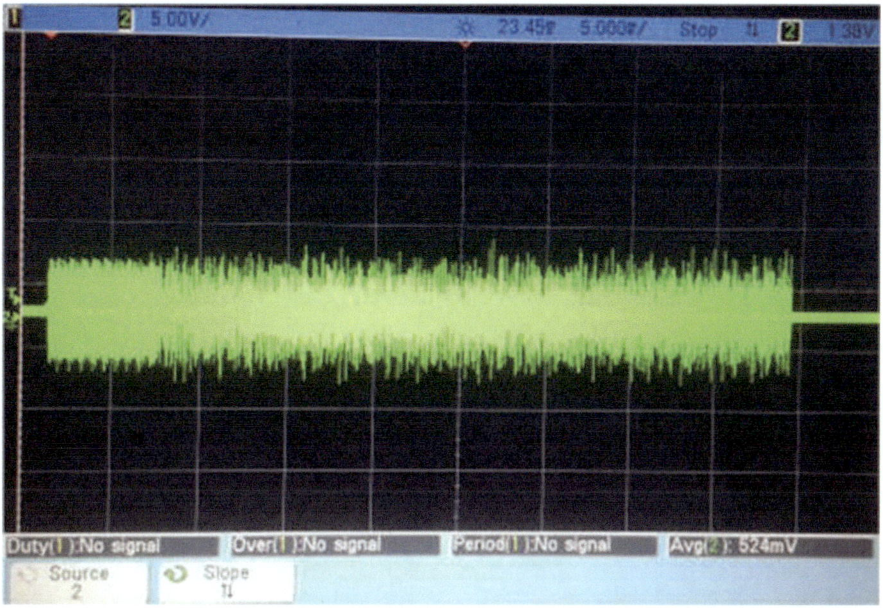

Fig. 4 G3-PLC Waveform captured at PHY Layer

controls the light and communicates to the central server based on user requirement. Figure 4 shows the waveforms of G3-PLC communication at physical layer. Figure 5 shows the G3-PLC controlled unit and LED driver which is positioned at individual electric pole.

5 Conclusion

In this paper, the evaluation and analysis of DLMS data transport in tiny intervals for smart meter, data concentrator unit, head end systems are estimated. Our study also helps with the estimation of scalability, managing and processing massive amounts of large volume of meter data and DLMS data transport over different last mile network and communication technology options. At the end the field deployment at STMicroelectronics for 20 smart street led lighting testbed as an example of this PLC application implemented at the boundary of campus. It is

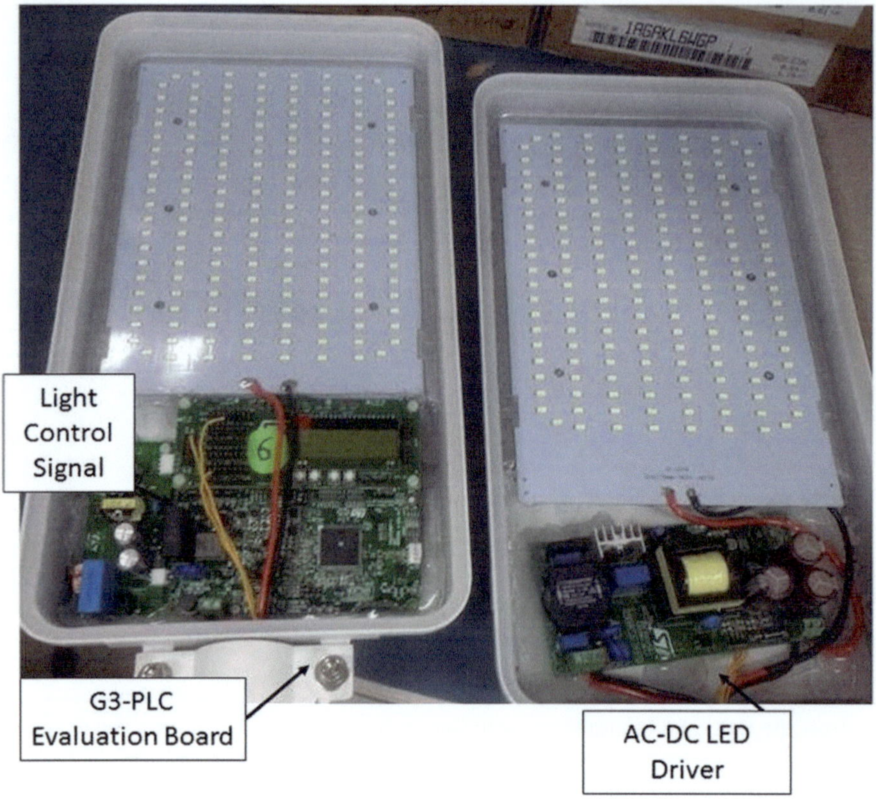

Fig. 5 G3-PLC based LED streetlight system

basically focused on G3-PLC system in comparison of other last mile communication options. In future this work may be extended to real time field implementation for smart meter and smart lighting using PLC technology options.

Acknowledgements This work is carried out as a part of IEEE PLC test bed working group, Bangalore and G3-PLC smart streetlight test bed deployment supported by STMicroelectronics Pvt. Ltd, Greater Noida.

References

1. Companion Specification for Energy Metering (2018, Mar) DLMS UA, EXCERPT FROM, COSEM Identification system and interface classes, 9th edn, (Online). http://dlms.com/documents/archive/Excerpt_BB9.pdf
2. https://standards.ieee.org/about/sasb/iccom/IC17-005-01_PLC_Test_Beds_in_India.pdf
3. ETD 18 (2018, Mar) IS 16444:2015 a.c. Static direct connected watthour smart meter class 1 and 2 specification (Online). http://www.bis.org.in/sf/etd/ET13_02082016.pdf02

4. India Smart Grid Knowledge Portal (2018, Mar) Smart grid updates pilot projects in India (Online). http://www.indiasmartgrid.org/pilot.php
5. AMI (2018, Mar) Advance metering Infrastructure (AMI) Rollout strategy for India (Online). http://www.indiasmartgrid.org/reports/ISGF%20White%20Paper%20on%20AMI%20Roll-Out%20Strategy%20for%20India.pdf
6. TEC Report (2018, Mar) Technical Report Communication technologies in M2 M/IoT domain (Online). http://tec.gov.in/pdf/M2M/Communication%20Technologies%20in%20IoT%20domain.pdf
7. IPv6 over low power wide area networks. https://datatracker.ietf.org/wg/lpwan/about/
8. National Smart Grid Mission, Ministry of Power, Government of India (2018, Mar) Status of Smart Grid Pilots & NSGM Smart Grid Projects (Online). http://www.nsgm.gov.in/sites/default/files/SG-Projects-Status-January2018.pdf
9. Jha A, Kumar M (2018) A wide range constant current LED driver with improved power quality and zero standby. In: 2018 IEEMA Engineer infinite conference, Accepted for Publication 13–14 Mar 2018
10. Jha A, Di Franco F (2018) Wide range constant current LED driver with high power factor, Low THD, and low standby power consumption. Arrow Times, USA. https://www.arrow.com/en/research-and-events/articles/st-micro-wide-range-constant-current-led-driver. 05 Jan 2018
11. Datasheet (2018, Feb) HVLED001A Offline controller for LED lighting with constant voltage primary-sensing and high power factor (Online). www.st.com/web/en/catalog/sense_power/FM142/CL1854/SC1572/SS1656/PF261186
12. Datasheet (2018, Feb) TSM101 voltage and current controller (Online). http://www.st.com/web/catalog/sense_power/FM142/CL1454/SC276/PF65510?s_searchtype=partnumber
13. Whitepaper (2018, Feb) G3-PLC power-line technology makes the smart grid smarter (Online). https://www.digikey.in/en/articles/techzone/2012/jun/g3-plc-power-line-technology-makes-the-smart-grid-smarter
14. Wikipedia (2018, Feb) Power-line communication (Online). https://en.wikipedia.org/wiki/Power-line_communication
15. STMicroelectronics PLC (2018, Feb) PowerLine transceivers (Online). http://www.st.com/contentinterfaces-and-/st_com/en/products/transceivers/power-line-transceivers/stcomet.html

Renewable and Microgrid

Evaluation of Technologies for Smart Microgrid Protection Schemes

Karthik Vagicharla and Harikrishna Muda

Abstract Due to the presence of distributed generation (DG) resources at different levels of the distribution network, the fault current seen by a relay in a microgrid is varied. Further, designing of a proper protection scheme for landing and grid connected modes is one serious challenge facing in a microgrid. Variation of fault current magnitude and direction makes the conventional methods such as over current and directional relays are inappropriate to provide complete protection for microgrids during the grid connected and island mode of operation. In this paper, various new adaptive protection techniques based on network reconfiguration, communication, non-communication for the microgrid system are analysed. These protection techniques are concentrated on addressing the solution to protection issues such as sympathetic tripping, protection blinding and protection coordination in both grid connected and islanded mode of operations.

Keywords Adaptive overcurrent protection schemes · Microgrids · The mode of operation · Protection issues

1 Introduction

The connection of distributed generation (DG) to the distribution network forms a microgrid which is capable of either operate in grid-connected or islanded mode [1]. The combination of several microgrids in the electric power grid establishes a new grid paradigm called microgrids, which is the corner stone of smart grid [2, 3]. Microgrid offers several advantages and benefits including increased power

K. Vagicharla (✉)
Regional Inspectorial Organization (West), Central Electricity Authority, Mumbai, India
e-mail: karthikbec@gmail.com

H. Muda
Sreenidhi Institute of Science and Technology, Hyderabad, India
e-mail: harikrishnam2@gmail.com

R. K. Pillai et al. (eds.), *ISGW 2018 Compendium of Technical Papers*, Lecture Notes in Electrical Engineering 580,
https://doi.org/10.1007/978-981-32-9119-5_10

generation capacity in each individual microgrids and increased reliability. However, as more microgrids form an integral part of the distribution network, a critical concern regarding protection coordination issues arise [4–6].

Protection coordination depends on the quick performance of the overcurrent relay as a primary operation, yet backed up by another overcurrent relay [7]. The operating modes of microgrid impose challenges to directional overcurrent relay (DOCR) scheme due to the change in current magnitudes, and alteration of power flow through the distribution lines. Due to the dual mode of operation of microgrids, penetration level, location, and type of DGs, the backup DOCR will sense high fault current compared to primary DOCR for a fault condition [8]. Therefore, traditional DOCR settings are affected and it results in blinding of line protection, sympathetic tripping, recloser failure and loss of protection coordination. With the inception of faults, the current contributed by the inverter based DG is in the range of 2 pu [9]. In such scenarios, the DOCRs find limitation in identifying the fault during inverter based DG operation.

In the literature, the proposed overcurrent protection techniques concentrated on the solution to a protection issues such as sympathetic tripping, protection blinding and protection coordination in a single microgrid. However, microgrids introduces a connection of DG, which may alter the prefault and fault current level seen by protective relay within adjacent microgrid. Further, the combination of many generating units in the form a microgrid connection can have significant impact on fault currents and affect the operation of overcurrent protection coordination.

In this paper, the pattern of fault current in different microgrids operation conditions is studied. Further, various new adaptive protection techniques based on network reconfiguration, communication, non-communication for the microgrid system are analysed. These protection techniques are concentrated on addressing the solution to protection issues such as sympathetic tripping, protection blinding and protection coordination in both grid connected and islanded mode of operations. A sequence fault current based interval linear programming technique is proposed to extenuate the impact of microgrids on operation of protection coordination.

2 Protection Coordination Issues with Microgrids

Due to the grid-connected and islanded operation of microgrids, the distribution system lost its radial nature and the power flow is bidirectional through distribution network [7]. Figure 1 shows the diagram of the microgrid structure having Grid, DGs, three individual microgrids of same configuration with switches and unbalanced loads at different buses. This model shown in Fig. 1 is adopted from [10]. The relays are considered to protect the distribution lines. Protection coordination issues associated with microgrid operation are described in the following sections.

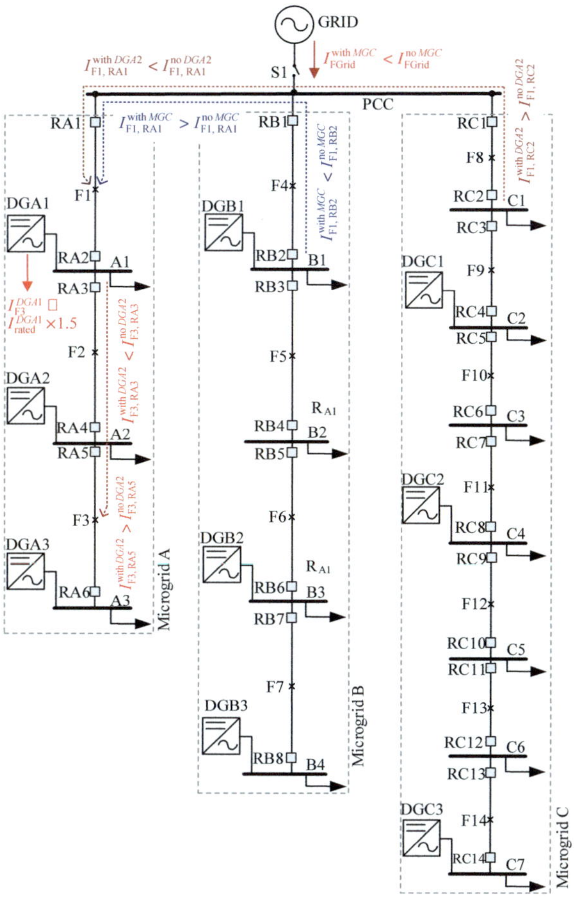

Fig. 1 The current passing through primary and backup DOCRs in microgrid for several fault conditions

2.1 *The Impact of Disconnection of DGs on Microgrid Protection*

The primary relay (PR) should operate as fast as possible, and coordinated with a backup relay (BR) for each fault event [11]. For fault event-F3, RA3 is backup DOCR for the primary DOCR-R5 as shown in Fig. 1. When the DGA2 is presented at bus-A2, fault current contribution from DGA2 may vary the total fault current seen by the PR and BRs. The fault current $\left(I_{\mathrm{F3,RA3}}^{with\ DGA2}\right)$ flowing through RA3 due to the association of DGA2 with system, is lower than the fault current $\left(I_{\mathrm{F3,RA3}}^{no\ DGA2}\right)$ obtained with the disconnection of DGA2. The fault current, $I_{\mathrm{F3,RA5}}^{no\ DGA2}$ seen by R5 is increased to $I_{\mathrm{F3,RA5}}^{with\ DGA2}$ with the connection of DGA2 to the microgrid.

Therefore, the operating times of PR RA5 and BR RA3 is decreased and increased respectively. This leads to the loss of protection coordination between RA5 and RA1 in a microgrid. Further, the fault current contribution from DGA1 $\left(I_{\mathrm{F}}^{DGA1}\right)$ is generally 1.5–2 times its rated value $\left(I_{rated}^{DGA1}\right)$ for such a situation [9]. The fault current seen by R3 may exceed the pickup current of it. As a result, the R3 may issue a trip signal and the healthy line may be disconnected unnecessarily. Such a phenomenon is known as sympathetic tripping [12]. Therefore, overcurrent relays must be integrated with directional elements in case of DGs integrated with DN to maintain selectivity.

2.2 The Effect of Disconnection of DG on Adjacent Microgrid

Reconnection of DG to the microgrid complies with the requirements of new interconnection standards [13]. The impact of connection of DG on adjacent microgird protection is shown in Fig. 1. A fault event at F1 is considered. The fault current $\left(I_{F1,RA1}^{with\ DGA2}\right)$ seen by PR RA1 is increased in comparison with the fault current $\left(I_{F1,RA1}^{no\ DGA2}\right)$ for disconnection of DGA2. For the same fault scenario, the fault current seen by BR RC2 is decreased in comparison with the connection of DGA2. When DGA2 exist in microgird, the protection reach of the BR (RC2) is reduced and the protection reach of the PR (RA1) is increased as DGA2 increased the equivalent impedance of the microgrid. The conventional DOCR settings are insufficient since there is prominent change in fault currents due to the disconnection of DG2 in adjacent microgrid.

2.3 The Impact of Disconnection of Microgird on Microgrid Protection

The microgrids provide improved reliability by sharing the power among microgrids, and results in the bidirectional power flow operation. Particularly, in case of microgrid operation. The level and direction of fault and prefault currents seen by PR and BRs are changed significantly during network disturbances.

A fault is created at point ‘F1’ as shown in Fig. 1. Before the integration of Microgrid C at PCC, the total fault current flowing through the DOCR, RA1 is $I_{F1,RA1}^{no\ MGC}$. When the Microgird C is connected to the PCC, the fault current $\left(I_{F1,RA1}^{with\ MGC}\right)$ which flows through relay, RA1 is reduced. It may result in protection blocking or delayed protection operation. For the same fault scenario, the fault current $\left(I_{F1,RB2}^{with\ MGC}\right)$ seen by Relay, RB2 in Microgrid B is increased to $\left(I_{F1,RB2}^{no\ MGC}\right)$

due to the fault current division between MGB and MGC. It can be seen that the primary (RA1) and backup (RB2) phase overcurrent relays fail to provide protection coordination measure for such a situation.

The 11 kV, 12MVA, 50 Hz, MMG architecture as shown in Fig. 1 is simulated using MALAB/Simulink platform. Detailed system data are provided in [21]. An ag-type of a single-phase-to-ground fault is created on fault location, F1 at time t = 0.06 s. The phase-a current available at RA1 (the PR) and RB2 (the BR) are shown in Fig. 2a, b respectively. Peak fault currents at RA1 and RB2 are found to be 8.87 A and 4.04 A respectively. The MGC is disconnected by keeping DOCR, RC1 opened at PCC. The fault currents available at RA1 and RB5 for the same fault situation are shown in Fig. 2c, d respectively. Peak fault currents at R5 and R1 are found to be 7.75 and 6.07 A respectively. The fault current seen by PR RA1 with the presence of MGC is increased in comparison with the disconnection of MGC.

To evaluate the impact of DG connection on adjacent microgrid, the prefault and fault current are obtained for An ag-type of a single-phase-to-ground fault at F1 location. The islanded mode of operation as shown in Fig. 1 is considered. With the connection of DGA2, the fault current seen by PR (RA1) and BR (RC2) is plotted in Fig. 3a, b respectively. The fault current seen by RA1 and RC2 are 8.87 and 18.02 A respectively. For the same fault event, results for PR and BRs are shown in Fig. 3c, d respectively. With disconnection of DGA2 in Microgrid A, the peak fault current at RA1 and RC2 are found to be 1.98 and 10.86 A respectively. The conventional DOCRs are insufficient since there is a wide variation in fault current in distribution networks with and without the presence of DG. Further, protection miscoordination is influenced due to the change MMG mode of operation.

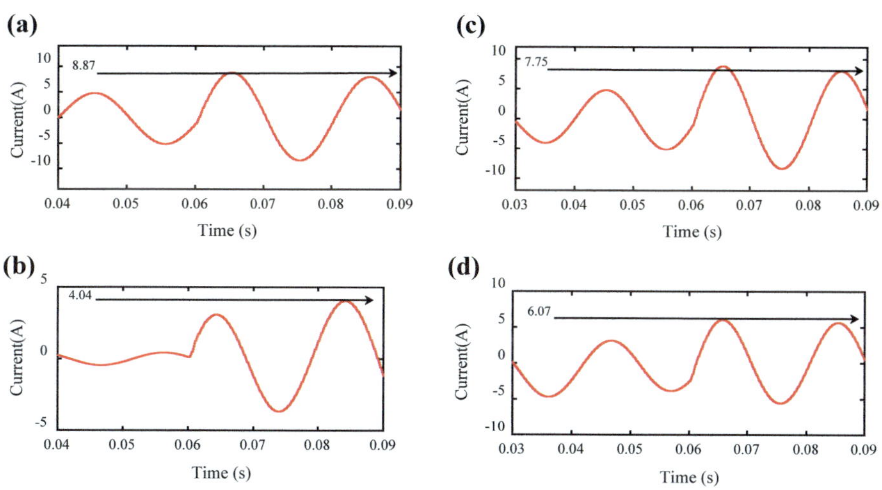

Fig. 2 The phase a relay current for fault F1 at PR and BR locations. **a** At primary DOCR, RA1 with the connection of MGC. **b** At backup DOCR, RB2 with the connection of MGC. **c** At primary DOCR, RA1 with the disconnection of MGC. **d** At backup DOCR, RB2 with the disconnection of MGC

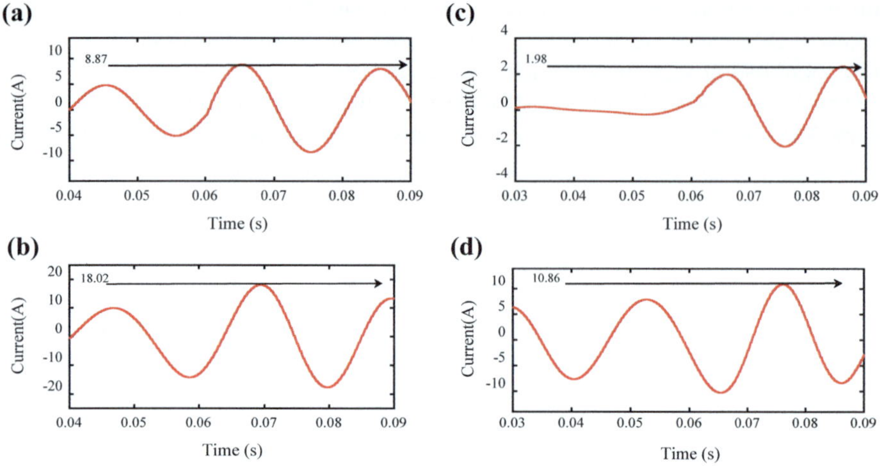

Fig. 3 The phase a relay current for fault F1 at primary and backup relay locations. **a** At primary DOCR, RA1 with the connection of DGA2. **b** At backup DOCR, RC2 with the connection of DGA2. **c** At primary DOCR, RA1 with the disconnection of DGA2. **d** At backup DOCR, RC2 with the disconnection of DGA2

It can be seen that the conventional DOCRs are insufficient since there is a wide variation in fault current in distribution networks with and without the presence of DG. Further, protection miscoordination is influenced due to the change microgrid mode of operation.

3 Recent Developments in Smart Microgrid Protection Schemes

To extenuate impact of DGs on protection schemes, the proposed methods by several researchers classified into

- Disconnection of DGs after fault detection [14],
- Deployment of fault current limiters [11, 15],
- Reconfiguration of network using protective devices [16],
- Communication [17, 18] and non-communication [19, 20] based adaptive protection techniques.

In [14], it is mentioned that DGs should be disconnected from the electric power system during faults. However, the disconnection of DGs may reduce system reliability. Fault current limiter is an effective method since the presence of fault current limiter in distribution system decreases fault current by the effect of its increase impedance during fault [15]. In [11], Inductive and resistive fault current limiters are used to restrict the fault current within the desired value and maintain the fixed relay settings for fixed DG capacities. Here, extra cost is involved in connecting the fault current limiters.

In order to overcome the DOCR protection coordination problems in microgrid environment, researchers have reconfigured the network using protective devices. In [16], whole DN in addition to DG are reconfigured using CB arrangement to avoid the fuse-fuse, fuse-relay and relay-relay co-ordination difficulties. Further, a relaying scheme is developed by dividing the microgrid into different zones and acting individually in islanded mode. However, these methods make protection process more difficult due to the requirement of utility participation. In addition, reconfiguration of the protection system is costly.

3.1 Non-communication Based Adaptive Protection Techniques

In [21], pre-calculated non-directional and DOCR settings are computed for the protection of DG connected system. In [17], the adaptive protection along with the conventional protection scheme is proposed using fuses, digital relays, and recloser. In this technique, online estimation of currents at all fuse, relay and recloser locations of distribution networks is necessary. The APA shift curves of the downstream protective devices to maintain the protection coordination depending on the relative current magnitudes. In [8], a microprocessor based recloser-fuse coordination strategy for typical distribution networks embedding DGs is proposed. In this technique, the ratio of fault current to relay current is considered to adapt the recloser characteristics. Online estimation of currents at all fuse, relay and recloser locations of distribution network is necessary. However, protection coordination may not exist due to the high fault currents at the downstream relay locations for increasing penetration of DGs. In [22], the adaptive relaying scheme is presented using multiple characteristics for the protection coordination. The relay characteristics are selected based on the ratio of fault currents measured at PR and BR locations. In [20], overcurrent relay is implemented for the protection of the distribution system equipped with synchronous based DGs. In this technique, the settings of protection scale that fit into the grid-connected mode of operation and the islanded mode of operation are described."

Figure 4 shows the coordination between the primary phase directional overcurrent relay, R3 and one backup phase directional overcurrent relay, R1 (dashed lines) with the help of a time-overcurrent plot for normal (disconnection of DGs), grid connected and the islanded mode, respectively. However, protection coordination is lost when grid connected and islanded modes were considered with traditional approach. Protection coordination between the DOCRs is guaranteed for any fault with adaptive approach for different operating modes as depicted in Fig. 5. The performance of the method is evaluated for limited system conditions.

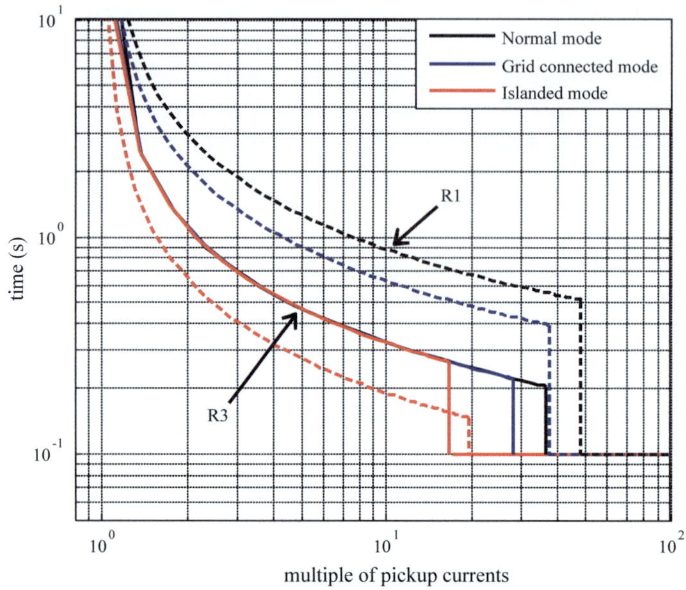

Fig. 4 Protection coordination IDMT characteristics for different operating modes with conventional approach

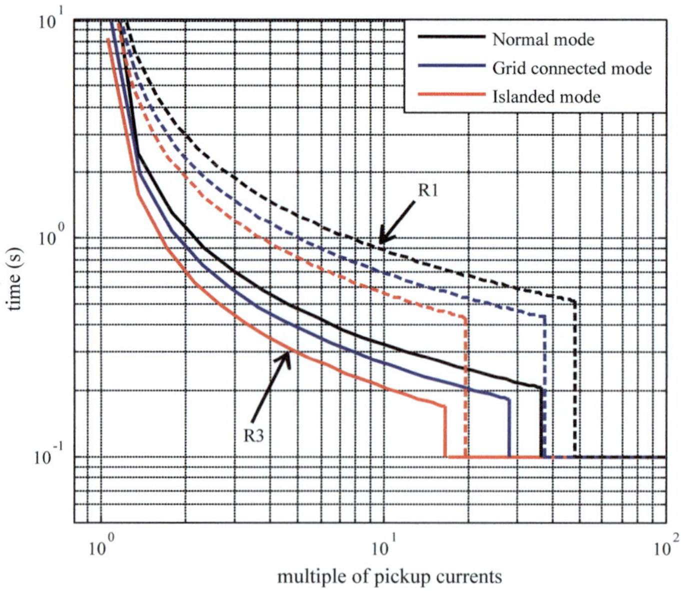

Fig. 5 Protection coordination IDMT characteristics for different operating modes with adaptive approach

3.2 Communication Based Adaptive Protection Techniques

The communication-enhanced coordination will support real-time protection time–current adjustments in the system. In [23], digital relays with a communication network for the protection of loop structured distribution networks are proposed. The proposed method can be used to protect the microgrid during the normal and high impedance fault periods. To locate and isolate a faulty segment in a distribution system with DGs, wavelet coefficients of the high-frequency transients using communication schemes is proposed [24]. Microgrid control center acts as a server and generates appropriate signals by fetching the status of the agents. The communications-based approach is shown in Fig. 6 is used to transmit the signal information to the central microgrid controller. to update the status of the DGs and grid, the communication is required between the microgrid control centre and DOCRs as shown in Fig. 6. It can be seen that the microgrid central controller with a communication network is provided to monitor the status of every DG in the microgrid. As the microgrid mode of operation changes, the settings in all relays are adapted to ensure protection coordination between DOCRs.

In [25], agent-based communication protection scheme is proposed to adapt prevailing system conditions. By using this technique, different types of relays are coordinated in the microgrid protection, and other equipment. In [26], communication-based central protection system monitors the microgrid mode operation and updates the overcurrent relay operating settings for safe operation. The fault current coefficient of inverter based DGs can be taken as 1.5 regardless of fault location and type of fault in microgrids. The adaptive method for overcurrent relays is demonstrated by considering the DG operation, network management and islanded mode in [27]. In this approach, relay settings are computed and verified for each mode of operation. The real-time digital simulator (RTDS) is used to test adaptive protection algorithm for active distribution networks. However, the directional element is needed to enhance the selectivity property of the PR and BRs.

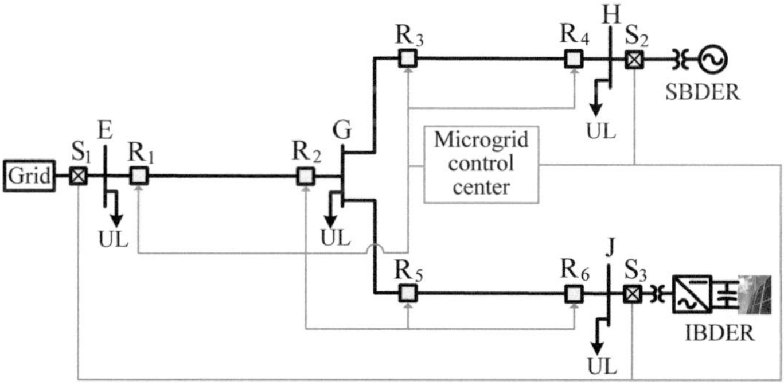

Fig. 6 Schematic diagram of the microgrid with communication infrastructure

4 Conclusion

The problem of protection coordination in microgrid environment is addressed. The advantages and disadvantages of the fault current limiter based, voltage, impedance, non-communication, communication-based adaptive, sequence, differential and optimization methods are summarized. The introduction of renewable electricity production needs for adaptive protection schemes in the grid. Most of the methods for microgrid protection are unable to meet the challenges of low fault current levels and bidirectional power flow from DGs, high impedance faults and connection/disconnection of DGs in microgrids. These protection techniques are concentrated on addressing the solution to protection issues such as sympathetic tripping, protection blinding and protection coordination in both grid connected and islanded mode of operations.

References

1. Najy AWK, Zeineldin HH, Woon LW (2013) Optimal protection coordination for microgrids with grid-connected and Islanded capability. Ind Electron IEEE Trans 60:1668–1677
2. Laaksonen JH (2010) Protection principles for future microgrids. IEEE Trans Power Electron 25
3. Gopalan SA, Sreeram V, Iu HHC (2014) A review of coordination strategies and protection schemes for microgrids. Renew Sustain Energy Rev 32:222–228
4. El-khattam W, Sidhu TS (2009) Resolving the impact of distributed renewable generation on directional overcurrent relay coordination: a case study. IET Renew Power Gener 3(4):415
5. Birla D, Maheshwari RP, Gupta HO (2006) A new nonlinear directional overcurrent relay coordination technique, and banes and boons of near-end faults based approach. IEEE Trans Power Deliv 21(3):1176–1182
6. Singh M, Panigrahi BKK, Abhyankar ARR, Das S (2014) Optimal coordination of directional over-current relays using informative differential evolution algorithm. J Comput Sci 5(2): 269–276
7. El-Saadany EF, Saleh KA, Al-Hinai A, Zeineldin HH (2015) Dual-setting characteristic for directional overcurrent relays considering multiple fault locations. IET Gener Transm Distrib 9(12):1332–1340
8. Zamani A, Sidhu T, Yazdani A (2010) A strategy for protection coordination in radial distribution networks with distributed generators. In: IEEE power and energy society general meeting, pp 1–8
9. Nimpitiwan N, Heydt TG, Ayyanar R, Suryanarayanan S (2007) Fault current contribution from synchronous machine and inverter based distributed generators. IEEE Trans Power Deliv 22
10. Kersting WH (1991) Radial distribution test feeders. IEEE Trans Power Syst 6(3):975–985
11. El-Khattam W, Sidhu TS (2008) Restoration of directional overcurrent relay coordination in distributed generation systems utilizing fault current limiter. IEEE Trans Power Deliv 23 (2):576–585
12. Muda H, Jena P (2017) Sequence currents based adaptive protection approach for DNs with distributed energy resources. IET Gen Trans Distrib 11(1):154–165
13. IEEE SA - 1547.6-2011 (2014) IEEE recommended practice for interconnecting distributed resources with electric power systems distribution secondary networks

14. Muda H, Jena P (2017) Superimposed adaptive sequence current based microgrid protection: a new technique. IEEE Trans Power Deliv 32(2):757–767
15. Sood VK, Alam S (2006) 3-phase fault current limiter for distribution systems. In: 2006 international conference on power electronic, drives and energy systems, pp 1–6
16. Brahma SM, Girgis AA (2004) Development of adaptive protection scheme for distribution systems with high penetration of distributed generation. IEEE Trans Power Deliv 19(1):56–63
17. Shah PH, Bhalja BR (2014) New adaptive digital relaying scheme to tackle recloser–fuse miscoordination during distributed generation. IET Gener Transm Distrib 8(4):682–688
18. Zamani MA, Yazdani A, Sidhu TS (2012) A communication-assisted protection strategy for inverter-based medium-voltage microgrids. IEEE Trans Smart Grid 3(4):2088–2099
19. Rajaei N, Salama MMA (2015) Management of fault current contribution of synchronous DGs using inverter-based DGs. IEEE Trans Smart Grid 6(6):3073–3081
20. Sharaf MH, Zeineldin HH, Ibrahim KD, EL-Zahab AEE-D (2015) A proposed coordination strategy for meshed distribution systems with DG considering user-defined characteristics of directional inverse time overcurrent relays. Int J Electr Power Energy Syst 65:49–58
21. Oudalov A, Fidigatti A (2009) Adaptive network protection in microgrids. Int J Distrib Energy Resour 5:201–226
22. Hussain B, Sharkh SM, Hussain S, Abusara MA (2013) An adaptive relaying scheme for fuse saving in distribution networks with distributed generation. IEEE Trans Power Deliv 28 (2):669–677
23. Sortomme E, Venkata SS, Mitra J (2010) Microgrid protection using communication-assisted digital relays. IEEE Trans Power Deliv 25(4):2789–2796
24. Perera N, Rajapakse AD, Buchholzer TE (2008) Isolation of faults in distribution networks with distributed generators. IEEE Trans Power Deliv 23(4):1
25. Xu Y et al (2014) Distributed subgradient-based coordination of multiple renewable generators in a microgrid. IEEE Trans Power Syst 29(1):23–33
26. Ustun TS, Ozansoy C, Zayegh A (2013) Fault current coefficient and time delay assignment for microgrid protection system with central protection unit. IEEE Trans Power Syst 28(2)
27. Coffele F, Booth C, Dyśko A (2015) An adaptive overcurrent protection scheme for distribution networks. IEEE Trans Power Deliv 30(2):561–568

Design of Hybrid DC Mini Grid for Educational Institution: Case Study

C. Sharmeela, P. Sivaraman and S. Balaji

Abstract An education institution campus consists of several academic departments, laboratories, class rooms, seminar halls, administration office, guest house etc. The major operating electrical loads (base loads) are lighting and HVAC loads, intermittent loads are laboratory equipments etc. The fluorescent, CFL lightings are mostly used for indoor applications and sodium vapour lamp for outdoor applications. The advantages of LED in illumination technologies take the attention to switchover from other illumination devices to LED. Presently most of the LED lightings are connected in 1ϕ AC supply, AC to DC conversion is required for LED operation resulting to additional energy loss in the converter and current harmonic injection to the grid. This paper proposes the design of hybrid (solar, battery energy storage and utility) 110 V DC mini grid for lighting, HVAC and corridor lighting application. Solar is used as a prime source to power the loads in day time, energy stored in the battery is used for night time, utility power supply is used when solar and battery energy is not available. During the non-working days and holidays, power output from the solar PV is exported into the grid.

Keywords DC · Mini grid · LED · BLDC · Solar PV

C. Sharmeela (✉)
Department of EEE, Anna University, Chennai, India
e-mail: sharmeela20@yahoo.com

P. Sivaraman
TECh Engineering Services, Chennai, India
e-mail: psivapse@gmail.com

S. Balaji
Anna University, Chennai, India
e-mail: balajisriram15@gmail.com

R. K. Pillai et al. (eds.), *ISGW 2018 Compendium of Technical Papers*,
Lecture Notes in Electrical Engineering 580,
https://doi.org/10.1007/978-981-32-9119-5_11

1 Introduction

In olden days, electric power is generated from power stations like thermal stations, hydro stations, nuclear stations etc. at remote locations and transferring the power through transmission system to the load center (where the actual load is located) [1]. Depletion of fossil fuels, energy crisis and energy loss are the major concern in every power system to think for the alternative sources and technology to enhance the power system operation with less energy loss [2]. Renewable Energy Sources (RES) like solar, wind etc. are the alternative energy sources to conventional energy sources. Distributed Generation (DG) is the methodology to integrate the energy sources into the grid locally to reduce the energy loss and enhance the voltage profile [3]. Distribution systems are highly affected by energy conversion (AC to DC and DC to AC), poor power quality problems like harmonics, voltage flicker etc. resulting in energy losses in the conversion and poor performance of the equipments [4–6]. In the recent era, power electronics based equipments are used for the sophisticated operation and control requires AC to DC or DC to AC conversion resulting in energy losses in the conversion [7]. On the other hand recent invention in low voltage DC distribution equipments, availability of solar PV output in direct DC supply drawing the attention of the people to looking into DC distribution [8].

An education institution campus consists of several academic departments, laboratories, class rooms, seminar halls, administration office, guest house etc. The major operating electrical loads (base loads) are lighting and HVAC loads, intermittent loads are laboratory equipments etc. The fluorescent, CFL lightings are mostly used for indoor applications and sodium vapour lamp for outdoor applications. In this paper solar PV, Battery Energy Storage (BES) and utility powered hybrid 110 V DC mini grid for academic institution. Fluorescent lamps, CFL and fans in the AC system is replaced with DC operated LEDs and BLDC based fans to save the energy for the same requirements. This paper presents the design and modelling of hybrid 110 V DC mini grid for one department building. The sizing of solar PV analysis is performed in PVsyst simulation software.

2 Data Collection

This building has the class rooms, staff rooms, a conference hall, office, laboratories and toilet. Presently fluorescent lamp of 40 W and ceiling fan 60 W are powered through 1ϕ, 240 V, 50 Hz AC supply. The ground floor and first floor load details are listed in Tables 1 and 2 respectively.

The total building load in AC system is listed in Table 3.

The power quality parameters of ground floor and first floor ceiling fan loads measured by Fluke 434 power quality analyzer is shown in Fig. 1.

Table 1 Ground floor load details

S. No	Location	No of fluorescent lamp	No of ceiling fan
1	Corridor	6	–
2	Class room	18	18
3	LAB	26	26
4	Toilet	4	4

Table 2 First floor load details

S. No	Location	No of fluorescent lamp	No of ceiling fan
1	Corridor	4	–
2	Conference hall	16	6
3	Staff cabins	16	16
4	Computer center	4	4
5	LAB	6	6
6	Class room	4	4
7	Toilet	2	–

Table 3 Total load in AC system

Floor	Total lighting load (kW)	Total fan load (kW)	Total (kW)
Ground floor	2.1	2.2	9.4
First floor	2.2	2.9	

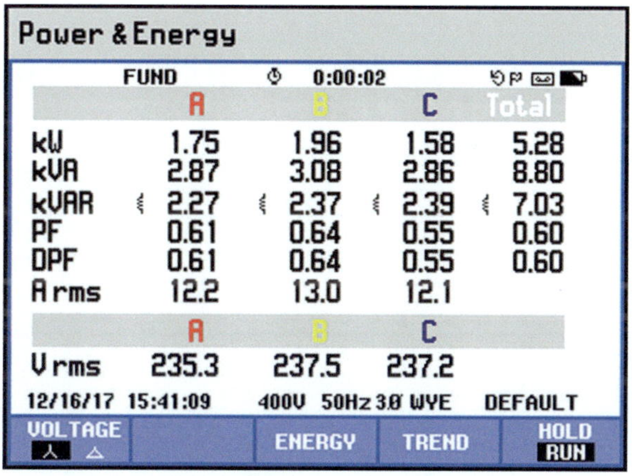

Fig. 1 Power quality parameters

3 Hybrid DC Mini Grid

The combination of solar photovoltaic, utility power supply and Battery Energy Storage (BES) are formed as a hybrid power source for DC mini grid is operating at 110 V DC. Solar PV is considered as the prime source to power the loads in the mini grid during day time. The excess solar power generated in the day time is stored in the BES. During the night time, energy stored in the battery is used to power the loads. If power output from the solar PV is lesser than the connected loads, energy stored in the BES will support the loads. When both solar PV and BES cannot support the connected load during cloudy/monsoon days, utility power supply is used to power the loads. During non-working days and holidays, literally there are no loads connected in the grid and solar PV power is exported to grid. The block diagram of electrical distribution is shown in Fig. 2.

4 Modelling of DC Mini Grid

The fluorescent lamp of 40 W and ceiling fan of 60 W, 1ϕ, 240 V, 50 Hz AC loads are replaced with 30 W LED lights and 30 W Brush Less DC (BLDC) motor based fan. The total building lighting loads and fan loads are listed in Table 4.

The sizing of PV capacity analysis is performed in PVsyst simulation software. In order to cater the present loads and future expected load growth 12 kW (250 nos of 60 Wp) solar panel is designed 25 strings of 10 modules. BES of 600 AH rating is considered to store the energy from the solar PV. When solar PV and BES cannot

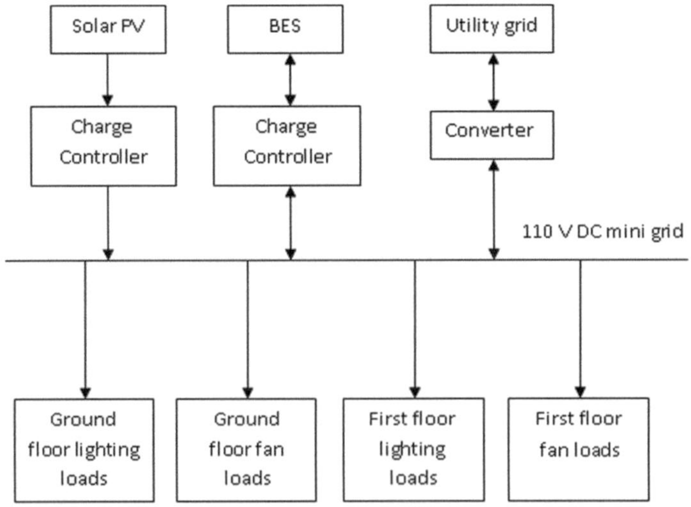

Fig. 2 Block diagram of electrical distribution

Table 4 Total load in DC system

Floor	Total LED load (kW)	Total fan load (kW)	Total (kW)
Ground floor	1.6	1.1	5.8
First floor	1.6	1.5	

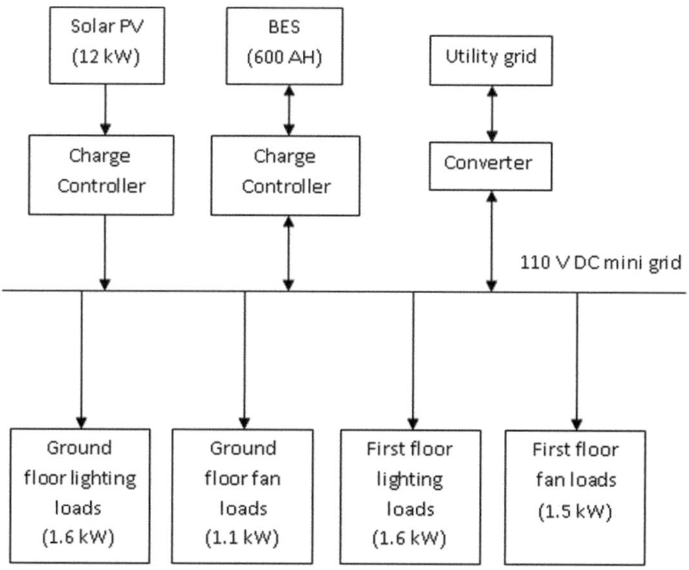

Fig. 3 Modelling of 110 V DC mini grid

support the loads, utility1ϕ, 240 V, 50 Hz AC supply is converted into DC supply in order to powering the DC loads. The layout of 110 V DC mini grid electrical distribution is shown in Fig. 3.

Solar PV is the preferred power source in day time, control logic is designed to prefer solar PV as a priority source for powering the load. It prevent the paralleling of solar PV output and utility output at 110 V DC mini grid by either solar PV and BES powering the loads or utility supply powering the load to avoid short circuit.

5 Result and Discussion

5.1 Energy Savings in LEDs and BLDC Fans

The energy consumption of ground floor, first floor with fluorescent lighting, fans for per day, per month and per year is listed in Table 5. Day is considered as 12 h, month is considered as 22 days and year is considered as 11 months for energy consumption details.

Table 5 Energy consumption of fluorescent lightings and fans

Energy consumption	Per day	Per month	Per year
Energy consumed by first floor (kWh)	51.6	1135.2	12487.2
Energy consumed by ground floor (kWh)	61.2	1346.4	14810.4

Table 6 Energy consumption of LEDs and BLDC fans

Energy consumption	Per day	Per month	Per year
Energy consumed by first floor (kWh)	32.4	712.8	7840.8
Energy consumed by ground floor (kWh)	37.2	818.4	9002.4

Table 7 Comparison of energy consumption of fluorescent & fans and LEDs & BLDC fans

Energy consumption per year (kWh)	Per day	Per month	Per year
Fluorescent and fan	112.8	2481.6	27297.6
LED & BLDC fan	69.6	1531.2	16843.2

Total energy consumption of ground floor, first floor loads are 112.8 kWh per day, 2481.6 kWh per month and 27297.6 kWh per year.

The energy consumption of ground floor, first floor with LED lighting, BLDC fans for per day, per month and per year is listed in Table 6.

Total energy consumption of ground floor, first floor loads are 69.6 kWh per day, 1531.2 kWh per month and 16843.2 kWh per year.

Comparison of energy consumption of fluorescent & fans and LEDs & BLDC fans per year listed in Table 7.

Energy savings by replacing the fluorescent lightings and induction motor based fans by LED lightings and BLDC based fan per year reduce the energy consumption of 19% of the total energy consumption. Figure 4 shows the energy savings per year.

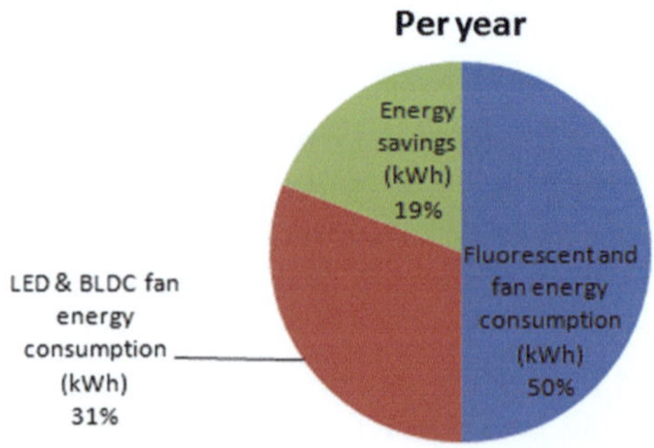

Fig. 4 Energy savings per year

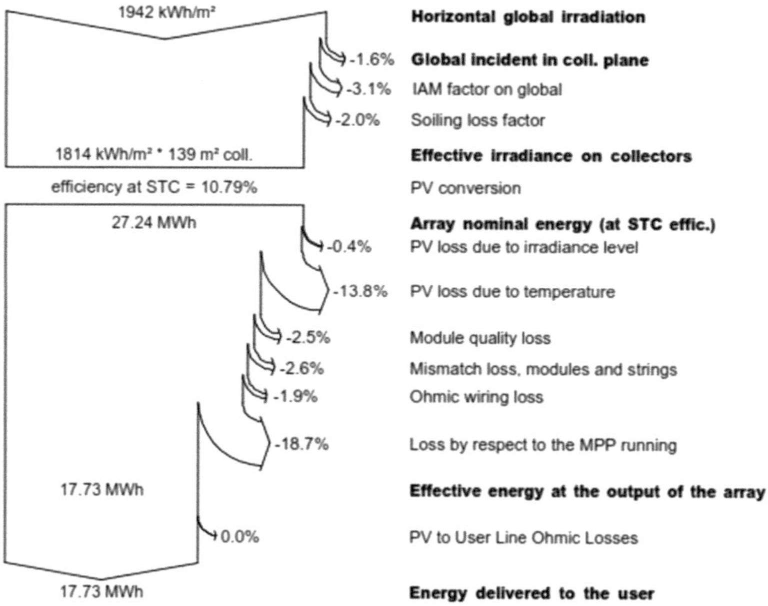

Fig. 5 Energy loss diagram per year

5.2 *Sizing of Solar PV Capacity*

The sizing of PV capacity analysis is performed in PVsyst simulation software. In order to cater the present loads (16843.2 kWh) and future expected load growth, 12 kW (250 nos of 60 Wp) solar panel is designed 25 strings of 10 modules. The energy produced by the 12 kW of solar PV installed capacity per year is 17.73 MWh or 17,770 kWh. The energy loss is shown in Fig. 5.

The produced useful energy per kWh/kWp/day is 3.24. Figure 6 shows the useful energy produced by solar PV system per day.

5.3 *Energy Exported to the Grid*

During non-working days and holidays of the institute, energy requirement for the building literally nil and energy from the solar PV is exported to the grid. The average energy generation of solar PV is 50 kWh per day.

Non-working days and holidays per month is 8 days except government holidays and local holidays. Non-working days and holidays per year is 118 days (88 days for 11 months and 30 days for one month) except government holidays and local holidays. The energy export from the solar PV for day, month and year is listed in Table 8.

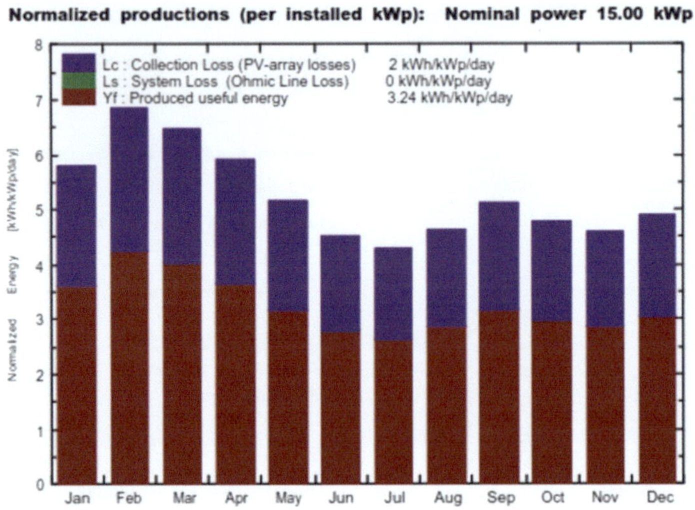

Fig. 6 Useful energy produced by solar PV per day

Table 8 Energy export to the grid from solar PV

Energy export	Per day	Per month	Per year
in kWh	38.88	311.04	4587.84

6 Conclusion

A hybrid 110 V DC mini grid is designed by combination of hybrid solar PV, BES and utility supply to power the lighting (LED) and cooling (fans) in the building of education institution campus. The major lightings load of fluorescent lamps is replaced with LEDs and induction motor based ceiling fan is replaced with BLDC based fans operated through 110 V DC mini grid. The LEDs and BLDC fans are directly connected and operated in 110 V DC supply by eliminating the AC to DC, DC to AC conversion resulting in reduced energy losses. Direct DC power supply is used to eliminate the AC to DC conversion for LED lightings. This paper discusses the design of hybrid (solar PV, BES and utility supply) 110 V DC mini grid for lighting and HVAC applications. Solar is used as prime power source to powering the loads in day time, energy stored in the battery is used for powering the loads during night time, utility supply is used to powering the loads when both solar and battery energy is not available. During the non-working days and holidays, power output from the solar PV is exported into the grid.

References

1. https://en.wikipedia.org/wiki/Electric_power_system
2. Nehrir MH, Wang C, Strunz K, Aki H, Ramakumar R, Bing J, Miao Z,, Salameh Z (2011) A review of hybrid renewable/Alternative energy systems for electric power generation: configurations, control, and applications. IEEE Trans Sustain Energy 2(4):392–403
3. https://en.wikipedia.org/wiki/Distributed_generation
4. Kurohane K, Senjyu T, Yona A, Urasaki N, Goya T, Funabashi T (2010) A hybrid smart AC/DC power system. IEEE Trans Smart Grid 1(2):199–204
5. Loh PC, Li D, Chai YK, Blaabjerg F (2013) Autonomous operation of hybrid microgrid with AC and DC subgrids. IEEE Trans Power Electron 28(5):2214–2223
6. https://en.wikipedia.org/wiki/Electric_power_quality
7. Lidula NWA, Rajapakse AD (2011) Microgrids research: a review of experimental microgrids and test systems. Renew Sustain Energy Rev 15(1):186–202
8. http://www.bis.org.in/sf/etd/ETD50(11294)_27012017.pdf
9. Tan KT, Peng XY, So PL, Chu YC, Chen MZQ (2012) Centralized control for parallel operation of distributed generation inverters in microgrids. IEEE Trans Smart Grid 3(4):1977–1987

Dr. C. Sharmeela (CS) holds a B.E. in Electrical and Electronics Engineering, M.E. in Power Systems Engineering from Annamalai University, Chidambaram and a Ph.D. in Electrical Engineering from Anna University, Chennai in 1999, 2000 & 2009 respectively. She did P.G Diploma in Electrical Energy Management and Energy Audit from Annamalai University, Chidambaram in 2010. At present, she holds the post of Associate Professor in department of EEE, CEG campus, Anna University, Chennai.

She has 16 years of teaching experience and taught various subjects to undergraduate and post graduate students. She did number of research projects and consultancy work in renewable energy, power quality & design of compensators for various industries. She published more than 30 research publications in peer reviewed journals, more than 50 research papers in international and national conferences in India and abroad. CS talks importance of power quality more than 10 years. She guides 14 Ph.D.s (two completed and 12 ongoing), 55 Master's level, 2 Undergraduate's level and 2 AMIE students. She co-ordinate and organized short term course on power quality for Tamilnadu Electricity Board Engineers in various years. She received best paper award in IEEE POWERCON 2004 conference held in Singapore. She is a Life-member of the Institution of Engineers (India), ISTE and SSI.

Er. P. Sivaraman was born in Vellalur, Madurai district, Tamilnadu. He completed schooling in Govt. Higher Secondary School, Vellalur, B.E. in Electrical and Electronics Engineering from PGP College of Engineering and Technology, Namakkal and M.E. in Power Systems Engineering from M. Kumarasamy College of Engineering (autonomous), Karur affiliated to Anna University, Chennai in 2012 & 2014 respectively. Presently, he is working in TECh Engineering Services, Chennai as sr.engineer. He has three plus years of experience in the field of renewable energy, energy storage, power quality, harmonic assessments, trouble shooting analysis, providing the techno economical solution to various power quality problems in across India. He had published many research papers in national and international conference.

Mr. S. Balaji graduated from SASTRA UNIVERSITY in the year 2016, HE holds a B. Tech degree in Electrical and Electronics Engineering, currently perusing M.E. in High Voltage Engineering in Anna University Guindy, An Electrical enthusiast with Hands-on Experience in Design of Electrical Insulations & Switch gear Equipment for Medium & High Voltage, PV Solar panels and Cooling systems for PV-panels. Design of Power factor correction and control mechanisms for BLDC motors. An Engineer with in-plant training in ABOND Strands pvt. LTD & an Internship at VI Micro Systems. He has also attended trainings on Electrical system design, Electrical Transients, Electrical Protection & Safety. He has lectured on the topics: Electrical system design using ETAP & Design and mitigation of issues in Micro Grids using PV-SYST. Additionally he has one year of professional experience, has Hands-on Experience with Wide range of Network and Cyber-security Products such as CISCO ASA, Fortigate, Juniper SRX& cloud based Proxy.

Integration of Rooftop Solar: A Step Towards 100% Renewable

Akshay Kumar Gera

Abstract Harvesting of Solar Energy in form of electricity experienced a phenomenal growth in past few years due to recent advancement in technology. Thus harvesting solar energy to its maximum extent would be a vital step towards achieving goal of 100% dependency on renewable. Integration of Rooftop solar is big step in this direction. This white paper discuss about the challenges for Roof Top Solar PV System and providing recommendation for Way forward on various aspects for Distributed Solar Power including Remote Supervision, Control and Network Operation, Power Evacuation and Grid Stability, Power Portfolio Management, Network Planning, Power Quality Management, Policy advocacy and regulatory approvals concepts.

Keywords Photo voltaic · Delhi electricity regulatory commission · Geographical interface system · Supervisory control and data acquisition · Advance demand response

Abbreviations

TPDDL	Tata Power Delhi Distribution Limited
DSM	Deviation settlement mechanism
SCADA	Supervisory control and data acquisition
ADR	Advance demand response
GIS	Geographical information system
AT&C	Aggregate technical and commercial
PV	Photo voltaic
DT	Distribution transformer
OD/UD	Over drawl/Under drawl

A. K. Gera (✉)
Power System Control, Tata Power-DDL, New Delhi, India
e-mail: akshay.gera@tatapower-ddl.com

R. K. Pillai et al. (eds.), *ISGW 2018 Compendium of Technical Papers*,
Lecture Notes in Electrical Engineering 580,
https://doi.org/10.1007/978-981-32-9119-5_12

1 Introduction

Tata Power Delhi Distribution is a joint venture between Tata Power and the Government of NCT of Delhi with the majority stake and management control being held by Tata Power (51%). Tata Power DDL distribute electricity in North & North-West parts of Delhi and serve a populace of over 60 lakhs. With a registered consumer base of approx. 14 lakhs and a peak load of around 1852 MW, our operations span across an area of 510 km^2. Today in our area of operations AT&C losses stand at 8.07% which is an unprecedented reduction of around 84.78% from an opening loss level of 53% in July 2002. Being a smart utility TPDDL always welcome implementation of new and advance technologies like Advance metering infrastructure, Advance Distribution Management system, MW scale energy storage technologies, Grid connected solar rooftop plants etc. (Table 1).

Delhi is blessed with almost 300 sunny days and the rooftop space available for solar panels is estimated to be 31 km^2, giving Delhi a solar energy potential of

Table 1 For Pan India. *Source* Wikipedia

Year	Installed solar capacity MW	Average Unit cost (Rs/unit)
2010	161	12.16
2011	461	7.5
2012	1205	7.1
2013	2319	6
2014	2632	5.6
2015	3744	5
2016	6763	4.44
2017	12,289	3.15 (Till April)

Fiscal Year	New Solar Energy (MW)	Cumulative Solar Energy (MW)	Annual Growth (%)	Percentage of peak grid load*1	Percentage of total electricity consumption*2
FY 16	30	35	700%	1%	0.15%
FY 17	84	119	240%	2%	0.56%
FY 18	193	312	162%	5%	1.43%
FY 19	294	606	94%	9%	2.66%
FY 20	385	991	63%	14%	4.16%
FY 21	285	1275	29%	17%	5.10%
FY 22	228	1503	18%	19%	5.73%
FY 23	187	1690	12%	20%	6.14%
FY 24	161	1850	10%	21%	6.40%
FY 25	145	1995	8%	21%	6.57%

Fig. 1 Year wise solar targets, as per Delhi Solar Energy Policy, 10th September 2015

2500 MW (annually approx. 3500 MU). Of this potential, 26% is in the government/public sector (650 MW), 25% in commercial/industrial sector (625 MW), and 49% in domestic sector (1225 MW).

While solar energy tariffs have, on average, fallen 6–8% per year since 1998 (solar panel prices have dropped 75% in the last six years), conventional energy tariffs in Delhi have risen 6.9% per year on average since 2007. After years of innovation and declining prices, solar energy tariffs in Delhi have become cheaper than conventional energy tariffs for the government the low and medium domestic segment as early as 2018 commercial-industrial, and Delhi Government has consequently established solar generation targets of 1 GW by 2020 (4.2% of energy consumed) and 2 GW by 2025 (6.6% of energy consumed) (Fig. 1).

2 Characteristics of Delhi Distribution System

Electricity Distribution business in Delhi is a regulated business governed by DERC and distribution is done by five different DISCOMS namely BRPL, BYPL, Tata Power DDL, MES and NDMC. Fixed area of operation is allotted to each DISCOM. Accordingly different assets are distributed and different DSM limits are set.

Considering technical point of view, primary distribution is at 66 kV or 33 kV & secondary distribution is at 11kV and 0.4 kV. There is both radial and ring type feeders in Delhi network. Urban area in Delhi is mainly ring type consisting RMU's and in sub urban parts, type of network is mainly radial consisting of autoreclosers and sectionalizes. Few issues like over voltage in winter and over loading of feeders in summers are frequently faced in system. Integrating PV to system may even help in alleviate some of these recurring issues.

Also by increasing localized generation amount of technical losses in distribution grid can be reduced. Total number of rooftop installations in TPDDL serving area have reached 179 numbers including high percentage of residential installations (Fig. 2).

Following interconnection voltage limits are recommended to connect different solar rooftop based on their capacity (Table 2).

3 Challenge with Distributed Solar Generation

Indian government have set solar targets for the whole country and distributed those to the states accordingly. Achieving those targets would mainly remain as a state subject and depends on each state's Government, electricity boards, distribution companies and state regulators. Significant progress has been made in past for setting the policy and regulatory frameworks right, however still several technical, financial as well as socioeconomic barriers exist in accelerated deployment of rooftop PV.

Major Challenges faced with distributed solar generation from a utility perspective are mentioned below:

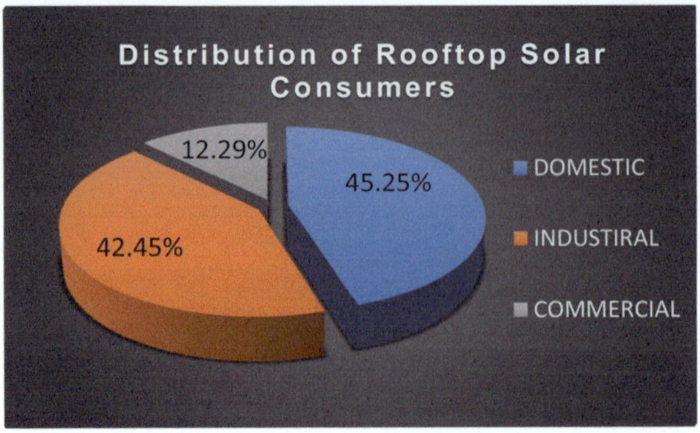

Fig. 2 Percentage wise distribution of rooftop PV consumers in Delhi

Table 2 Volatage limits recommended as per DERC for PV connections

Capacity range	Connecting voltage
Upto 10 kW	240 V-single phase or 415 V-three phase at the option of the consumer
Above 10 kW and upto 100 kW	415 V—three phase
Above 100 kW	At HT/EHT level

- This source of power is unpredictable in nature and complete dependency cannot be achieved unless a backup mechanism for maintaining reliability is available.
- Monitoring, control and management of large number rooftop solar plants to ensure safety and grid stability.
- The power generated from solar is DC and distributed, it's conversion to AC put a challenge to control harmonics in the grid.
- Inspecting and Maintaining quality of the power injected into grid.
- Sudden Overloading and frequent over and under voltage issues.
- Ensuring proper isolation and preventing reverse power flow in case of fault/ outage.
- Ensuring safety for utility workforce.

4 Recommendation for Way Forward Regarding Rooftop Solar Aspects

To achieve the desired targets set and overcome the possible challenges separate analysis has been done and following recommendations are concluded for below mentioned domains:

- Remote Supervision, Control and Network Operation
- Power Evacuation and Grid Stability
- Power Portfolio Management
- Network Planning
- Power Quality Management
- Policy advocacy and regulatory approvals concepts.

5 Remote Supervision, Control and Network Operation

As the cost of solar PV module decrease more and more consumers will opt for Rooftop solar generation and providing remote monitoring and controlling to each of these consumers is not economically feasible.

Also as per current DERC net metering regulation [2]: "*The Distribution Licensee shall have the right to disconnect the Renewable Energy System at any time in the event of possible threat/damage, from such Renewable Energy System to its distribution system, to prevent an accident or damage*". Thus controlling is only available in case of above mentioned scenarios.

Taking all these factors and recommendation provided in IEEE 1547/IEC 62116 into consideration, it is recommended to provide mandatory remote monitoring for plants above 10 kW and enable controlling mechanism for PV generation plants above 50 kW capacity. Smart metering should be provided for PV plants up to 10 kW capacity. Cost of required equipment for performing desired functions should not be more than 1% of total capital cost of plant. This capacity limit set shall vary as the quantum of rooftop solar increases in near future.

Proposed quantum growth and corresponding monitoring and control quantum is as follow (Table 3).

To ensure effective network management and control, mapping of rooftop consumer in GIS as well as in SCADA is necessary as these consumer can act as both load and generation source.

6 Power Evacuation and Grid Stability

To maintain Grid stability and Frequency within range, specific deviation limits are set for each states. For Delhi, deviation within range of ±150 MW is allowed and correspondingly ±38 MW for TPDDL.

Nature of PV generation source is quite variable as it highly depends on temperature, weather conditions, cloud cover, irradiance etc. As quantum of PV generation increases, dependency on this source increases which can lead to protean OD/UD (Figs. 3 and 4).

As per Delhi solar policy we can expect 2 GW of rooftop solar by 2025. This quantum is highly unreliable as it depends on weather condition, cloud cover etc. In case of under-generation, this high quantum as predicted will not be available. Following options can be explored in case of variation in solar generation from predicted:

Table 3 Expected potential growth of PV generation in TPDDL distribution area and recommended quantum for monitoring & control

Roof Top Solar Generation potential for TPDDL			
FY	Potential Growth in Solar (MW)	Min Quantum	
		Monitoring (kW)	Control (kW)
2017–18	14	10	50
2018–19	43	10	50
2019–20	83	10	30
2020–21	128	10	30
2021–22	178	5	10
2022–23	238	5	10
2023–24	318	5	10
2024–25	418	5	10

- Large scale energy storage can be used to supply power in such case, also energy storage can be used in case of over-generation from rooftop solar to prevent curtailment.
- Opting ADR is another option recommended. As per ADR process consumers are encouraged to take part, incentives to be provided to such consumers. During OD beyond limits, non critical load of these consumers can be switched off.
- OD up to 150 MW for Delhi and subsequently 38 MW for TPDDL is allowed.
- Dynamic Electricity pricing, by varying the cost of electricity and encouraging consumers to either increase or decrease their consumption depending on electricity rates.
- Fast EV charging is another option available in case of over generation to limit under drawl.

7 Power Portfolio Management

As per currently followed process each distribution utility should submit their requirement from each generating station to associated State Load Dispatch Centre on day ahead basis. For predicting the accurate demand, proper forecasting is done.

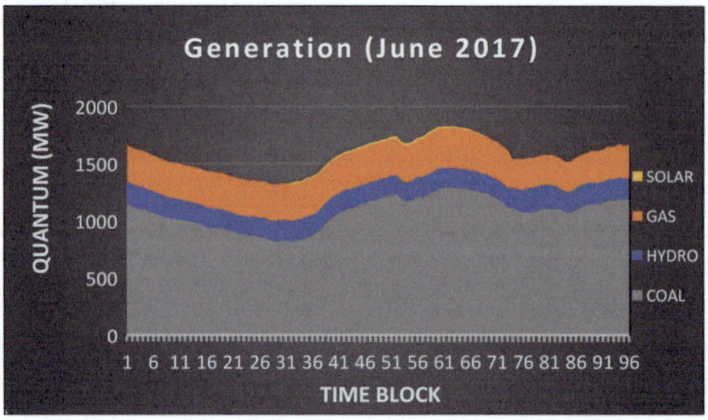

Fig. 3 Meeting TPDDL Peak demand on 6th June 2017 from different energy sources

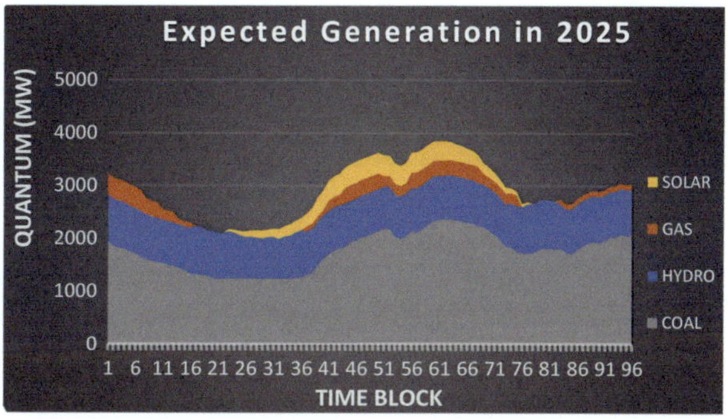

Fig. 4 Expected generation quantum available to meet TPDDL demand in 2025

As per current scenario rooftop solar generation is still not consider while predicting the demand but in near future quantum from rooftop solar generation cannot be neglected. Thus it is recommended to have proper solar forecasting tools which will use day ahead weather and irradiance data and predict the generation.

8 Network Planning

As per current scenario limit on PV penetration is set only at DT level. For Delhi maximum PV penetration limit on a DT is 20% of its rated capacity. It is observed that most of the DTs are operating well within limits and it is recommended that this limit can be increased to 50% without introducing overloading and voltage fluctuations issues. As the quantum of PV generation plants increase there will be need of setting penetration limits feeder wise. These limits will be highly variable and depends on network of a particular state.

8.1 Reverse Power Flow Protection in Case of Fault/Outage

It is noticed that all solar plants does not work till they get grid reference voltage from utility network through net metering. However, in time to come, consumers may be having other sources like battery storage, DG set etc. to operate their load. For connecting all such sources, consumers have to connect through Hybrid inverter having anti islanding protection which will ensure that no energy gets exported to utility network unless reference voltage is available in utility network.

Following network connectivity is recommended to prevent reverse power flow into fault point area or in case of outage by PV generating system (Figs. 5, 6 and 7).

9 Power Quality Management

Invertor injects harmonics into the grid via distribution system, thus it is necessary to approve only those solar connection that maintain standards and harmonics within limits as per IEEE STD 519-2014.

To limit the voltage rise issue which is caused due to injection of active power at connection point of a PV, inverters can be set to operate at non-unity power factor. Thus it will draw reactive current and in turn reduces the voltage.

Harmonic limits as per IEEE519-2014 (Tables 4 and 5):

9.1 Policy Advocacy & Regulatory Approvals

As the penetration of the PV increases network will face new challenges and hence policy advocacy is imminent. Following are the areas in which regulatory amendments are required:

- Control of consumer inverters to the utility beyond 50 kW.
- Increase of threshold limit of PV penetration on DT.
- Authorizing utilities to levy network augmentation charges in case of any requirement of DT capacity upgradation or laying of new network for accommodating solar plants.
- Review of long term PPA's in view of solar installation.
- Review of solar power tariff
- Proposal for requirement of harmonic filters.
- Proposal of large battery storage plants.

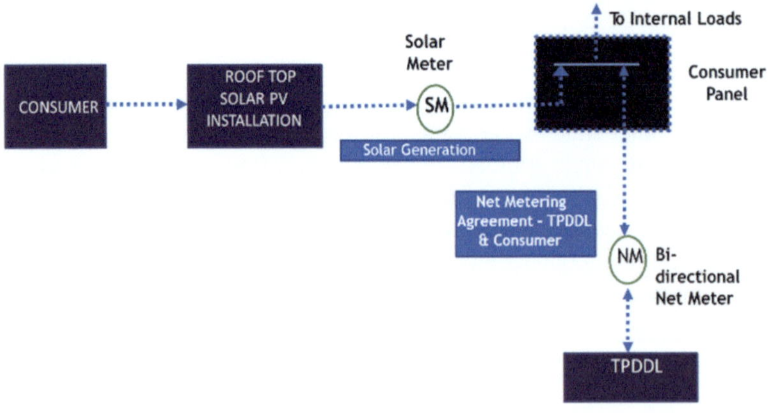

Fig. 5 Existing connectivity of the net meter

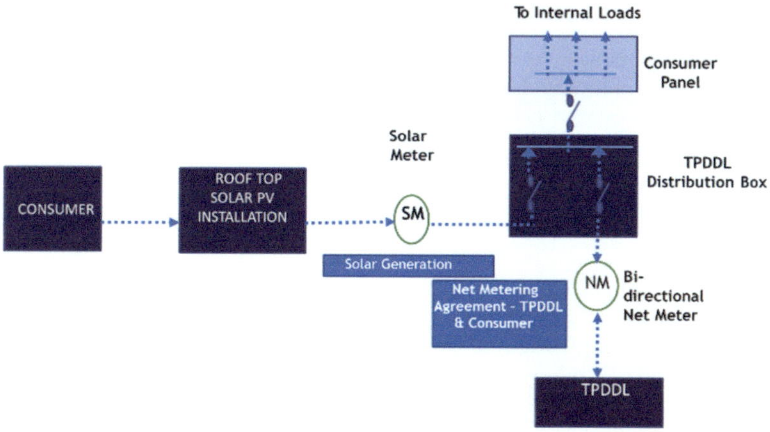

Fig. 6 Proposed connectivity of the net meter without DG set

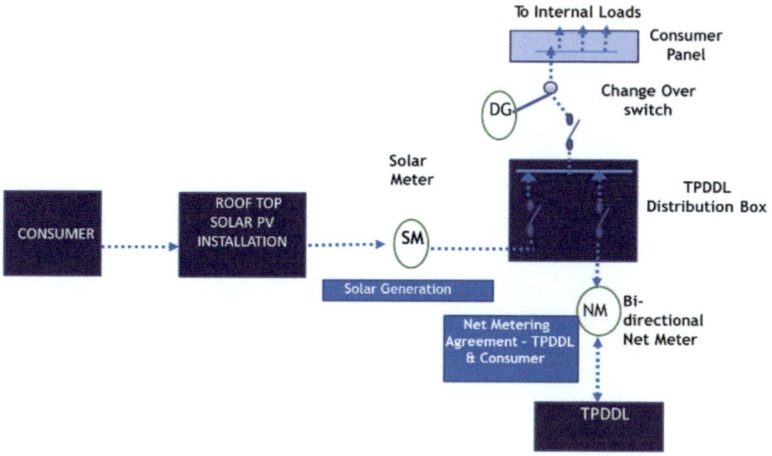

Fig. 7 Proposed connectivity of the net meter with DG set through changeover switch

Table 4 Voltage distortion limits

Bus voltage V at PCC	Individual Harmonic (%)	Total harmonic Distortion THD (%)
$V \leq 1.0$ kV	5	8
1 kV < $V \leq$ 69 kV	3	5
69 kV < $V \leq$ 161 kV	1.5	2.5
161 kV < V	1	1.5[a]

[a]High-voltage systems can have up to 2.0% THD where the cause is an HVDC terminal whose effects will have attenuated at points in the network where future users may be connected

Table 5 Current distortion limits for system rated 120 V to 69 kV

Maximum harmonic current distortion in percent of I_L						
Individual harmonic order (odd harmonics)[a, b]						
I_{SC}/I_L	$3 \leq h < 11$	$11 \leq h < 17$	$17 \leq h < 23$	$23 \leq h < 35$	$35 \leq h \leq 50$	TDD
<20[c]	4	2	1.5	0.6	0.3	5
20 < 50	7	3.5	2.5	1	0.5	8
50 < 100	10	4.5	4	1.5	0.7	12
100 < 1000	12	5.5	5	2	1	15
>1000	15	7	6	2.5	1.4	20

[a]Even harmonics are limited to 25% of the odd harmonic limits above
[b]Current distortions that result in a dc offset, e.g., half-wave converters, are not allowed
[c]All power generation equipment is limited to these values of current distortion
I_{sc}/I_L. Where, I_{sc} = maximum short-circuit current at PCC and I_L = maximum demand load current (fundamental frequency component) at the PCC under normal load operating conditions

10 Conclusion

Amid stride to achieve the vision of 100% renewable both government and utilities need to work together. An attempt has been made in this white paper to identify the key challenges which will be faced and provide possible recommendations to overcome them.

As per current regulations control and monitoring is mandatory for solar plants above 1 MW but it will be necessary to monitor distributed solar generation sources as they will be the main contributor to the total generation. Similarly penetration limit on DT is set to be 20% but it can safely increase to 50% as per normal DT loading. Although this study is performed considering Delhi distribution network, the challenges and recommendation provided will be nearly same for any distribution utility of India. On some issues policy advocacy is required as regulatory framework is still under constant development in India.

References

1. Delhi Solar Energy Policy (2015). Retrived from: http://ipgcl-ppcl.gov.in/documents/renewable/2016_08_03_6_Delhi_Solar_Policy.pdf
2. Delhi Electricity Regulatory Commission (Net Metering for Renewable Energy) Regulations (2014). Retrived from:https://mnre.gov.in/file-manager/UserFiles/Grid-Connected-Solar-Rooftop-policy/DERC_Renewable_Energy_Net_Metering_Regulations_2014.pdf
3. IEC 61836: (1997) Solar photovoltaic energy systems—terms and symbols
4. Best practices guide: implementation of state-level solar rooftop photovoltaic programs in India (2016). Retrived from: https://mnre.gov.in/file-manager/UserFiles/Best-Practices-Guide-on-State-Level-Solar-Rooftop-Photovoltaic-Programs.pdf
5. Government of India—Ministry of New and Renewable Energy, Jawaharlal Nehru National Solar Mission (2017). Retrived from: https://mnre.gov.in/sites/default/files/uploads/mission_document_JNNSM.pdf

Hybrid Microgrids for Diesel Consumption Reduction in Remote Military Bases of India

Soham Chakraborty, Sarasij Das and Manoj Negi

Abstract Grid connection is often not available at remote locations. As a result, diesel generators are commonly used in Indian remote military bases to generate electricity locally. Transportation of diesel to these remote locations is often difficult and expensive task. Communication and surveillance equipments along with border fences and illumination need reliable electricity supply. Any disruption in diesel supply can severely impact the operation of the military base. Fuel transportation routes are often cut off during extreme weather conditions. In addition, diesel generators are prone to failure and cause significant air pollutions. Renewable energy sources are often available at these remote locations. These renewable energy sources can be used to substitute the diesel based generation partially. In this paper, hybrid microgrids consisting of both renewable and diesel generators are proposed for remote military bases. Use of local renewable sources reduces the dependence on external diesel supply. In addition, the consumption of diesel is reduced due to the use of renewable energy sources. Detailed Cost Benefit analysis has been done to show the effectiveness of the hybrid microgrid. It is shown that this hybrid microgrid can be a possible solution for reduction of diesel consumption in remote military bases of India.

Keywords Cost-benefit analysis · Diesel generators · Remote/forward military base · Renewable energy sources · Microgrid

---

S. Chakraborty · S. Das (✉) · M. Negi
Department of Electrical Engineering, Indian Institute of Science, Bangalore, India
e-mail: sarasij@iisc.ac.in

S. Chakraborty
e-mail: sohamchak92@gmail.com

M. Negi
e-mail: manojnegi@iisc.ac.in

R. K. Pillai et al. (eds.), *ISGW 2018 Compendium of Technical Papers*,
Lecture Notes in Electrical Engineering 580,
https://doi.org/10.1007/978-981-32-9119-5_13

1 Introduction

The changing global geo-political dynamics presents the Nation with multiple security challenges. While constantly reviewing its operational preparedness to meet the perceived security challenges, the Indian Army is committed to the defence of the country from external and internal threats across the entire spectrum of warfare. Over the years, there have been constant efforts from Indian government in modernization of infrastructures and technologies for Indian Army to keep pace with technology transition. One of the trending initiatives entrusted to Military Engineering Service (MES) is committed on the field of energy conservation using renewable energy and green technology. The ultimate aim is to enhance energy security for Indian armed forces, especially in remote military bases of India. Across the country's border, Indian Army has several remote bases such as Siachen base, Dras base in Jammu and Kashmir, Tawang base in Arunachal Pradesh, Munabao base in Rajasthan, Dantewada base in Chhattisgarh, bases in Assam's forests etc. These remote bases often employ temporary Forward Operating Bases (FOBs) which are designed to support combat operations in austere environment. Communication and surveillance equipments along with border fences with electrification and illumination are some of the critical loads required by these bases which need reliable and secure electricity. A reliable power system is critical to these remote army outposts in the rough weather conditions and deployed environments to protect and deliver sufficient energy to meet operational needs. Loss of power in these bases can have crippling effects which could put soldier's lives in danger.

Difficult terrains are the main hindrance of getting power from grid for these remote military bases. Grid connection is often not available at the bases situated in remote locations. However, military operating bases have several loads that need power supply. Usually the loads in military bases are categorized as Critical loads i.e. those electrical devices that must be powered at all times to ensure safe mission operations and as Non-Critical loads i.e. those loads that may be briefly turned off without causing a major disruption to safety or operational requirements (Table 1). Therefore, employments of numerous diesel generators which are sized to handle any power contingency along with meeting all load demand are the only way out for these bases. A typical remote military operating base along with their own FOBs and corresponding diesel supply chain is illustrated in Fig. 1. For maintenance and refuelling the diesel generators, soldiers need to put a great deal of efforts in transportation using defence logistics. The cost of fuel can be significantly higher in a deployed environment primary due to the cost of security and transportation through hostile and sometimes difficult terrain. Extreme weather conditions often cut off the fuel transportation route which poses a significant risk to armed forces employed in the remote bases. Apart from that, diesel generators are usually operated in underrated conditions with own set of loads independent from other generators in these remote bases. Therefore most of the diesel generators operate inefficiently which results significant amount of energy loss [1–5]. Moreover, under-rated diesel generators consume more fuel than at rated condition as per fuel consumption chart of any diesel generator [6]. Diesel generators often fail to start or even prone to failure due to run by fuel contaminated by sludges, water and asphaltenes [6].

This burdened cost of diesel and corresponding issues incentivize the utilization of renewable energy sources for these remote military bases. Many researchers have started proposing optimized energy management systems (EMSs) for remote military bases with diesel generators, renewable energy sources and battery based energy storage systems [1–5]. With the joint initiatives of Department of Defense (DoD), Department of Energy (DOE), Department of Homeland Security (DHS), and individual military services (Army, Marines, Navy, and Air Force) of USA, several proposal have been made to employ a cyber-secure microgrid with integration of smart grid technologies, distributed and renewable generation and energy storage on military installations for enhanced mission assurance. There are several remote military bases that have already employed EMSs controlled by locally generated renewable energy resources along with diesel generations [7–10]. Researchers are still working on optimizing the EMS for reducing diesel consumption and enhancing energy security for the operation of remote military bases.

India has immense geographical advantages to avail various types of renewable energy sources such as solar panel, wind farms, hydro plants, geothermal plants, etc. for remote military bases. MES has already tendered projects to develop a

Table 1 Possible loads presents in remote military operating bases

Types of load	Loads
Surveillance system	Electrification of border electric fences
	Illumination of fences
	Target acquisition radars
	Mobile artillery monitoring battlefield radar
	Unmanned aerial vehicles (UAVs)
	Global positioning systems (GPSs)
	Thermal imaging scopes, satellite phones
	Counterintelligence and human intelligence
Logistics systems	Terrain transport vehicles
	Heavy equipment transporter
	Demountable rack off load and pickup
	Support vehicles
Weapon systems	Ballistics and cruises
	Artilleries
	Armored combat vehicles
Other loads	Lighting loads for military base camps
	Heating loads for military base camps
	Loads in hospital facilities
	Pumping systems

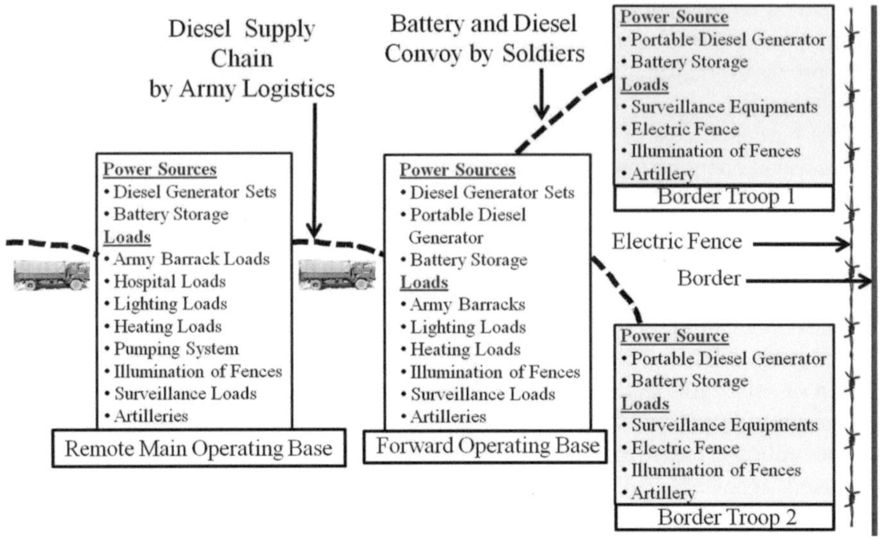

Fig. 1 Overview of power generation and diesel supply chain for remote military operating bases

2 MW solar plant at Jodhpur, a 3 MW solar plant at Jaisalmer in Rajasthan and is going to tender a 1 MW solar project to be developed at Binnaguri Cantonment, a 1 MW solar project at 5 BRD AFS in Sulur and a 3 MW solar project in Ezhimala, Kerala [11]. However, these projects are tendered mostly in urban or sub-urban areas where grid is available. Therefore, these projects are not beneficent for remote military bases where grid is not available. Employment of renewable sources in the remote military bases surely reduces the burden of fuel cost by substituting the diesel generator based electricity generation. However, these renewable resources have high dependency on weather conditions which is uncertain and variable. To ensure reliable power generation, both renewable sources and diesel generation should be employed. Therefore this paper is proposing a hybrid microgrid with both available renewable energy farms and diesel generators for remote military bases. Several possible renewable resources available in remote regions of India are discussed that can come handy in the proposed hybrid microgrid for remote military bases. A Cost Benefit Analysis has also been done to show the effectiveness of the hybrid microgrid in reduction of diesel consumption leads to reduction in running cost of military bases and in lowering the burden imposed by diesel dependency of military bases. It is shown that a hybrid microgrid can be a possible solution for reduction of diesel consumption in remote military bases of India.

2 Proposed Hybrid Microgrid for Electrification of Remote Military Bases of India

In this section, a hybrid microgrid consists of local renewable energy sources along with conventional diesel generators are proposed for electrification of remote military bases of India.

The main objective of the proposal is to achieve reduction in diesel consumption leads to reduction in running cost of military bases, lowering the burden imposed by diesel dependency of military bases. Figure 2 shows the foundation of the proposal to employ hybrid microgrid for remote military bases of India. It shows the hierarchy of the Indian military force structure which starts from army commands (usually situated at urban areas) to the lowest components i.e. platoons (usually situated in remote areas in deployed environments). Usually the commands/corps/ divisions or even brigades sections have access to the grid connectivity. The components such as battalions, companies or platoons are mostly situated in remote regions of India where grid connectivity is not possible. However, due to locational advantages, these remote regions are endowed with several renewable energy resources which can come handy for electrification for the bases. In the next section, several possible renewable resources available in remote regions of India are discussed.

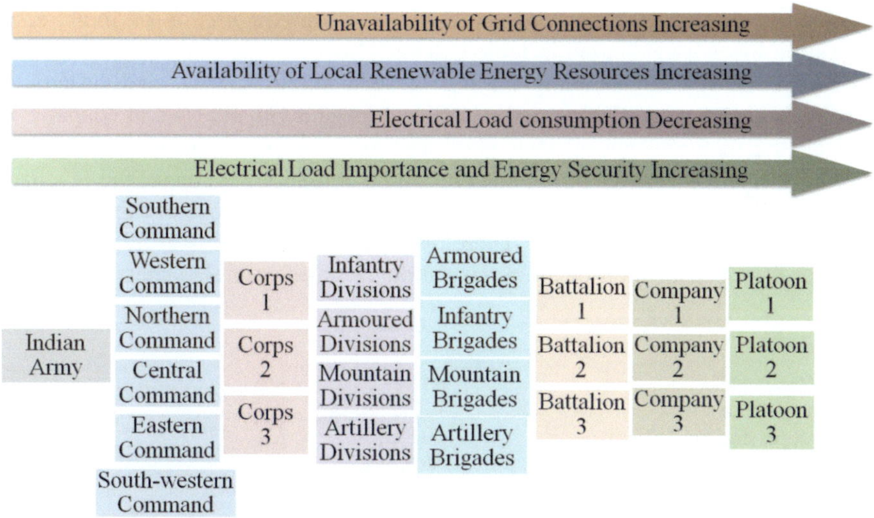

Fig. 2 Indian army structure

2.1 Possibilities of Renewable Energy Integration for Remote Military Bases of India

With geographical advantages of India, it is possible to avail various types of renewable energy sources for electrification of remote military bases. Remote locations of India are endowed with rich renewable energy resources such as solar panel, wind farms, hydro plants, geothermal plants, etc. Figure 3 shows the geographical overview of majorly available renewable energy resources in India [12]. Major remote military bases of India are situated in the Indo-Pak border area in Rajasthan and J&K states and Indo-China border in J&K State and north east region of India. These remote military bases can therefore utilize the renewable energy resources available nearby for integrating into hybrid microgrid as proposed in this paper. The details regarding the current scenario and available potentials for utilizing the renewable resources in these remote places are discussed in the following section.

2.1.1 Solar Energy Integration

According to [13], Rajasthan and Gujarat are India's most solar-developed states, with its total photovoltaic capacity reaching 2156 and 1262 MW respectively by the end of 2017. Availability of clear sunny days in a year and vacant spaces are the two major requirements for developing a solar energy farms. In Himalayan regions of India, especially in Jammu and Kashmir or north-east section of India, solar integration has not yet started in large extents. However, the Ministry of New and Renewable Energy is promoting the use of solar energy devices and systems in those regions. Remote military bases in Rajasthan and Gujarat state are capable to avail all the requirements of solar plant and generate power independently. The increasing interest in solar power integration by Indian government, the levelized cost of solar PV electricity becomes cheaper than fuel cost of any pit head coal-based power plants in India. Therefore, solar energy seems to have immense potential for electrification of the remote military bases in these states.

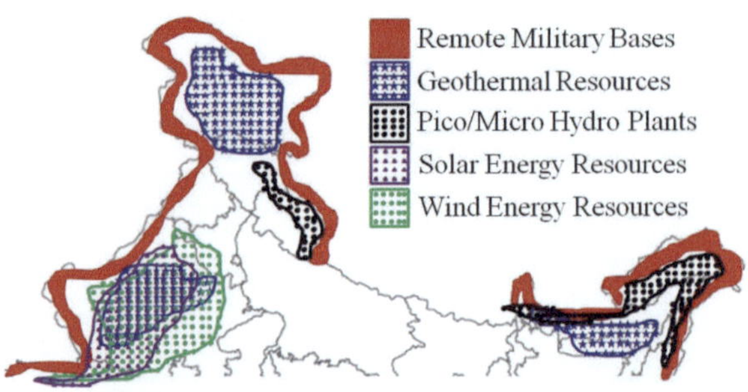

Fig. 3 Geographical overview of India's remote military bases and possible renewable resources

2.1.2 Wind Energy Integration

Wind power has become one of the prominent power generation technologies in India. This potential is distributed mainly in the states of Tamil Nadu, Andhra Pradesh, Karnataka, Gujarat, Maharashtra, Rajasthan and J&K [14]. Free land without any major obstacles for high wind flow rate is the main requirement for developing a wind energy farm. The remote regions of Jammu and Kashmir State are potential wind energy areas, which are yet to be exploited [15]. Especially, the Kargil, Ladakh, Gilgit of J&K state are potential wind energy areas where wind speeds are higher during the winter months, complimentary to the hydro power available during the summer months from the snow melt water. The levelised tariff of wind power reached a record low of ₹2.43 per kWh in 2017 [14]. Therefore, wind energy farm seems to have immense potential for electrification of the remote military bases in these states.

2.1.3 Pico/Micro Hydro Power Integration

In India, pico and micro hydro plants are categorized by the power generation capabilities (pico: 1–10 kW, micro: 10–100 kW). India's economically exploitable and viable hydroelectric potential is increasing at a rapid rate by integration of these pico/micro hydro plants. Northern and north-eastern regions of India are mostly endowed with rich hydro-plants [16]. The hilly and rough terrains in these regions pose high current in small/medium rivers which in turn facilitate the hydropower generation. Therefore, remote military bases in these regions have immense possibilities to employ small hydro plants for local power generation.

2.1.4 Geothermal Energy Integration

Ministry of New and Renewable Energy contemplates major initiative in RDD&D of Geothermal technology for harnessing the geothermal energy in the country. There are seven geothermal provinces in India e.g. the Himalayas, Sohana, West coast, Cambay, Son-Narmada-Tapi (SONATA), Godavari, and Mahanadi. Northern region of India, especially Jammu and Kashmir State is reach in geothermal energy resources. Several projects are going on aiming to utilize geothermal energy sources efficiently [17, 18]. Therefore, remote military bases in Himalayan regions have immense possibilities to employ several either flash steam plants or binary plants for power generation locally.

2.2 *Proposed Hybrid Microgrid for Remote Military Bases*

An overview of proposed hybrid microgrid for remote military bases is shown in Figure 4. It consists of all possible renewable energy sources integrated along with conventional diesel generator with battery energy storage system (BESS). Due to unavailability of grid connection, the microgrid network will mainly operate in island mode. Unlike the common practices of military network, the primary sources for the proposed network are the available renewable energy sources depending on the geographical environment as discussed in previous section. All the critical and non-critical loads as well as FOBs are supplied through a distribution network. A BESS is usually employed in microgrid to manage the surplus or deficit in power generation by renewable energy sources. The unavailability of enough generation by renewable sources in severe weather conditions can be a possible situation. Therefore, the diesel generators are employed for backup operation in these austere conditions to ensure sustained power generation at all time. The possible aspect in development of distribution network integrated with all the sources are important to be mentioned briefly. For remote military bases, the choice in network arrangement will be determined by the condition of terrain where the base is situated and the geographical spread of the operating bases. The choice in allocation of local generations will be determined primarily by the available locations, challenges in employment of the renewable farms in the terrain etc. Apart from that, a properly functioning EMS is required for the proposed hybrid microgrid with the following guiding principles.

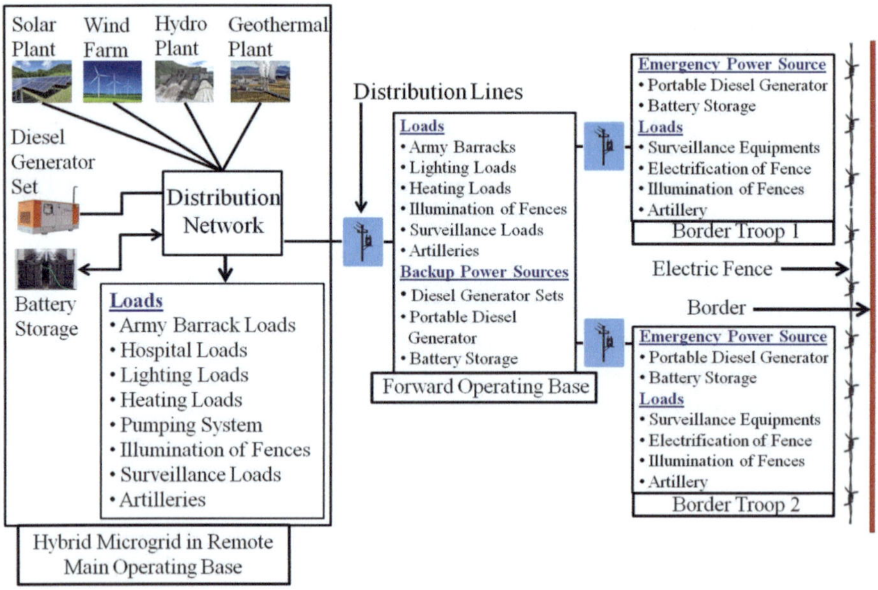

Fig. 4 Proposed hybrid microgrid for electrification of remote military operating bases of India

- Maximum utilization of renewable energy sources to supply all critical and non-critical loads to achieve maximum fuel saving of diesel generator sets.
- Maintaining uninterrupted power supply to critical loads especially loads in FOBs in austere conditions considering the unavailability of renewable sources.

3 Cost-Benefit Analysis of Proposed Hybrid Microgrid

The cost-benefit analysis of the proposed hybrid microgrid is discussed by considering a general remote military base. The base contains both critical and non-critical loads connected into the distribution network. With the available information, the load profile is selected from [1, 2] as shown in Fig. 6. Two different cases are selected for the cost benefit analysis.

- Case-1: The practiced system that contains only diesel generator sets and BESS as shown in Fig. 5a.
- Case-2: The proposed hybrid microgrid that contains solar PV plant, diesel generator sets and BESS as shown in Fig. 5b.

For both the cases, the loads are assumed to be same as mentioned previously. The details regarding the chosen diesel generator, PV panel, BESS are furnished in Table 2 [19, 20]. This information is required for the cost-benefit analysis for both the cases. The cost benefit analysis will fundamentally calculate the Initial Capital Cost and the Running Cost per year. The fundamental cost equations [21] for diesel generator set, battery storage system and PV panel are provided below.

- Cost Function for Diesel Generator Unit

$$C_T^D(P_G^D, t) = C_U^D(P_G^D) + C_{O\&M}^D(P_G^D, t) + C_F^D(P_G^D, t) \quad (1)$$

- Cost Function for PV Panel

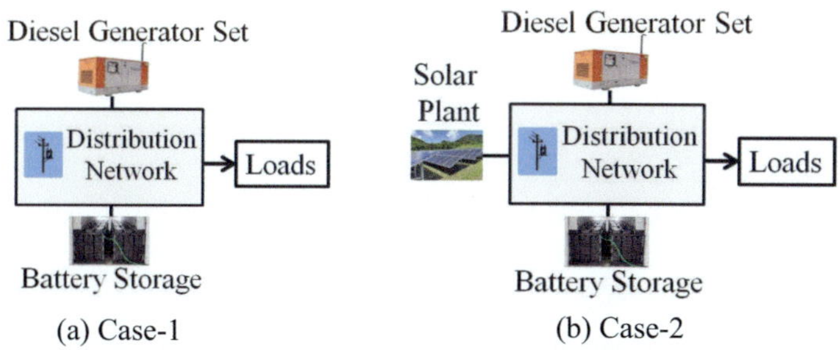

Fig. 5 **a** Practiced network (Case-1) and **b** proposed hybrid microgrid (Case-2) for remote military operating bases

Table 2 Model description for cost-benefit analysis

Sources	Details
Diesel generator unit [20]	TATA diesel generator RT-75-697TC50
	kVA rating: 75 kVA
	kW rating: 60 kW (0.8 pf)
	Efficiency curve: shown in Fig. 7b
	Fuel consumption chart: shown in Fig. 7a
	Lifetime: 25,000 h
	Minimum loading factor: 10%
PV panel [19]	Kenbrook solar 300 kW off-grid plant
	Capacity of plant: 300 kW
	Lifetime: 25 years
	De-rating: 80%
	Degradation (1–10 years): 0.05%
	Degradation (11–25 tears): 0.67%
BESS [25]	EATON 9355
	Rating: 24 kVA/kW

$$C_T^{PV}(P_G^{PV}, t) = C_U^{PV}(P_G^{PV}) + C_{O\&M}^{PV}(P_G^{PV}, t) \quad (2)$$

- Cost Function for BESS

$$C_T^{BESS}(P_G^{BESS}, t) = C_U^{BESS}(P_G^{BESS}) + C_{O\&M}^{BESS}(P_G^{BESS}, t) \quad (3)$$

where, $C_T^D, C_T^{PV}, C_T^{BESS}$ are total cost for diesel generator set, PV panel and BESS respectively. $C_U^D, C_U^{PV}, C_U^{BESS}$ are initial capital cost of diesel generator set, PV panel and BESS respectively. $C_{O\&M}^D, C_{O\&M}^{PV}, C_{O\&M}^{BESS}$ are operation and maintenance cost of diesel generator set, PV panel and BESS respectively. C_F^D is the fuel consumption cost of diesel generator. t is time in hour and P_G^D, P_G^{PV} and P_G^{BESS} are the power output of diesel generator set, PV panel and BESS respectively. The information regarding cost benefit analysis for both the cases is discussed next.

3.1 Cost Analysis

3.1.1 Cost Analysis for Case-1

In Case-1, the procedure of cost analysis is mentioned below.

- The total demand is supplied by diesel generator units along with BESS (Fig. 6a) ensured by employed EMS.

- The cost of diesel unit is considered ₹2.3 lacs/unit [20]. As per generator sizing and peak load demand, total number of generators is calculated as 9. C_U^D of (1) is calculated as ₹20.7 lacs.
- In calculation of running cost, $C_{O\&M}^D$ and C_F^D are important to be considered. Based on the lifetime and minimum loading factor as mentioned in Table 2, O&M rate is considered as ₹1.92 per hour per kW [22]. Therefore, $C_{O\&M}^D$ is calculated as ₹8.41 lacs per year.
- The efficiency and fuel consumption of a diesel generator shown in Fig. 7a and Fig. 7b are used to calculate the fuel consumption [20, 23]. The cost of diesel per litre is considered as ₹60 [24]. Therefore, C_F^D is coming around ₹16.73 lacs per year.
- C_F^{BESS} is calculated from the information of [25] which is coming around ₹1.75 lacs.
- $C_{O\&M}^{BESS}$ depends on Depth of Discharge, charging/discharging cycle etc. [2]. ₹2.5 lacs/year is assumed in this analysis by considering the operation of the BESS as mentioned in [2].
- The result is showing in Fig. 8.

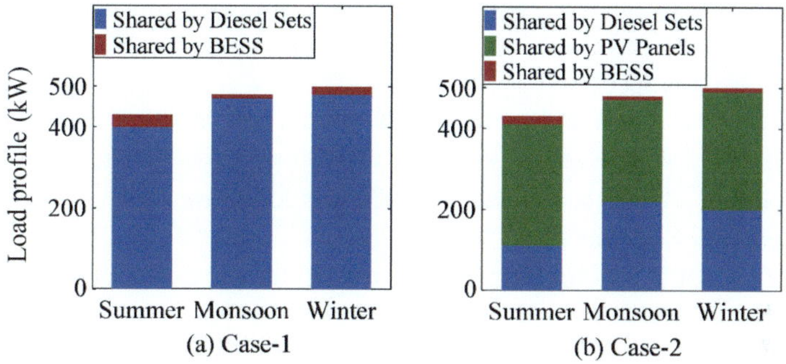

Fig. 6 Load profile of the military operating base for cost analysis

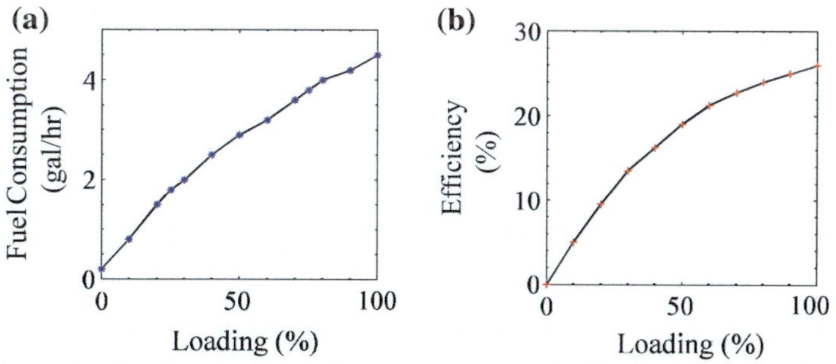

Fig. 7 **a** Fuel consumption curve, **b** efficiency curve of diesel generator

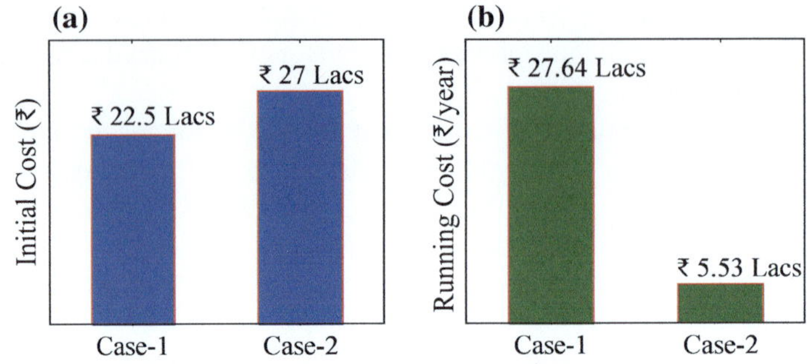

Fig. 8 Results of cost analysis. **a** Initial cost, b running cost

3.1.2 Cost Analysis for Case-2

Cost analysis for Case-2: In Case-2, the procedure of cost analysis is mentioned below.

- The total demand is supplied by PV panel, diesel generator units along with BESS (Fig. 6b) ensured by employed EMS.
- The shared load by PV panel is determined primarily on the available solar irradiation and temperature on the installed area [26]. The BESS is assumed to be same here as in Case-1.
- The employed diesel units and BESS system are assumed to be present in Case-2 also to ensure safest operation of military bases. Therefore, C_{U}^{D}, $C_{O\&M}^{D}$, C_{U}^{BESS} and $C_{O\&M}^{BESS}$ are calculated as in the previous case.
- C_{F}^{D} is coming down to ₹1.86 lacs per year.
- The initial capital cost for PV plant consists of module cost and civil engineering cost. Module prices are considered at $0.48/W [19]. Civil and general works cost is ₹35 lacs per MW which includes mounting structures and cost of power cable, transformers [19]. Therefore, C_{U}^{PV} is coming around ₹4.5 lacs.
- The running cost for PV panel includes only module degradation cost. Module degradation cost is considered as ₹640 per kW per year which depends on the de-rating factor and environment of the installed area [22]. Therefore, $C_{O\&M}^{PV}$ is coming around ₹1.92 lacs.
- The result is showing in Fig. 8b.

3.2 *Discussions on Benefits of Proposed Hybrid Microgrid*

The cost of setting up a solar power project has reduced significantly over the past years. With Indian government's initiatives, more renewable integration is being planned in future. Therefore, the initial capital cost for installation of new

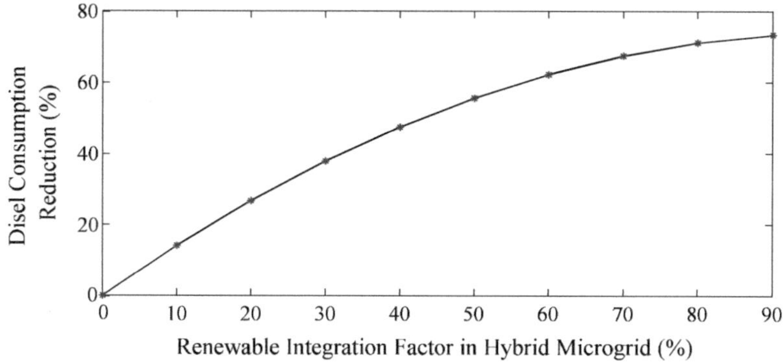

Fig. 9 Diesel consumption reduction versus renewable integration

renewable plant is expected to be reduced in future. Figure 8a shows that the initial capital cost for the hybrid microgrid is fairly comparable with the initial cost required for installation of new diesel generator set. The capital cost for diesel generator set based network is ₹22.5 lacs and for proposed hybrid microgrid (integrated with solar) is ₹27 lacs. Whereas, the benefit of renewable sources over diesel generator sets is the significantly less running cost. Figure 8b clearly shows that the integration of PV based hybrid microgrid reduces the running cost significantly. The running cost for diesel generator set based network is ₹27.64 lacs per year and for proposed hybrid microgrid is ₹5.53 lacs per year. Figure 9 shows that for the application of renewable sources, there is increasing profile in the amount of diesel consumption reduction. Also, Fig. 10 shows that there is diminishing profile in the energy loss by diesel consumption generation. This likely is occurring because increase in renewable energy sources offsets the load demand supplied by diesel generators. Therefore, it can be concluded that a hybrid microgrid powering remote military bases of India has some clear advantages over a purely diesel generator powered system.

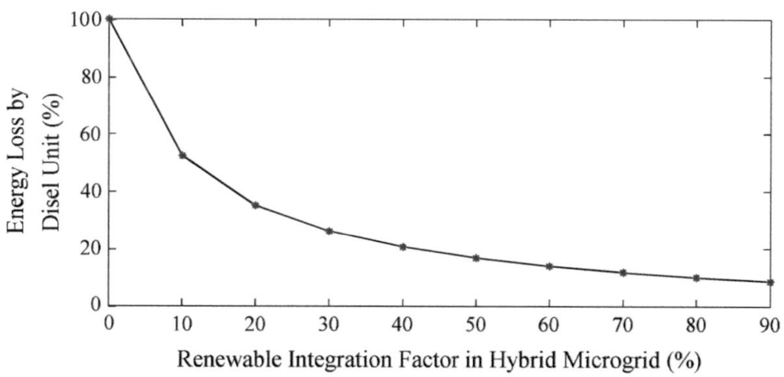

Fig. 10 Energy loss by diesel generation versus renewable integration

4 Conclusion

Grid connection is often not available at remote locations. As a result, diesel generators are commonly used in Indian remote military bases to generate electricity locally. Transportation of diesel to these remote locations is often difficult and expensive task. Any disruption in diesel supply can severely impact the operation of the military base. This paper proposes a hybrid microgrids consisting of both renewable and diesel generators for remote military bases. Use of local renewable sources reduces the dependence on external diesel supply. In addition, the fuel consumption and energy loss by diesel is reduced significantly by the use of renewable energy sources. Cost Benefit analysis shows that the running cost reduces significantly by the proposed hybrid microgrid with fairly comparable initial cost. Therefore, the proposed hybrid microgrid can definitely be a possible solution for reduction of diesel consumption in remote military bases of India.

Acknowledgements This work was supported by the Science and Engineering Research Board of Government of India under Grant ECR/2015/000504.

References

1. Kelly RL et al (2013) Reducing fuel consumption at a remote military base: Introducing an energy management system. IEEE Electrification Mag 1(2):30–37
2. Anglani N et al (2017) Optimized energy management system to reduce fuel consumption in remote military microgrids. IEEE Trans. Ind. Appl. 53(6):5777–5785
3. Tyner J et al (2011) The design of a portable and deployable solar energy system for deployed military applications. In: IEEE systems and information engineering design symposium, Charlottesville, VA, pp 50–53
4. Skowronska-Kurec AG et al (2012) Demonstration of microgrid technology at a military installation. In: 2012 IEEE power and energy society general meeting, San Diego, CA, pp 1–2
5. Severson B et al (2013) Feasibility study of photovoltaic panels in military temporary housing structures. In: 2013 IEEE green technologies conference (GreenTech), Denver, CO, pp 78–84
6. http://www.dieselserviceandsupply.com/Fuel_Management.aspx
7. Hall TL, Fsicher CH (2014) Microgrid enabled distributed energy solutions (MEDES)—Fort Bliss Military Reservation
8. SPIDERS Phase 2 Fort Carson Technology transition public report, joint capability technology demonstration (JCTD) version 1.1, Naval Facilities Engineering Command, 9 October 2014
9. Paquette A, Harley R, Bhavaraju V, Krstic S, Theisen P (2014) Design of the Fort Sill microgrid. In: 2014 IEEE energy conversion congress and exposition (ECCE), Pittsburgh, PA, pp 4640–4646
10. Fort Hunter Liggett, CA—1 MW solar microgrid. USACE Department of the Army/ Sacramento District. http://www.bothman.com/projects/federal-government/fort-hunter-liggett-phase-1-1-mw-solar
11. Military Engineering Services. https://mercomindia.com/
12. http://www.indiaenergyportal.org
13. State wise installed solar power capacity. Ministry of New and Renewable Energy, Govt. of India, 31 October 2017. 24 November 2017

14. Indian wind energy—projects, companies, research, data, statistics—energy alternatives india. In: EAI. Retrieved 27 Nov 2010
15. Estimation of installable wind power potential at 80 m level in India. Retrieved 16 May 2015
16. Status of hydro electric potential development in India. Retrieved 17
17. Harinarayana T et al (2006) Exploration of geothermal structure in Puga geothermal field, Ladakh Himalayas, India by magnetotelluric studies. J Appl Geophys 58(4):280–295
18. Harinarayana T et al (2004) Magnetotelluric studies in Puga valley geothermal field, NW Himalaya, Jammu and Kashmir, India. J Volcanol Geoth Res 138(3):405–424
19. Benchmark capital cost of solar PV project 2016–2017. http://www.ezysolare.com/blog/knowledge-center
20. http://www.rishabhengineering.com
21. Agalgaonkar AP, Dobariya, CV, Kanabar, MG, Khaparde SA, Kulkarni SV (2006) Optimal sizing of distributed generators in microgrid. In: 2006 IEEE power India conference, New Delhi
22. Omogoye OS, Ogundare AB, Akanji IO (2015) Development of a cost-effective solar/diesel independent power plant for a remote station. J Energy
23. Otasowie PO, Ezomo PI Life cycle cost analysis of diesel generator set and national grid in Nigeria. J Emerg Trends Eng Appl Sci (JETEAS) 5(5): 363–367
24. https://www.iocl.com/TotalProductList.aspx
25. http://powerquality.eaton.in/India/Products-services/Services/IN-Large-UPS-Services.asp?cx=36
26. Region-wise monthly solar power generation, CEA. Retrieved 30 Nov 2016

Microgrid for the Purpose of the Development of Net-Zero Energy Green Buildings and Communities Towards 100% Renewables Vision: Microgrids as the Buildingblocks

Abilash E. T. Nair and PMP®

Abstract The usage of energy in the world has greatly increased, including those from renewable resources. The usage of renewable energy, especially in the form of distributed generation, is beneficial for the environment. The delicate ecosystem needs effective systems for the better utilization of the natural energies. Solar and Wind power lead the types of renewable energy that are integrated into electrical power system, thanks to major advancements in conversion technologies like solar panels and wind turbines. However, wind and solar are intermittent and hence need complex control systems in the power systems to address such variations. State-of-the-art solutions are available today that enable the usage of these variable energies without losing the full capacity of generation and also stabilize the overall grid on a modular level to provide the maximum efficiency for use with the loads. These are made possible due to Microgrid control systems. For a (net-zero energy) green building requirement, the control of power through such automated systems make available the best quality power and also utilize any generation and storage sources within the building. These hybrid–like systems make possible optimum utilization of the grid's sources and varied loads to give the most stable system with quality power control.

Keywords Microgrid · Multiple source · Distributed generation · Energy storage

1 Introduction

With the invention of control technologies, there has been considerable increase in the effective use of natural energy resources in the world. In this manner the delicate ecosystem needs effective systems for the better utilization of the natural energies like Solar and Wind. While optimizing the utilization of energy resources by

A. E. T. Nair (✉)
Power & EV Charging, ABB India Ltd., Bangalore, India
e-mail: Abilash.e.t@gmail.com

R. K. Pillai et al. (eds.), *ISGW 2018 Compendium of Technical Papers*, Lecture Notes in Electrical Engineering 580,
https://doi.org/10.1007/978-981-32-9119-5_14

improving the efficiency is expected to bring about some improvements, using renewable energies like solar and wind would significantly address the above eco-system issue. Such renewable energy sources are not completely in our control and may not closely match with the variable loads that may be present in a building or a community. However, this can be largely overcome by combining such resources and adding a reservoir (Energy storage). Such blocks of Microgrid with high levels of renewable energy penetration adding to full visibility, controllability also can be combined together to provide highly resilient and intelligent power grid. Such solutions inertly give rise to green buildings and communities with Net Zero Energy. Energy-balanced localized systems reduce the need for transmitting and distributing power over longer distances, resulting in inherently smaller distribution losses, making the whole power system greener. Such ability to generate, store, and utilize nature-based (renewable) electricity locally from clean and sustainable energy sources without the need for transmission enables transformational possibilities for the electric utility and transportation infrastructures. Every technology into smart Microgrid ecosystem (like Solar, Wind, Energy Storage, etc.) complements each other to provide betterment at every transformational stage of the energy system for the future. The use of a hybrid system enhances system-wide efficiency and reduces the overall lifecycle cost and saves the environment from further damages. These hybrid systems can also provide emergency power in geographical areas that are disaster-prone. Emerging Economies with non-existent, non-reliable, or weak grid infrastructure can best use Microgrid for better utilization of renewable energy resources. With this there is very high utilization of concepts like cluster Microgrid, building-level Microgrid and renewable integration to move towards a 100% renewable energy-based world.

It is well recognized that the conventional forms of energy from petroleum, coal, natural gas, etc. are harmful to the environment, not only because of emission of green house and noxious gases into the atmosphere but also because of major mining activities, both on land as well as offshore, disturbing fragile eco-systems. With the advancement of technologies, there has been considerable increase in the effective use of renewable energy resources in the world. Such systems result in better utilization of the natural energies, typically from solar and wind, which help maintain the delicate balance of our eco-system. While optimizing the utilization of energy resources by improving the efficiency is expected to bring about some improvements, using renewable energies like solar and wind would significantly address the above eco-system issue. Renewable energy sources, however, are not fully in our control and may not closely match with the variable loads that may be present in a building or a community. This can be largely overcome by combining such resources along with energy storage, resulting in a Microgrid which could be either operated in conjunction with the utility supply or independently if need be. Microgrid with high levels of renewable energy penetration with full visibility and controllability can be combined together to provide highly resilient and intelligent power grid. These solutions inherently can give rise to *Net Zero Energy* (also known as Zero Net Energy—ZNE) buildings and communities. Energy-balanced localized systems reduce the need for transmitting and distributing power over longer distances, resulting in inherently smaller distribution losses, making the whole power system greener. This is especially relevant in Indian scenario with one of the highest distribution system losses in the world. Ability to generate, store, and utilize electricity locally from clean and sustainable energy sources without the need for transmission enables transformational possibilities for the electric utility and transportation infrastructures. Various technologies used in smart Microgrid ecosystem (including solar, wind, energy storage, etc.) complement each other to provide betterment at every transformational stage of the energy system for the future. The use of a hybrid system enhances system-wide efficiency and reduces the overall lifecycle cost and saves the environment from further damages. These hybrid systems can also provide emergency power in geographical areas that are disaster-prone. Emerging economies with non-existent, non-reliable, or weak grid infrastructure can constitute best use cases of Microgrid for better utilization of renewable energy resources.

2 Renewable Energy-Related Problems

Mother Nature has endowed humanity with abundance of energy. This includes various forms, only bound by the rule that "Energy can neither be created nor destroyed"; One flexibility, however, is that energy can be changed from one form to another.

2.1 Intermittence in Energy Sources - a need for solution

Fossil fuels (the origin of which is actually solar) and nuclear fuels, are forms of stored energy, that can be used in a power plant to derive steady energy. These have been the main stay of major power grids. Microgrid, often called off-grids, too have been using typically diesel fuel with minimum control systems. On the contrary, energy from wind, solar etc., are dependent on nature, constituting a rapidly variable, but an important parameter to be considered while running a power system. While wind could be highly seasonal, solar is affected by clouds and shades. Both could be affected during extreme weather conditions such as during storms. Another important consideration with solar energy is that it is available only during daytime, peaking around noon, while the maximum load in the grid is often around evening hours. Solar eclipse condition, though rare, could be another aspect to be considered while integrating large scale solar into grid systems. While availability of more economical energy from nature like solar and wind are changing the scenario rapidly, we have not yet reached a stage when we can outright shut down base fossil or nuclear power plants and substitute them with wind and solar plants. So long as renewables in a grid system form a small percentage, the overall grid stability is not threatened. However, when the penetration levels of renewables reach a critical level one needs to add necessary control systems to do a multitude of controls including load control, energy storage systems to absorb excess generation from the renewables or augment them during minimum or nil generating conditions so as to stabilize the grid system. The control system should watch out for not only energy balancing to have absolute stability of frequency but also the reactive power flow to provide adequate voltage support to enable quality power in the system. One important functionality of the control system would of course be to optimize the resources which would result in minimum capital and operating expenses. Such systems are commercially feasible for small building and small community levels by way of Microgrid technology.

2.2 Indian Power Grid Scenario

The energy mix in developing is an ever changing chart that specifies the amount of implementation that the world expects from developing countries like India and China. The Oil mix is considerably larger and the 56% mixture using petroleum products make the largest countries in terms of population to be largest in need for these powers from these sources. This in-turn generates the largest amount of pollution and Smog in the atmosphere. With renewable energies like Solar and Wind increase the nature energy consumption, there is greater need to increase the supply-demand ratio for the same with better and efficient systems and increase the energy mix contribution to more than the current 25%. The Indian grid as a whole has contributed to the indentation of solar into the electrical grid and in-turn cause the current de-stabilization of the grid scenario in India. These figures in the long run will give India the needed stability with generation and consumption reaching the tip-off point. The complex distribution will need evolution for increasing the efficiency of the system and in the long run make the possible of 100% renewable consumption.

The power grid in India is considered as one of the most complex of all the grids across the world. While the major portions of the grid system have been strengthened, thanks to a number of 765, 400 kV and HVDC lines, the grid is still evolving rapidly, including plans to add 1200 kV AC system. The rapid addition of significant renewable power into the grid could pose major challenges unless controls are put into place at distribution levels themselves down to building and small community levels by way of Microgrid. The present scenario of installed capacity in India are indicated in Fig. 1. The government has ambitious plans to install more than 175,000 MW of renewables by 2022AD, which would surely mean more penetration of renewables into grid. With Distributed generation and consumption with Electric Vehicles, the dream for 100% renewable energy is possible as the system evolves for better supply-demand ratio. The possibility of this scenario in the Indian grid is quite near and just past the horizon with several implementations planned (Fig. 2).

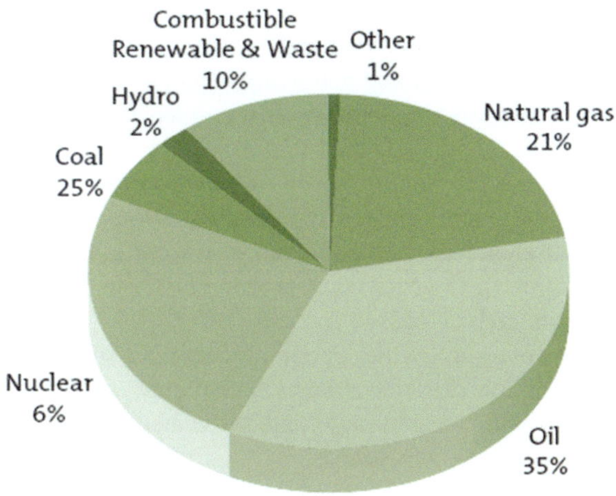

Fig. 1 Energy mix in developing countries

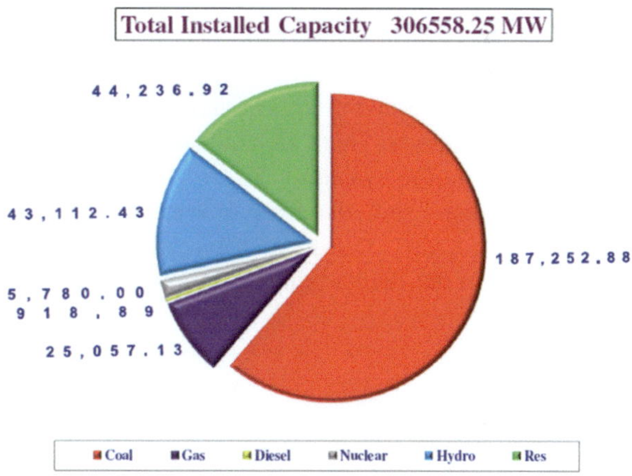

Fig. 2 Installed capacity in MW. *Source* CEA

2.3 *Maintaining the Integrity of the Power Flow*

There are numerous fluctuations and variations caused by the grid. This on the whole is caused by various electrical systems, the parameter changes and the effective balancing the grid tries to perform. The variations in the electrical system are caused by the non-availability of the expected demand and supply at the expected place at the expected time. Grid parameters change when a load wants the power but the sources are not able to provide the same at that instant of time. In steady state scenario, such unbalance may cause unforeseen power flow across the systems, causing both frequency and voltage stability of the system, causing brown outs, frequency variations and occasional blackouts.

In depth power system analysis, both static and dynamic, experienced operating personnel, accurate load and generation forecasts, high speed communication, protection, control and automation system etc. are all essential to maintain integrity of a major grid system. With reference to renewables, predictions over a larger geographical area could be reasonably done, and may not pose operational issues so long as the penetration levels are not too high. It is not out of place to mention that even after adjusting such prediction errors, a certain amount of redundancy of critical elements such as transmission corridors, spinning reserves etc. need to be provided while operating a power system to account for loss of at least one such element such as caused by a fault.

However, in smaller geographical areas, such as a small community or building, the variation of renewables is a challenge and hence the system needs to be augmented with local storage and associated control system for guaranteeing quality and resilient power source.

2.4 Components Causing Issues

The issues that were mentioned earlier are in real time and hence any solutions need to be in real time too. There are several combinations that work together to provide the balancing in the system.

On the other hand, there are several factors that cause the power grid to be unstable, and cause ‘nuisance’ trips and affect power flow. Some of the major contributors to the fluctuations in the modern power grid are described below.

2.5 Sources (Solar and Wind)

The intermittent form of electricity may cause the system to become unstable due to the unplanned fluctuations in the system. The most prominent sources of renewable energy in the system are solar and wind (Fig. 3).

Solar Energy is one of the most intermittent and is available only a maximum of 12 h a day. These are the major disruptors in the grid and demand special measures for stability and efficiency.

Wind energy is also quite intermittent in nature and the electrical system that handles these changes needs to be able to cater to the disruptions provided. This being said, the wind might be available throughout the day and cannot be predicted over a small geographical area. This causes the destabilization of a weak grid system.

Apart from Wind and solar, there are several different renewable energy systems like small hydro, biomass, etc. that have their own patterns and intermittent characteristics.

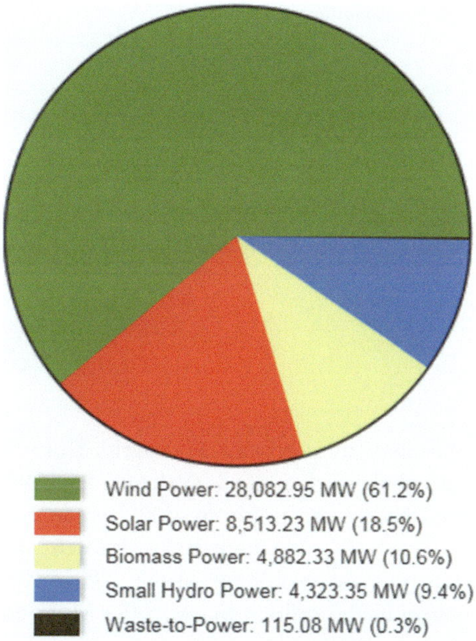

Fig. 3 Renewable energy installed capacity. *Source* CEA

The integration of wind and the ability of the combined renewable sources may give steadier energy as many of the renewable energy systems compensate each other. This eliminates some of the intermittence and deficiencies from the individual power systems. A smoother power generation may be expected when there are multiple sources in the power system. The sources need to be as varied as possible, somewhat sized to reduce and optimize the net variations at the bus level. This aspect can be one of the design features of modern day Microgrid.

2.6 Sources (DG Set)

There are several sources of electricity including renewable and non-renewable. But the most prominent of all the systems is the diesel generator set (DG). The DG set burns fossil fuel like diesel to give the desired electricity. The electricity and the necessary real and reactive power can be catered by the DG set as it can work on the requirements based on the speed and the sequence of synchronization (Fig. 4).

Even though the DG set is a conventional source used in buildings, the usage of them are not green as they pollute the environment the most.

There are different types of DG sets that cater to the requirements of the loads or customer. This might typically be a building, community or a factory that need the same. There are several reasons for the inclusion of DG sets into the Microgrid. In some cases, for the purpose of stabilizing and adding power factor correction,

Fig. 4 Conventional DG set

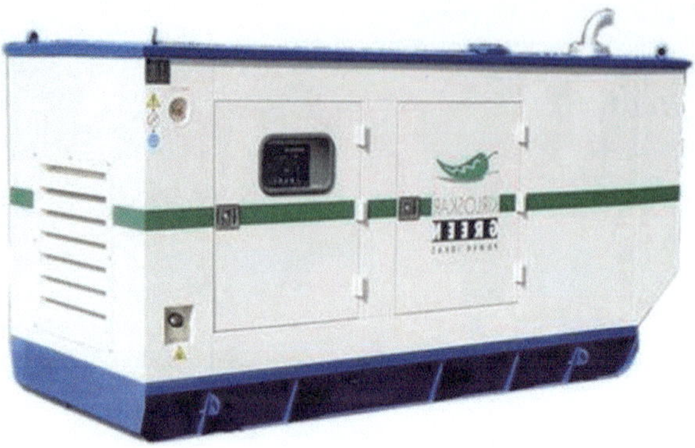

DG sets were used as a Microgrid. But with the introduction of energy storage and state-of-the art control systems, the DG set can almost be replaced for such requirements, resulting in totally green power system. HFO DG sets are heavy fuel oil DG sets, also sometimes known as special oil DG sets, are also used in industries for prolonged usage. These typically form the 2nd or 3rd level of protection to the critical loads in the Microgrid.

These DG sets are also some source of problem as they do not cater to the full load requirement upon immediate termination of the primary source i.e., from utility or Power Grid. The start and stop functionality does not start immediately and it takes a controller a few minutes to reach base load requirements.

2.7 *Variable Loads*

Highly varying loads could be one reason for power fluctuations in a system. The system may not be able to maintain the balance of demand and supply, both with respect real and reactive power, which may hence lead to frequency and voltage instability. Sometimes local equipment may not be suitable for the expected power flow, causing serious congestion and overload, resulting in protection tripping, which might in turn cause further overloads in adjacent system, leading to brown or blackouts. Inadequately tuned power systems may cause continuous oscillations in the power systems, leading to ultimate instability. Weak systems, in the absence of adequate reactive support may collapse. Electro-mechanical resonances, serious harmonics especially introduced by non-linear loads and inverters/rectifiers all need to be addressed by a mature power system.

There are several types of loads that are present in the power grid. Table 1 summarizes some of the loads that are typical depending on the application.

Table 1 Types of loads

Legend
No demand ○ Moderate demand (yellow)
Low demand (green) High demand (red)

Sl No.	Type of Load	Base demand	Peak demand	Power requirement duration	Micro Grid solution improves
1	Industrial	High demand	High demand	◕	Peak power consumption, power charges, High reliability, Quality power
2	Domestic	Moderate demand	High demand	●	Base & peak power consumption, Lower power charges, High reliability, Quality power
3	Domestic Community	Moderate demand	Moderate demand	◔	Base & peak power consumption, Lower power charges, High reliability, Quality power
4	Commercial	High demand	Moderate demand	◕	Base & peak power consumption, Lower power charges, High reliability, Quality power, less downtime
5	Industrial Backup	No demand	High demand	◔	Peak power consumption, Lower power charges, High reliability, Quality power, less downtime
6	Community backup	No demand	Moderate demand	◑	Peak power consumption, Lower power charges, High reliability, Quality power, less downtime
7	Rural load	Low demand	Low demand	●	Base & peak power consumption, Lower power charges, High reliability, Quality power

2.7.1 Industries

These loads are very intermittant. The base load requirements for these sites are very crucial. The need for uninterrupted power supply is one of the primary drivers that keeps the increased power demand for prolonged period of time. The importance of uninterrupted power supply is a major requirement as peak power consumption at the power charges due to high reliability and quality power become th primary reasons. The industrial microgrids are the best solution for this type of requirement.

2.7.2 Domestic

These loads are comparatively smaller as the base load characteristics are very good. The base load requirements for these sites are very crucial. The need for uninterrupted power supply is fairly important but the requirement will be fr the full peak load requirement of the loads. Thus, the energy and power requirement to maintain the peak load and at the same time the improvements like base and peak load consumption, lower power charges, high-reliability and quality power are all independent requirements of the electrical microgrid.

2.7.3 Domestic Community

These loads are based on a community of domestic loads that cater to the initial load requirements. They include the overall experience of having a cluster of communities that unforgivably are accustomed to the green buildings and their requirements. The requirement needs to cater the moderate base demand and peak demand in places that includes the necessities of the green buildings and communities that require the microgrid. These microgrids are more efficient in terms of small power for short periods of time as they manage and care to their microgrid requirements like base demand and consumption along with high reliability and very good quality power. There are several cases like green buildings and remote communities that have this kind of requirements and are very much in requirement of these types of solutions.

2.7.4 Commercial

Commercial buildings are huge load centers that have a very huge base load requirements that the microgrid or the control system has to cater to for the majority of time the total system is in need. These type of commercial buildings have best suitability to have more renewable energy due to the space available for solar, wind and other integrations into the building loads. Roof top solar and wind power generation capabilities need to be assessed before the microgrid is designed with respect to the level of penetration of renewables into such systems.

2.7.5 Industrial and Community Back-up

These loads are in the category of lowest of the demands and the requirement and duration of the different loads vary with respect to the application. In industries the back up required is directly dependent on the base load and the requirement is upon the maximum demand for short period of time whereas in communities, the moderate demand is for the base load for a little higher time before the DG set is turned ON. This in variably creates the time lag and helps synchronization of the DG set connected.

2.8 Control Technologies

Different philosophies exist as to the structure and architecture of control system, depending on price and performance levels, with either centralized control system or distributed architecture.

Battery being one of the important constituents of a microgrid system, the battery control is an important element of the microgrid. When combined with a suitable battery system, such a controller becomes a versatile grid energy storage and stabilization system. It can be a lead-acid, Li-ion or a fly-wheel type of a set up.

The main purposes of such a controller are:

- Stabilize grid voltage and frequency
- Provide spinning reserve
- Store and release energy to resolve asset constraints
- Form the reference frequency for an isolated or islanded grid
- Enable a MicroGrid to transition from grid-connected to islanded operation and back to grid connected.

Grid stabilization can require highly dynamic power injection and absorption for short amounts of time, while other applications require more energy and charge/discharge over longer time periods. The battery energy storage system should ideally combine solid state power inverters, flexible energy management and control and LV-switchgear in standardized switchboard arrangements to reliably and safely provide a range of energy storage-based grid functionalities.

3 Virtual Generators in Disaster Zones

Virtual Generators with latest Power Electronic technologies that can either absorb power into the storage device or alternatively generate power emulating the DG sets, have a great role and can enable a wide variety of applications including the creation of sustainable Green Buildings and Communities. Hybrid systems by integrating Solar, Wind, Geo-thermal, battery systems and other types of electricity generation techniques are available commercially across the globe. Dual combination offerings such as PV Solar & DG, etc. open up unlimited possibilities for effective utilization of natural resources with other numerous applications for power optimization. These hybrid systems can also provide emergency power in geographical areas that are disaster-prone. Emerging Economies with non-existent, non-reliable, or weak grid infrastructure can best use Microgrid for better utilization of renewable energy resources.

The inverter/rectifier with an intelligent control is the heart of the system (Fig. 5). This in the actual sense is a reference of the grid's ability to outperform the other inverters to give the best possible power quality while managing all the different sources and loads (Fig. 6).

It helps maximizing the use of energy storage system to deliver exceptional return on investment. Typically, it has modular design with advanced controls to give maximum availability, value and performance of large or small energy storage systems for a variety of applications.

The Microgrid control system maintains and manages the control system through the architecture and the Microgrid controllers that form the hardware of the control and automation system. It provides many functions like reporting and it is versatile to design for each requirement.

Fig. 5 Hybrid and High-performance DG set

Fig. 6 Typical container solution with battery energy storage system, associated power controller

Typically, a distributed architecture may be applied with independent controllers for specific application with pre-programmed controllers for different applications like Load, Solar, Wind, EB, etc. These work together in sync to create the Microgrid controller network.

3.1 Energy Storage

There are several types of storage that are available or have been tried in the industry ranging from pumped hydro for bulk storage at grid level down to electro-chemistry, such as lead acid and Lithium ion (Li-ion). The latter technology has rapidly evolved in the last few years that it is the mainstay storage option today in electric cars. This technology is marching its way into small and medium scale storage option in electrical power systems as well. The battery technology has matured and is well suited for the versatile applications in green buildings and is well suited towards achieving net zero-emission in building and small communities. Li-ion type of batteries is modular and are very effective with this type of systems. The Li-ion cells are connected in required series and parallel combinations to achieve required storage level in a power system. Adding storage option is attractive for a lot of buildings today that independent of being used to smooth out variations of renewable power inputs, they can be charged during off peak hours and the energy so used can be used during peak hours. Peak shaving technique as this not only reduces the tariff with respect to peak demand charges but can also be used to mitigate the effects of differential pricing that a utility may impose on its customers. Lead-acid (SMF) battery systems conventionally used in UPS is also a solution that is feasible for the short term enabling and the test-run for the Microgrid system.

4 Solution for Grid Integrity

Grid Integrity is very important for green buildings and the factor respective to the control and maintenance of grid integrity. The grid integrity is maintained through the concept of Microgrid that enables the monitoring and control of the various sources and loads through one or more control units. The architecture of the control system could be such that it provides adequate security of power in green buildings even during disasters.

The functionalities of a Microgrid are performed real-time, through state-of-the-art techniques like peak shaving, power factor correction, power application, STATCOM, etc. (Table 2).

4.1 Utility-in-a-Box Solution

Consumers most likely have the technology utilized by the utility scale system. Companies like ABB have developed the cluster type systems that are most effective at smaller scales limited to the buildings. At about 150 kW the 'MGS100' type systems will revolutionize the market and the smart grid to effectively increase the efficiency hit-rate for the Microgrid (Fig. 7).

Table 2 Microgrid functionalities

Functionality	Description
Islanded operation	Acts as a grid forming voltage and frequency source
Seamless	Allowing seamless transition between grid-connected ad islanded (and return) as required
Stabilization	Continuously stabilizes voltage and frequency
STATCOM	Provide reactive power to perform pf correction
Spinning reserve	Provide spinning reserve to the microgrid
Smoothing	Provide power in response to generation intermittency
Shaping	Provide peak looping to manage overloading
Shifting	Shifting of energy generation to meet specific schedules

Fig. 7 ABB's MGS100 microgrid system

Having multiple buildings enables a tandem connection among the peers to create a Microgrid inside a Microgrid that has limitless systems connected to form the network/chain of load based nodal point system that are smart and are control points in a smart system. These will integrate with multiple sources and the grid to effectively give the world a self-sustaining system with very good utilization and mix of renewable energy (Fig. 8).

These systems are on the whole on a new level of technology that work together with the intelligence and logic to form the greater grid. The "*Utility-in-a-box*' type of solutions will give the electrical system the much needed stability and work at unity power factor. The uniqueness of this system at smaller power levels makes it an ideal candidature for the building block of the smart grid. The ecosystem can be

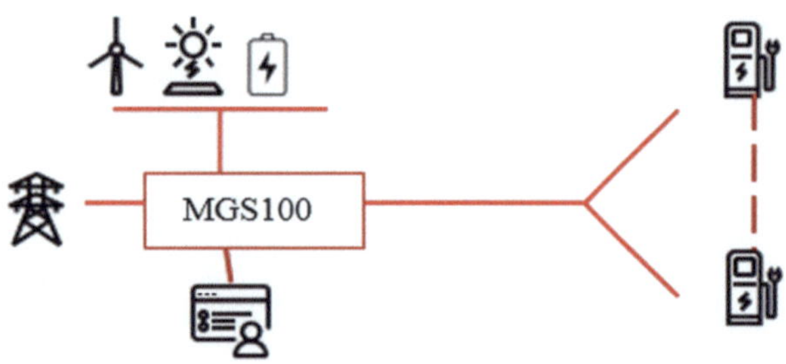

Fig. 8 Multi-source microgrid system

enabled by having these king of units with their own protection and communication panel. With this the entire incomer panel can be replaced with these systems that will make possible the usage of 100% renewables without compromising for the fixed and variable parts. These in-turn give us the much needed system level design for the building block.

4.2 *Microgrid and Distributed Generation*

Distributed Generation refers to "*Distributed Energy or Power generation at the consumer's end.*" By generating power onsite, the complexity of the system is eliminated and the cost of generation along with efficiency increases. The most effective distributed generations have better efficiency that are further enhanced by the automation and control feature induced by the control system.

The unique feature of a Microgrid is that the controls works with the battery in sync to provide the best quality power. This Microgrid addresses the problems of quality power, power factor correction, voltage and frequency regulation and energy management. These automation techniques provide the best possible Microgrid that can manage and maintain the sources and loads within the definite area to give the best quality power.

With the Microgrid established as a context, the current technologies enable the possibility of having distributed generation along various buildings in a campus with unique requirements. The possibility of a cluster Microgrid as shown in Fig. 9 makes it possible to use the 100% renewable energy by constituting the power in a cluster with the much needed control of power between these clusters (Fig. 10).

These on the overall sense can create a world that is intelligent and it can create the world as we see it today. Even 100% utilization like EV Charging can be enabled with these smart systems that are self-sufficient with their own generation sources that power their regional consumption make this efficiency possible.

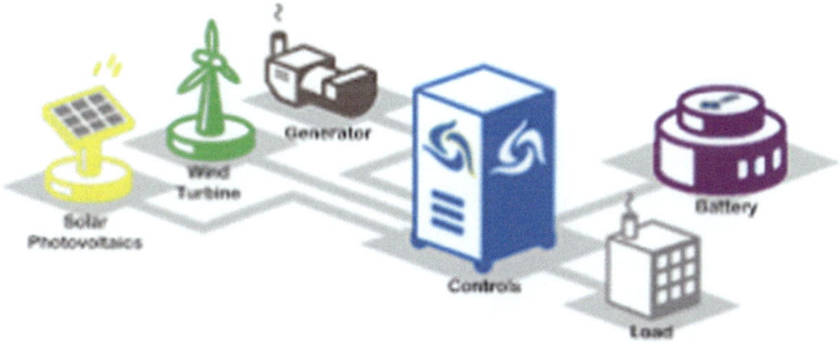

Fig. 9 Microgrid concept

Fig. 10 Nested microgrid

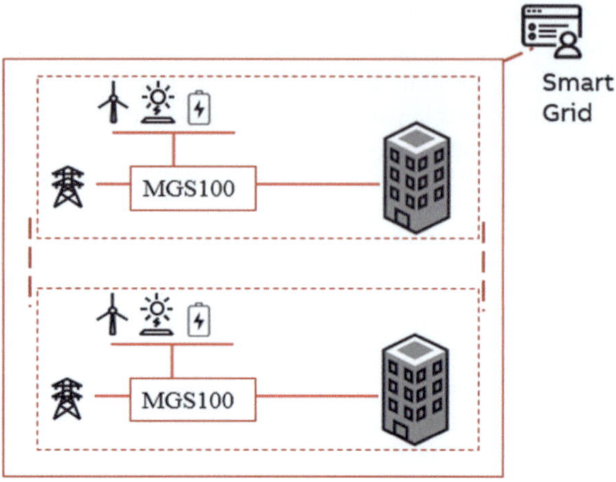

Concepts like '*Generation at the point of Consumption*' *and* '*Consumption at the point of Generation*' are two ways the grid will make the high-efficiency and distributed generation possible.

4.3 Distribution Automation

The Power Grid and the utilization of various activities that cater to the green build requirement of net-zero emission constitutes the usage of all the above mentioned techniques along with the usage of automation at the substation to generate the best quality power. There are several scenarios that the substation automation system along with the power control techniques and dedicated protocols will help protect the system from exterior hacking. The Smart grid is a logic that needs these small systems to be building blocks and also the nodal points for them to work.

5 Conclusion

Electricity is a valuable asset to humans. It has become a link to get essential items in life including food, water and shelter. Since the times of Nicola Tesla and Thomas Alva Edison, when commercialization of AC & DC power systems started, electricity has been essentially a link to connect up humanity with natural resources available with Mother Nature. With the exponential growth of population, coupled with increased economic activity in the last few decades, we have used up resources which took billions of years to form and in the process we have already disturbed the delicate eco-system balance of our planet. Microgrid control systems offer very versatile system of power control system with possibility to integrate major renewable energy, minimizing and eliminating fossil fuel in green buildings. They achieve this through a combination of control, automation system with adequate energy storage component to bridge the intermittent renewable energy inputs. The storage elements incidentally can eliminate the need for UPS systems often used in mission critical applications. The microgrid system has the capability to provide the best quality power without the use of DG sets for full integration of the renewable energies for a building or community. Technology is available to today to have 100% renewable integration, making a building totally green without power from the grid by time-based battery control and automation that defines the net-zero emission design for green buildings.

References

1. Bush SF (2014) SMART GRID: communication-enabled intelligence for the electric power grid, 1st edn. IEEE Press, pp 113–321
2. Battery/diesel grid-connected microgrids: a large-scale industry-based case study of future microgrid capabilities. ABB Ltd., 2015. 28 July, 2016. http://www300.abb.com/global/gad/gad01025.nsf/0/098d1a625d9368d085257f77004e4256/$file/microgrids_primer+for+policymakers_12-04-15.pdf
3. Microgrids for disaster preparedness and recovery with electricity continuity plans and systems. ABB Ltd. March 2014. 8 Aug. 2016. https://library.e.abb.com/public/c537a8400f074fb1bb1bb947614bc291/IEC%20Microgrids%20for%20disaster%20preparadness%20and%20recovery%202014%20White%20Paper.pdf
4. Ahuja D, Tatsunani M (2018) Sustainable energy for developing countries. http://journals.openedition.org/sapiens/823

Technologies for Integration of Large-Scale Distributed Generation and Volatile Loads in Distribution Grids: Technologies for Integration of Large-Scale Renewables Multi-slot EV Charging Stations in Distribution Grids

Jagdish Achara, Jean-Yves Le Boudec and Mario Paolone

Abstract As fossil fuel reserves are limited in stock and there is an urgent call to reduce the carbon footprint, distribution grids urgently need to move towards heavy use of local and distributed generation of electricity using renewable energy sources. Another promising prospect for the future of the planet is wide adoption of electric cars. However, large-scale integration of these highly volatile resources in distribution grids is challenging: distribution grids may face power quality problems and fuel-based generators may be needed to compensate for high volatility (which defeats the original purpose). Additionally, with the large penetration of solar and wind energy, the grid becomes inverter-dominated (has little inertia) and therefore traditional methods for controlling the frequency, voltage, and congestion of lines are no longer sufficient. In this article, we present different activities carried out in our research groups to tackle above-mentioned challenges in large- scale integration of such volatile resources. These activities range from advance planning to real-time monitoring and operation of distribution grids. We have also developed a software testbed, called T-RECS, for testing software agents before deploying them in the field. Finally, as the reliability and robustness is very crucial in such critical infrastructures, we have developed some reliability solutions that are suitable for use case scenarios typical to distribution grids.

J. Achara (✉) · J.-Y. Le Boudec
LCA2, EPFL, Lausanne, Switzerland
e-mail: jagdish.achara@epfl.ch

J.-Y. Le Boudec
e-mail: jean-yves.leboudec@epfl.ch

M. Paolone
DESL, EPFL, Lausanne, Switzerland
e-mail: mario.paolone@epfl.ch

R. K. Pillai et al. (eds.), *ISGW 2018 Compendium of Technical Papers*,
Lecture Notes in Electrical Engineering 580,
https://doi.org/10.1007/978-981-32-9119-5_15

Keywords Renewables · Electric cars · Distribution grids · Advance planning · Real-time monitoring and operation

1 Introduction

Solar and wind energy is certainly one promising prospect to reduce the carbon footprint. Contrary to energy generated from burning fossil fuels, solar and wind energy is decentralized, intermittent, variable, and less predictable. Therefore, adding a large amount of such energy in electrical grids has undesirable consequences on the operation of electrical grids: large voltage deviations may happen and line currents may surpass their limits. The addition of EV charging stations will make the problem even worse as the distribution grids are typically not designed for such large and sudden deviations in the load.

Above-mentioned power quality problems can be addressed by grid re-enforcement but the cost is often prohibitive. Dynamic control of resources is an alternative but it has to be economic, reliable, and scalable. Moreover, such control should be pseudo real-time if we need to manage resources with large variability.

For example, the most extreme case of such resources with large variability is that of solar panels which, during the passage of clouds, can lose or regain most of their power in a few seconds (Fig. 1).

A second type of problem caused by the high penetration of decentralized electric power generation concerns the transmission systems. Unlike traditional generators, extreme volatility of decentralized generation is associated with a lack of support to transmission networks for system services (primary and secondary control of frequency and voltage). This leads to an increase in tuning needs for national networks. They have to find ways to compensate for this volatility by

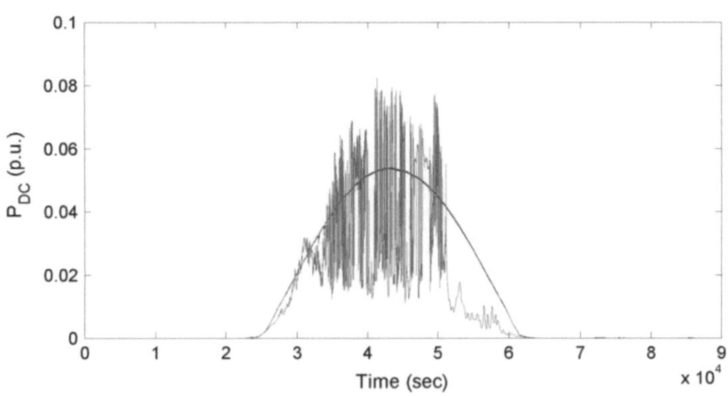

Fig. 1 Power injected by solar arrays at the roof of the DESL, EPFL building on one day in the month of November, 2013. *Source* Prof. Mario Paolone's lab (DESL, EPFL)

employing conventional generators with high flexibility. These generators are typically fossil fuel generators with a negative impact on the environment (CO_2 emissions).

Traditional methods of frequency control are no longer sufficient to manage high variability because frequency is very global quantity. It therefore becomes necessary to control directly the active and reactive power of resources in the distribution grids themselves. We already start to see the development of such systems: voltage control in low-voltage network by controlling the active and reactive power of solar panels [1], and controlling the active power of residential thermal loads using "Demand Response" by an operator [2].

The trends are now rapidly changing not only in terms of distributed generation of electricity but also in terms of load/consumption with the arrival of electric vehicles. As per various studies, the number of electric vehicles (EVs) will significantly rise in next few years. As a result, we will soon see the deployment of large multi-slot charging stations in public or parking places where many EVs (in the order of hundreds or more) can be charged simultaneously. Such a charging station will definitely be a huge load for the local distribution grid. As our current electrical grids were not dimensioned for them, power quality problems may arise if such charging stations are plugged at some arbitrary place in the grid. To deal with these power quality problems, the most environment-friendly solution is to deploy a local generation (PV panels) and storage system (Batteries) along with the EV charging station.

With the arrival of large multi-slot EV charging stations, we will also see more peaks in the consumption trends. While charging EVs at such charging stations, we should be able to avoid as much as possible the charging of EVs at peak times of the load in the grid. Moreover, at times when there is large production from renewable energy sources, the charging station should consume more power (for example, by following OpenADR or similar signals). For example, it is probably better to charge the EVs at mid-day when the sun shines the most or during those times when the wind blows. Here it is probably worth to mention that such renewable sources of energy production can vary in terms of their predictability. For example, solar energy is generally more predictable than wind energy. As advance planning is one of the crucial things in integrating large volatile loads/generation, predictability of a resource plays a key role.

In our research groups, we are building solutions for smooth inclusion of such distributed generation and heavy unpredictable loads such as large multi-slot EV charging stations. These solutions range from advance planning, monitoring, and real-time operation of distribution grids. For testing our software agents in silico before their actual deployment in the field, we have also developed a software testbed, called T-RECS. As reliability and robustness are also crucial elements in such critical infrastructures, we have developed some reliability solutions that are suitable in the context of real-time operation of distribution grids. Next sections follow a detailed description of our activities in each of these directions. Last section concludes the paper and gives some future perspectives.

2 Advance Planning

Advance planning is usually the first step in making conventional grids smart. The objectives usually consist in forecasting (usually day-ahead) power consumption/ generation for resources with uncertainties. In case of solar generation, forecasting can be done in many ways using various information like historical data, weather forecasts, and real-time pictures of sky. In the same fashion, power consumption can be forecasted by finding patterns in historical data. As another example of uncertainties in load, large multi-slot EV charging stations typically do not know the exact arrival and departure time of EVs, type of EVs, and their energy demand at arrival time.

After forecasting, the second step usually is to optimally manage a given set of resources. For example, in case of large multi-slot EV charging stations, an environment-friendly solution is to deploy such charging stations with local generation and storage solutions. By optimally managing all these resources, this solution can also be made economical. In this setting, there can be many optimization objectives: minimize the required capacity of electricity storage, minimize the total import of electricity from the main grid, or simply minimize the cost of electricity purchased from the grid depending on the pricing-scheme. For example, in the above scenario, we can compute the day-ahead optimal power trajectory at the local connection bus where the objective is to minimize the import of electricity from the main grid. The day-ahead power trajectory can be computed periodically (such as every 15 min) by taking into account the uncertainties of EV charging station and local generation.

3 Real-Time Monitoring

Having the current state of the grid is a requirement for many ensuing tasks like fault-detection, fault-prevention, and real- time operation of the grid. Therefore, we target that monitoring should be as fast-paced and accurate as possible. This is achieved by our current state-of-the-art monitoring infrastructure, which consists of phasor measurement units (PMUs), a communication network, a phasor data concentrator (PDC), and a state estimator (SE) unit (Fig. 2). More details about this real-time state estimation can be found in [4].

4 Real-Time Operation (Commelec Framework)

With the monitoring infrastructure, we have a real-time (with a reactivity of the order of a fraction of a second) operation framework, called Commelec. It sends active or reactive power setpoints to the resources involved in balancing the

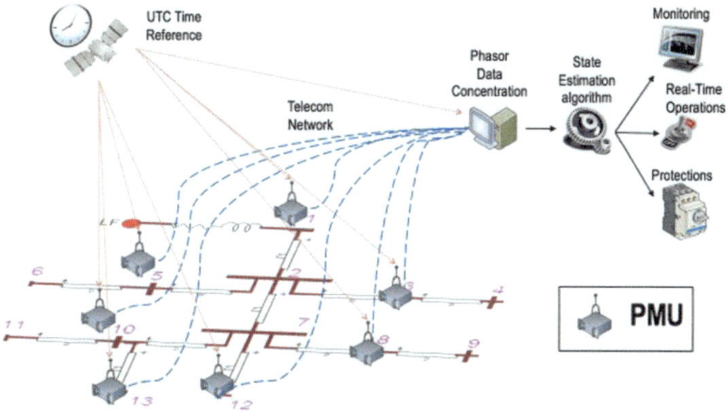

Fig. 2 Monitoring infrastructure developed at EPFL campus based on the use of PMUs. Real-time state estimation results are available at [3]

network. As, in most cases, it is not enough to control only the reactive power but also the active power, it is necessary to have some form of energy storage. These storage solutions can be: supercapacitors for the very short-term storage (of the order of a second), batteries for the short term (hours/days), and fuel cells/hydrolyzers for the longer term (weeks/month). It should be noted here that the cheapest storage in the minutes/hours dynamic is the manipulation of demand, mainly thermal loads (heating, hot water, refrigeration, air conditioning) which offer a huge virtual storage capacity.

We have developed a real-time operation framework, called "Commelec", with the support from "SNSF—NRP 70" Energy Turnaround Project [5] and "SCCER-FURIES" project [6]. With the Commelec framework [7, 8], a grid controller (Grid Agent or GA), implemented on a microcontroller, is responsible for managing an electrical network, for example a building or a district. It receives information on the status of different resources (a solar panel controller, a building or a battery) under control by their respective resource agents (RAs) (Fig. 3). GA sees the state

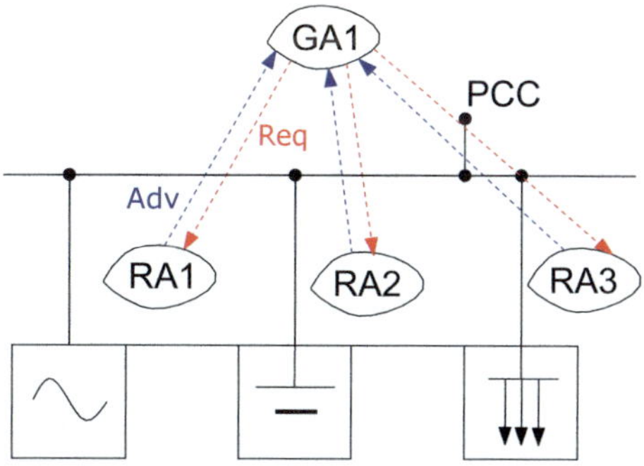

Fig. 3 Structure and principle of Commelec protocol. Resource agents (RAs) send advertisement containing the state of resources under control to the grid agent (GA) and GA sends power setpoint requests to the RAs

of the electrical grid for which it is responsible and sends power setpoints every 100 ms to resource agents, so as to maximize overall utility while keeping the grid in a stable and secure state.

4.1 Simplified Functionality Due to Its Universal Protocol

The first essential feature of Commelec is the use of a universal protocol, which is independent of the hardware. Thus, the battery agent BA, when it informs its network controller GA1 of its internal state, does so using an abstract (virtual) cost function and does not use a battery specific language (Fig. 4). When the battery is almost full, e.g., for a charging state equivalent to 0.9, the cost function indicates a preference to generate power (positive power domain, right part of Fig. 4) rather than to consume (negative power domain, left part of Fig. 4) and conversely when the state of charge is low. It is therefore not necessary to expose to the grid controller all the internal characteristics and variables of each battery or resource.

Another system, e.g., the Intelligent Building Agent, sends similar messages. These messages differ only in the value of the cost function, which can also change for each message every 100 ms. The work of the network controller is therefore always the same: to calculate the total cost gradient including a penalty term to keep the electrical network in a safe state, to perform a step of minimization with a gradient descent, and send the corresponding power setpoints to different agents.

4.2 Separation of Concerns

In particular, all network controllers are identical and there is one code to develop for all power grids, regardless of their size and specificity. In addition, this code is small and can be validated by rigorous development methods. Only agents, such as the battery agent, for example, must be specific to the nature of the system they represent. But, in return, they perform simpler functions.

Fig. 4 Virtual cost function for the active power used by battery agent as a function of state of charge (SoC)

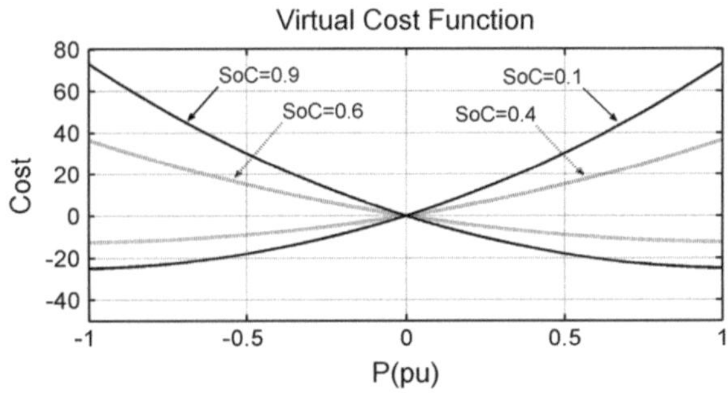

4.3 Composability

Composability is the second essential feature of Commelec: the GA1 network controller in Fig. 3, which, for example, controls an EPFL building, is itself controlled by a higher-level network controller, for example, the EPFL campus controller. In this interaction, the controller GA1 and the entire network of the building it controls appear as a single resource, somehow a battery of a special kind. The campus controller thus sees only a small number of systems, which keeps it simple and small (and therefore reliable). This composability property can be repeated at multiple levels, making it easy to manage systems of any size.

This property has another positive effect: it simply allows to realize the much-desired concept of "dispatchable feeder" with and without frequency control and primary/secondary reserves. In fact, the campus controller can send a power instruction to the building controller GA1, for example by asking him to reduce his consumption by × kW. The controller GA1 will incorporate this setpoint into its optimization function and, depending on the state of the systems it controls, will try to approach the setpoint. For example, the reduction will be possible if the heating of the buildings can be delayed for a few moments or if a battery has a sufficient state of charge. The building can therefore appear as a controllable electrical resource; and by composability, the entire campus too. In extreme cases, the software makes it possible to automate the islanding operation, that is to say the disconnection/reconnection to the electrical grid from the main grid. In the long run, it is a city or a portion of the country as a whole that can enjoy this flexibility of the electricity grid—a breakthrough that will allow for a strong penetration of renewable energies.

4.4 Commelec API

In practice, the deployment of Commelec requires that certain critical resources, such as heating for a building or a battery, respond to the instructions of an electrical network controller. Developers of building management systems or electric cars however do not need to know the details of Commelec, thanks to its programming interface (API) which is freely available [9]. They can simply interface with the control system of electrical networks (Fig. 5). In a way, the Commelec API provides the "operating system" of an active power grid.

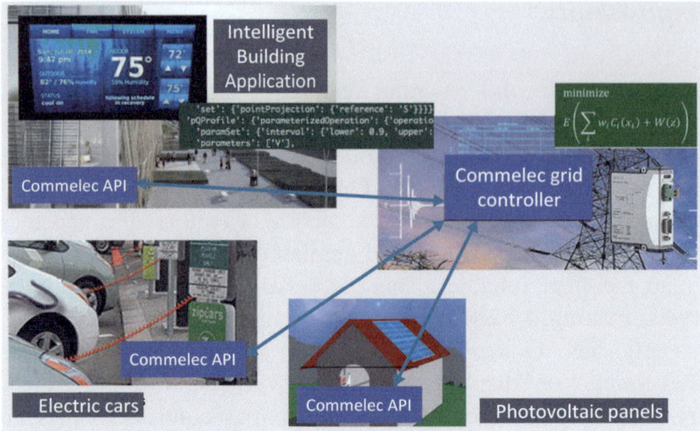

Fig. 5 Application programming interface (API) of Commelec allows to simply connect an intelligent building control system, an EV charging station, and photovoltaic panels; all these systems therefore can actively contribute to the management of electric distribution networks

4.5 Implementation/Demonstration

The Commelec control framework has been tested on the experimental electrical network of EPFL's Distributed Electrical Systems Laboratory (DESL). In a first step, it reproduces a Cigré1 benchmark of low voltage electrical distribution network. It incorporates:

- 40 kW of solar panels;
- Leclanche Li-titanate battery of 25 kW and 25 kWh;
- 75 kW and 2 kWh supercapacitors;
- 15 kW fuel cell and 6 kW hydrolyser connected to 30 bar hydrogen and oxygen storage system with a capacity of 2.5 MWh;
- a heat pump;
- and a building emulator connected to the hot water network with a heat output of 55 kW.

The first tests for the automatic control of this test network by the Commelec software took place in the autumn of 2015. In a second step, the test network is extended to few buildings of EPFL's electrical department (medium voltage grid). With Commelec, we have also successfully tested the islanding (automatic disconnection/reconnection) of DESL microgrid to the main grid. Outside EPFL network, Commelec is successfully tested at EMPA NEST building [10] and is now being deployed by few utilities in Switzerland and outside.

5 T-RECS Testbed

A recurrent requirement for software-based real-time monitoring and operation software agents is to test them as they are in silica before their actual deployment in the field. Monitoring and operation performance of such software-agent based solutions is influenced by software non-idealities such as crashes and delays, and message losses and delays due to the underlying communication network. Therefore, to study the effect of these non-idealities, we developed T-RECS: an open-source virtual commissioning system or software testbed.

They key feature of T-RECS is that it allows to test existing software without modification by running them in software containers. The communication network among these software containers is emulated using Mininet framework. As the communication network is emulated, it allows for real packets being exchanged between software agents as is the case in the real-world. The electric resources in the grid are simulated and the grid is modeled in the phasor domain.

We are currently using T-RECS to test Commelec software agents before deploying them in electric grids of different utilities we are working with. This allows us to study if Commelec software agents are bug-free and if they are able to achieve their desired objectives. More details about T-RECS and its design can be found in [11].

6 Reliability

To achieve a control system that takes into account electrical networks, buildings, loads, and distributed generation systems such as solar panels, etc., the challenge becomes, as for any process control system on a very large scale, to ensure its reliability and robustness. Such systems must withstand hardware failures and software bugs.

To tackle reliability in communication networks with stringent delay constraints, multiple fail-independent paths are necessary. However, existing solutions such as parallel redundancy protocol (PRP) only work for local area networks. Such a limitation on scalability, coupled with lack of security, and diagnostic inability, renders PRP unsuitable for reliable data delivery in smart grids. To address this issue, we developed a transport-layer design: IP parallel redundancy protocol (iPRP). Besides unicast, iPRP supports multicast, which is widely using in smart grid networks [12].

For Commelec style controllers where setpoints have real-time constraints (implementing a setpoint after its deadline, or not receiving setpoints within a deadline, can cause failure), delay faults can cause setpoints to violate their real-time constraints. To address these delay faults, we developed a fault-tolerance protocol, called Axo, that guarantees safety and improves availability [13].

7 Conclusion

The formidable challenges posed by strong penetration of decentralized production and large multi-slot EV charging stations can be solved by a combination of the following activities: advance planning, real-time monitoring, and real-time operation of electric grids. In our two laboratories, we are working on each of these directions of activities. We have already built a set of technologies that can help integrating large volatile distributed renewable energy sources (solar, wind) and loads (multi-slot EV charging stations). We continue working in these research directions and now actively trying to make a positive impact by bringing our research work to the industry.

References

1. Georges D, Coutarel A (2014) Voltage regulation in LV networks with PV installations (Régulation de la tension dans les réseaux BT avec installations PV). Bull SEV/AES 5(2014):29–32
2. For example: Swisscom Energy Solutions. www.tiko.ch
3. https://smartgrid.epfl.ch
4. Pignati M et al (2015) Real-time state estimation of the EPFL-campus medium-voltage grid by using PMUs. In: Innovative Smart Grid Technologies Conference (ISGT), IEEE PES
5. NRP70 Research Program. http://www.nrp70.ch/en
6. Swiss Competence Center for Energy Research—Future Swiss Electrical Infrastructure. http://sccer-furies.epfl.ch
7. Bernstein A, Reyes Chamorro LE, Le Boudec J-Y, Paolone M (2015) A composable method for real-time control of active distribution networks with explicit power set points. Part I: framework. Electr Power Syst Res 125:254–264
8. Reyes Chamorro LE, Bernstein A, Le Boudec J-Y, Paolone M (2015) A composable method for real-time control of active distribution networks with explicit power set points. Part II: implementation and validation. Electr Power Syst Res 125:265–280
9. See https://github.com/LCA2-EPFL/commelec-api and https://smartgrid.epfl.ch/?q=node/14
10. https://smartgrid.epfl.ch/?q=node/22
11. Achara J, Mohiuddin M, Saab W, Rudnik R, Le Boudec J-Y (2017) T-RECS: a software testbed for multi-agent real-time control of electric grids. In: 22nd IEEE international conference on emerging technologies and factory automation. IEEE
12. Popovic M, Maaz MM, Tomozei D-C, Le Boudec J-Y (2014) IPRP: parallel redundancy protocol for IPv6 networks. Technical report. EPFL
13. Mohiuddin M, Saab W, Bliudze S, Le Boudec J-Y (2016) Axo: masking delay faults in real-time control systems. In: Industrial electronics society, IECON 2016—42nd annual conference of the IEEE. IEEE, pp 4933–4940

Disruptive Technologies

IoT for Indian Power Sector

Rakesh Kumar Goyal

Abstract The Internet of Things (IoT) has the potential to significantly transform the industrial sector. IoT is considered as fourth industrial revolution. General Electric (GE) predicts that $1.3 trillion of value can be captured in the electricity value chain from 2016 to 2025 by IoT. SCADA, PLC, smart meters and smart devices are few old examples of IoT. What is new that would create so much value in power sector is deployment of smart devices, advance analytics, advance dash board, cloud computing and the integration of several services into a common platform due to drastic reduction in prices of Wi-Fi chipsets and availability of variety of sensors like accelerometers, cameras, tactile, sound, temperatures and others. The three global digital themes for electricity industry from generation to consumption are; (1) Asset Management, (2) Operation Optimization, and (3) Enhanced Customer Services. The Indian power sector is facing several challenges such as high AT&C losses, poor reliability, public and political demand to keep tariff low, integration of high RE capacity into the grid and aging infrastructure. IoT can support resolution of most. This paper will attempt to identify, IoT applications for India power sector, Challenges and way forward.

Keywords IoT · Power · Asset · Optimaisation · Grid

1 Introduction

Internet of things (IoT) is new buzz word and being considered as fourth industrial revolution. Its impact on power sector is no different than on other industrial sectors. IoT is revolutionizing the power sector by addressing some of the most critical issues industry is facing such as optimization of asset utilization, asset management to increase reliability of power supply by avoiding failure of network and equipment's, providing customer with load of information to play the role of

R. K. Goyal (✉)
Tetra Tech ES India Private Limited, New Delhi, India
e-mail: Rakesh.goyal@tetratech.com

R. K. Pillai et al. (eds.), *ISGW 2018 Compendium of Technical Papers*,
Lecture Notes in Electrical Engineering 580,
https://doi.org/10.1007/978-981-32-9119-5_16

prosumer and moving the retail and bulk electricity pricing based on economic analysis. The ball has already started rolling. Smart meter is no longer a meter to record electricity consumption and transfer the data to central severer. It is now hub of many activities from recording electricity consumption to calling ambulance and police in case of emergency and smart meter has become advance metering infrastructure (AMI). General Electric (GE) predicts that $1.3 trillion of value can be captured in the electricity value chain from 2016 to 2025 by IoT [1]. Indian power sector has lot more to gain as currently there is limited use of technology by traditional utilities.

Before moving further let us briefly understand what is IoT. The Institute of Electrical and Electronics Engineers (IEEE) developed a definition of IoT based on review of the several definitions created by various organizations and individuals; "An IoT is a network that connects uniquely identifiable 'Things' to the Internet. The 'Things' have sensing/actuation and potential programmability capabilities. Through the exploitation of unique identification and sensing, information about the 'Thing' can be collected and the state of the 'Thing' can be changed from anywhere, anytime, by anything."

It has three main components; (1) Uniquely identified things, (2) Sending and actuation and (3) anywhere anytime. For easy understating of the concept of IoT (Fig. 1).

The main applications identified for each component of the value chain of the Indian power sector is listed below.

Generation: Machine efficiency, load control, voltage control, health of the machine and generation dispatch to maximize machine efficiency and revenue.

Transmission: Differential control using phasor measurement leading to fault anticipation and avoidance, loss reduction by load balancing on lines and transformers, health monitoring of transformers, breakers and switchgears.

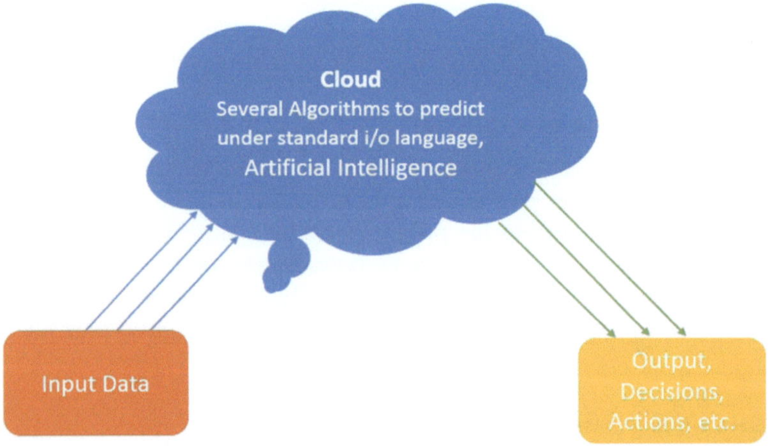

Fig. 1 Graphical representation of IoT

Distribution: Loss reduction, time varying pricing, peak load management, and customer services etc.

The global experience for use of IoT is very encouraging. The experience can be classified in three categories for estimating its usefulness Indian Power Sector.

2 Asset Management (AM)

Few examples of GE IoT solutions implemented globally are presented in Table 1.

Table 1 Asset management by IoT solution in convention power plant

Plant	Solution	Benefits
Bord Gais, Ireland 445 MW combine d-cycle gas power plant	**Goal**: Continuous operation with no unplanned downtime • 141 total sensors around the plant to monitor the condition of assets in the plant • Early warning of failure mechanism resulting in efficient outage management	Reduce plant downtime and reduced BoP operations cost €2.28 million positive financial impact in year 1 from cost savings and cost avoidance
Scottish Southern Energy, United Kingdom Thermal generation fleet at 11 locations	**Goal**: Increase plant availability through early detection of potential failure and prevent past failures from reoccurring • Created an equipment performance center to improve the reliability of its thermal fleet of generators • It continuously monitors combustion dynamics, turbine vibration, boiler temperature, creep and others at 11 different locations and 1026 assets • Predictive analytics algorithms are deployed	• Early failure detection has resulted in savings of approximately £3 million per year • Overall insurance costs have been reduced (£7.5 million per year) and with improved maintenance, a greater control over capital expenditures is expected • Savings of £100,000 in repair costs by not running the generator into a failed state
Salt River Project, Phoenix, Arizona, United States 11 power plant locations (Coal, gas, nuclear and renewable)	**Goal**: Integrate data across multiple plants for outage management, to optimize maintenance strategies and to understand where production issues might occur next	High level of asset, plant, and fleet reliability. Able to see problems before they happen—toward no unplanned downtime and improving fleet reliability $0.5 million per year savings

Source ADB White paper no 48 August 2017 "The Internet of Things in the Power Sector Opportunities in Asia and the Pacific-Arun Ramamurthy and Pramod Jain

Presently, AM is most popular application of IoT. It is more relevant for Indian power sector as assets in India are highly inefficient and unreliable due to poor maintenance. There are more cases of assets are being over loaded or under loaded rather than optimally loaded. From past few years, due to resource crunch there is trend in the utilities for reduction of maintenance budget leading to only breakdown maintenance from preventive maintenance. IoT helps in doing the maintenance of the asset when it is needed most based on the health of the asset, predicted time to failure and risk of failure. Thus maintenance can be shifted from preventive/break down to need based. It has several advantages such as improved reliability and availability of the asset, reduced cost of maintenance, lower inventory of spare parts and higher productivity. The asset which can be covered here are generator, turbine, transformers, breakers, switchgears from higher voltage to lower depending upon resources and priority for reliable power supply.

3 Operation Optimization

Unlike AM application, operation optimization application of IoT has not been so much exploited. Although it offers more varieties than AM. The range includes from higher production to grid optimization. Few examples for generation are: It is possible to optimize generation based on renewable energy sources (RE) by using IoT. Wind turbine manufacturers are providing several IoT solutions based on measurement of wind speed, direction, pitch angle, yaw angle to increase generation. Same can be used for solar generation. This will be of great use to India as India has target of 175 GW of RE by 2022. GE has implemented IoT solution in NRG, USA power plant to bid additional capacity during peak period by balance peak firing and digital twin model. Several distribution transformers in Indian power sector are either overloaded or under loaded. IoT can help in optimizing their loadings and also in balancing the load among three phases. This is one of the potential source of high technical losses in Indian distribution system.

The grid operation is getting more complex due to variability in demand and generation and high infusion of RE at central and at discreet level. This variability in generation and demand causes variation in voltage and in flow of active/reactive power at various nodes of the grid. Optimizing this dynamic grid is impossible without real time data of current and voltages of all elements of the network and high level of analytics. IoT can play an important role in this regard—digital power plants, digital substations, and other data collection systems can fill this gap. According to GE, a new digital 500-MW power plant can save $230 million over the lifetime of the plant, and an existing power plant of the same size can save $50 million. This savings is from 3% higher fuel efficiency, 2% higher output, 20% less fuel on starts, 6–9% reduction in carbon dioxide emissions, 10% reduction in nitrogen oxides, 5% reduction in unplanned downtime, and 25% reduction in O&M costs [2].

4 Enhanced Customer Services

The most visible application of IoT in Indian power sector is in AMI or in smart meters. The AMI with the help of IoT is facilitating not just recording of electricity consumption but several other applications in the power sector and outside power sector. The popular examples in power sector are: time varying electricity tariff leading to modest reduction in usage and significant reduction in peak consumption, remote connection/disconnection based on contract load or on nonpayment, net metering, consumer turning prosumer, outage deduction, line voltage management etc. The application likely to come up outside power sector are ambulance calling, police calling, disaster warning, remote control of house hold devices (such as air conditioner, water heater/pumps, washing machines etc.) house surveillance, burglar alarm etc.

More and more applications are getting developed. Home Energy Management System is another emerging application similar to building management solution. Such an application has complex algorithms to integrate, power drawl from grid and power that can be supplied to grid from solar PV, storage, electric vehicle and guiding customer about buying and selling power to maximize his revenue or minimize his utility bill.

SCADA is another popular example of IoT application. Table 2 provides few examples of global experience [1].

Table 2 Global experience

Location	Description	Benefits
Glendale, California (population: 194,000)	46,000 users with AMI and customer portal to monitor home energy usage	Total customer savings: 5777 MWh (2–4%) Peak load reduction of 4.1% overall $24 million in positive value for an IRR of 11.5%
Burbank California (population: 108,000)	50,000 users with AMI, smart thermostat, distribution automation at 100 feeders and load management	Customer saving 4800 MWh (about 1–2%) field service request reduced from 2500 to 300 per month SAIFI 0.34–0.24I from 27.8 to 9.5
Danvers Massachusetts (population: 26,000)	13,000 smart meters, net metering and time of day rates	Deferral of $3million investment in distribution network capacity Service call decreased by 75%

5 Challenges

Investment: Indian utilities are suffering from financial crisis and has no money to invest in IoT. Luckily most applications of IoT are profit driven by optimizing performance or reducing life cycle cost. Thus a business model where no upfront investment is done by the utility and vendor is paid back from derived benefit need to be developed and explored. Such models are already popular in energy efficiency using Energy Service Company (ESCO). Indian IT companies such as Infosys, Tata Consulting, Wipro have large cash surplus. Thus they should welcome this model. However, this will require sound accounting practice and true measurements of the gain from IoT.

Standards: It is important to define the standards at the beginning to avoid any interoperability issues and difficulties India faced with electronic meters. Although IoT standards even globally at nascent stage. Some of the bodies which are trying to develop standards are IEEE, Open Connectivity Foundation, Open ADR Alliance, Zigbee, Industrial Internet Consortium etc. India may like to get associated with them and develop standards for IoT applications which can be sued seamlessly with the equipment's manufactured by different manufacturers, and flow of data can take place at various IT platforms.

Capacity Building: IoT is likely to automate labor intensive tasks such as meter reading, connect, disconnect, outage detection etc. According to a United States Department of Energy report: "The electricity system of the 21st century will require an adaptable and flexible workforce with additional areas of expertise and capabilities than the current workforce [3]. The integration of variable renewable sources, storage systems, smart grid, and demand management will require new training and skillsets India being hub of IT activities, it should not be difficult. What is important is structured and targeted approach for capacity building. IT Companies along with IITs and IIITs should lead the way by developing course content, training material and train the trainer workshops for utilities training institutes faculties.

Cyber Security: When such a large amount of data flows and mostly through open protocols to address interoperability issue, cyber security becomes critical. Partially it can be addressed by having good standards and partly I feel it is cultural issue which requires a more long term measures of improvement in education, income levels and law and order. Nevertheless I will suggest due care but cyber security should not be allowed to become impediment in development and usage of IoT.

6 Way Forward

My suggestions are Government should facilitate development of more and more applications of IoT involving technological institutes, IT companies and utilities. The utilization of all IoT applications should be based on cost benefit only with the end objective of reducing customer electricity bill. This will not require any regulatory approval and will not put any financial burden on the utility.

India is already leading the world in ICT sector, innovate/develop IoT be the new sector for India to lead the world.

References

1. ADB White paper no 48 August 2017. The internet of things in the power sector opportunities in Asia and the Pacific-Arun Ramamurthy and Pramod Jain
2. General Electric (2016) GE power digital. https://www.ge.com/digital/sites/default/files/Operations%20Optimization%20Solution%20Sheet.pdf. Accessed on 8 Jan 2018
3. US Department of Energy (2017) Chapter V: the electricity workforce: changing needs, new opportunities. https://energy.gov/sites/prod/files/2017/01/f34/Chapter%205%20The%20Electricity%20Workforce%20Changing%20Needs,%20New%20Opportunities_0.pdf
4. The internet of things: mapping the value beyond the hype, June 2015—Mckinsey Global Institute report
5. IoT in Indian electricity transmission & distribution sectors—Somesh Kumar E&Y

Solving Duck Curve Problem Due to Solar Integration Using Blockchain Technology

Aurabind Pal

Abstract The high penetration of renewables, particularly rooftop based solar plants, into electrical grid has resulted in dip in net demand of power during day hours while a sudden surge in net demand in the evening when sun sets. The duck shaped load curve thus generated poses the challenge to the power utilities to match the demand and supply. Various demand response (DR) method like dynamic tariff etc. can solve the issue but only by curtailing the growth of renewables. Electric Vehicle (EV) charging can come as a rescue to this problem supported by blockchain technology. The model proposed is prosumers with roof tops based solar plants along with EV charging station shall be registered in a blockchain based program, a decentralized system, designed for measurement, registration, transaction and settlement of clean energy. The charging station having pre-arranged its vehicles to get charged shall get into a dynamic auction in a mobile based app, to get the cheapest power from the neighbourhood rooftop solar plants. The prosumers and the EV charging stations shall be equipped with net meters and a blockchain based program running in background. The energy thus traded shall be accounted effectively and securely in the blockchain program. The prosumer having sold its clean power can earn solar currencies or any wallet money that can be redeemed or even traded. The prosumers and the charging stations shall be backed up with the conventional grid in tandem. In the above proposed model a parallel power vertical shall be maintained that shall galvanise not only the surge in demand of power due to Electric Vehicles but dip in requirement due to integration of rooftop solar plants into the grid. Rooftop solar and Electric Vehicles will grow together complimenting each other, encouraging clean form of energy.

Keywords Duck curve · Solar · EV charging · Blockchain

A. Pal (✉)
Rural Electrification Corporation, Bhubaneswar, India
e-mail: apal@recl.in

R. K. Pillai et al. (eds.), *ISGW 2018 Compendium of Technical Papers*,
Lecture Notes in Electrical Engineering 580,
https://doi.org/10.1007/978-981-32-9119-5_17

1 Introduction

World is adopting solar energy like never before. Electricity regulators have started creating conducive environment for grid connected rooftop based solar power plant. India has set a ambitious target to generate 100 GW solar power before 2022 with 40 MW as rooftop solar plants [1]. Rooftop solar shows up in a completely different way than large-scale utility solar. Instead of being counted as supply, it's adjusted from customers' loads, yielding a value known as average system load – how much electricity customers are using, minus whatever they're generating behind the meter. With such penetration there is all probability of over generation of solar power. In 2013, California Independent Operator came out with a "duck curve" [2] (Fig. 1).

In the graph, all lines represents the net load. The "belly" of the curve represent the period in which solar power generation is at peak. This reflects a significant drop in mid-day net load as grid connected solar photovoltaic are added to system. And then as the sun sets there is a steep increase in net load. With increase in penetration of solar power the dip shall deepen and the gradient of slope shall be steeper.

Now managing the demand and supply is critical for utilities. The intermittency in load curve due to integration of solar power into grid shall call for great flexibility of grid. Power plants cannot respond quickly enough to manage the sharp evening increase in usage and coal fired power plant running in low PLF during day. Under such circumstances utilities shall have few options to flatten the curve. The most pessimistic approach shall be curtailment of solar power. The utilities with inertia to change in policies and adoption of state of art technology shall adhere to such methods. This shall be catastrophic for the environment and

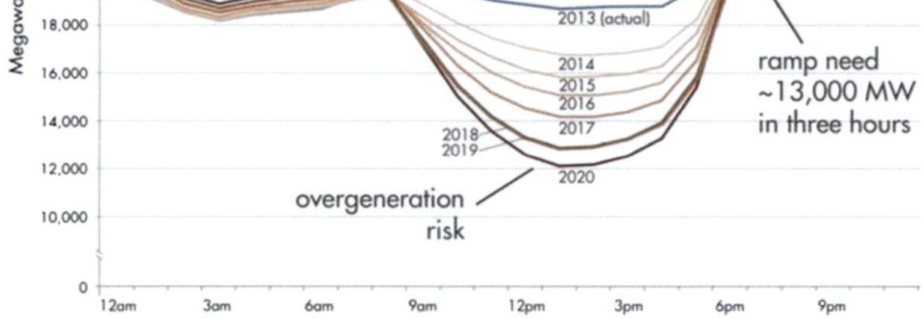

Fig. 1 Duck chart. *Source* CAISO 2013 [2]

sustainability. The other very clever approach shall be to use dynamic tariff or Time of Use (TOU) based tariff [3]. In the belly part of curve tariff shall be low, thus encouraging use of more power and in the neck side of curve, tariff shall shoot up. This approach shall be neither be financially beneficial for utility or consumers. Solar coupled with storage technologies can also alleviate over generation issue and curtailment of solar power will not be needed. But use of batteries is a costly affair and large storage technology is yet to mature.

Considering above, an apt solution shall be to arrange some dynamic load that will compensate the over generation and flatten the curve. Load of electric vehicle charging fits very well with the description of load required. Arranging this load in the correct slot of time shall be driven by blockchain technology.

2 Electric Vehicle Charging

Electric Vehicle (EV) is coming in a big way in India and they are here to stay. Indian Government have started framing policies and standardizing charging infrastructure [4].In the near future, electric vehicles (EV) are expected to dominate the vehicle market. However, the success of EV technology depends on the availability of EV charging stations. To meet this demand, utilities have to install EV charging stations at residential and commercial locations.

Different kind of chargers as categorized by the standard [4] are as follows (Fig. 2).

The home private chargers are generally used with 230 V/15 A single phase plug which can deliver a maximum of up to about 2.5 KW of power. A standard EV takes overnight to get charged. The billing for the power is part of home-metering. The batteries in EVs need DC power to get charged. So EVs have on-board chargers/converters that convert AC power to DC and get the batteries charged. But in India they are limited to 2.5–3 KW rating. So for fast charging, DC

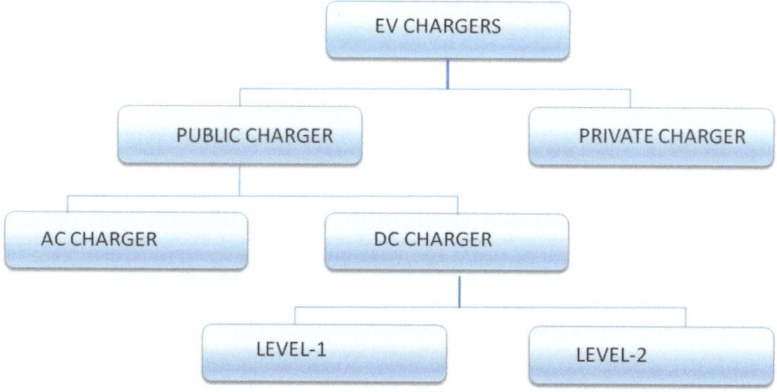

Fig. 2 Classification of EV chargers

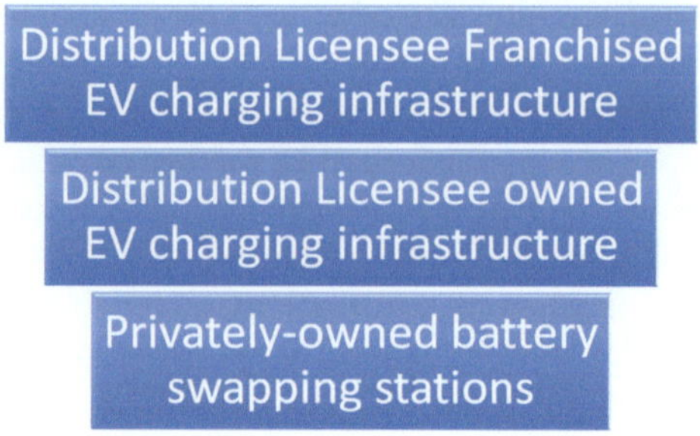

Fig. 3 Business models for EV charging

source is to be pumped into the vehicle directly. The standard has categorized DC offboard charger into level-1 and level-2 DC chargers. The level-1 DC public charger has output voltage of 48 V/72 V, with power outputs of 10 kW/15 kW with maximum current of up to 200 A. While Level-2 DC public charger has output voltage up to 1000 V, with power outputs of 30 kW/150 kW. The public chargers shall be billed separately and developing a sustainable business model, which shall adhere to the Electricity Act is primarily important.

The charging of EVs involve distribution of electricity. The Electricity Act, 2003 of India bestows this power only to the distribution licensor. There might be amendment in the future to suit the EV but with present legal framework the Central Electricity Regulatory Commission (CERC) of India has recently identified three business models for electric vehicle charging within the framework of the Electricity Act (Fig. 3).

The first business model suggested by CERC is distribution licensee selling power directly to the EV consumers as per rate fixed by SERC. And the second model is franchisee, deputed by distribution company (DISCOMS) could sell power to the EV consumers. The franchisee in turn could buy power at a rate determined by SERC or can buy power from the open market or prosumers without cross subsidy surcharge. The third model is an interestingly enterprising model. The drained out batteries driving the electric model are swapped by a charged battery at a battery swapping stations (BSS). The BSS can very well be owned by a private organization as the product being traded are not electricity but batteries. The battery owner can buy power from DISCOMS at a special category tariff or open access. In order to increase the penetration of electric vehicles, EV need to appeal to customers as much as their conventional fuel equivalents. However, major concerns include long battery charging times and range anxiety. These concerns can be minimized if the customers have access to BSS, where swapping of battery takes as much time as refilling the fuel in a gas station. On top of these bidirectional flow of power i:e grid to battery and battery to grid shall add to grid stability. They act as storage device and are capable of participating in the power market transaction.

The promise of clean transportation and getting rid of fossil fuel has motivated the federal Governments to adopt Electric Vehicles aggressively. While Govt. of India has aligned its policy to 100% EV by 2030 other state governments have started framing the policies for EVs. In the near future, electric vehicles (EVs) are

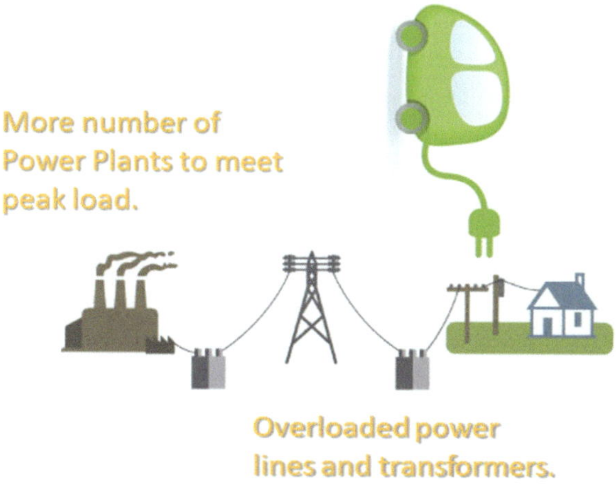

Fig. 4 Impact of EV charging on grid

expected to dominate the vehicle market. However, the success of EV technology depends on the availability of EV charging stations. To meet this demand, utilities are installing EV charging stations at residential and commercial locations. But such widespread adoption of EVs calls for proper planning of charging infrastructure, otherwise it is speculated that it will lead to significant increase in substation load demand. Thus setting more power plants and strengthening the grid infrastructure. Furthermore, the increased peak load demand due to EV charging load may overload service transformers. Certain measures have been recommended to mitigate the peak load (Fig. 4).

A very useful approach is utilities control EV charging using Time-of-Use (TOU) pricing. The reduced off-peak electricity rates in a TOU pricing scenario motivates EV owners to charge their vehicles during off-peak hours. This method can be really handful in reducing the peak load and relieve the stressed infrastructure. But this approach is not very convenient to the consumers and to certain extend this may curtail the adoption of EVs.At the same time if TOU pricing is not scheduled properly, it might lead to surge in charging of vehicles and a spike in load curve. Unlike the above approach, utilities instead of indirect control can directly control EV charging rates and charging start time using smart charging algorithms. This approach shall be symbiotic to both utility and consumers. This paper discussed a source of power, whose over generation is difficult to manage and a developing sink of power whose over-drawal can be headache to utilities. So if they are made to shake hands, the deal is sustainable and greener.

3 Blockchain the Deal Maker

The Indian Electricity Act has provision of obtaining power from open access. With little amendments in act, a charging station or battery swapping station can obtain power from neighbourhood prosumers who have power to trade. In the times to come when solar power shall constituent an appreciable percentage of power pool, solar power feed to grid shall not be lucrative. And with increasing penetration of EVs, buying power from utilities during peak hours shall be costlier. Power transaction between the two shall be win-win situation for all. The sustainability of these transactions demands a decentralized, fast system that deals with the financial transactions swiftly, accurately and safely. The block chain platform offers the solution where the energy transactions are recorded, distributed and generated in a decentralized manner with highest accuracy and speed. In conventional power distribution, power utility are responsible for energy exchange and trading. But blockchain is a tool that gives power to trade their own energy (Fig. 5).

A blockchain is a shared and encrypted ledger that is hosted, updated, and validated by several peer "nodes," rather than by a single centralized authority [5]. This eliminates the central authority and having immutable transaction records that are validated by several peers. The nodes here are network of computers. Once the sender and receiver agree on to enter in transaction, the datasets of transaction are set by specifying the sender, receiver, quantity of energy and other quality terms. In the transaction of power between solar power producers and EV charging station, energy transaction and other parameters recorded in meters are send directly to the local node to form part of datablock. The datablock is broadcasted to a P2P network of computers. So each transaction is encrypted and distributed to many individual computers (peer-to-peer), each of which stores the data locally. The members of the network automatically confirm (verify) the transactions stored on the individual computers. The block stored in the decentralized network gets verified by known

Fig. 5 Conventional vertical method of power distribution

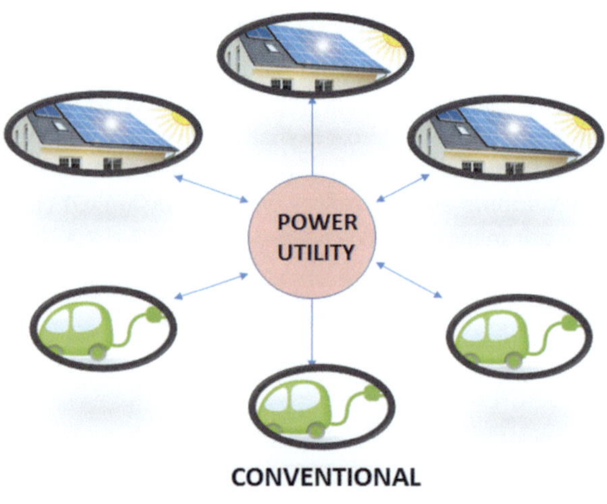

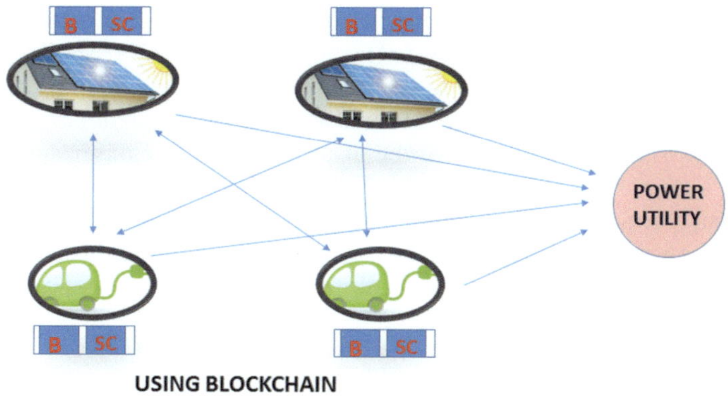

Fig. 6 Power distribution by bloackchain

algorithm. Once verified the block is combined with an existing blockchain, thus creating a continuously growing blockchain. The verification process ensures that all members can add to the blockchain but no subsequent revisions are possible. This enables direct, peer-to- peer transactions between persons or organizations that used to require the services of an intermediary in order for their transactions to be legitimately recorded. So as apparent, transactions are carried out between sellers and customers with no involvement of centralized authority. Apart from handling the energy transaction, blockchain is capable of handling smart contracts. Smart contract is a term used to describe a computer program code that is capable of facilitating, executing and enforcing the performance and negotiating terms of an agreement using blockchain technology. In the present context it is possible to create a fully automated smart contract between an energy producer (prosumer) and EV charging station that autonomously and securely regulates both supply and payment. These smart contracts shall replace the short term power purchase agreement (PPA) (Fig. 6).

4 Chronological Approach

In this model EV charging stations(CS) are also integrated battery swapping stations (BSS). They get themselves register in a blockchain based program and also get the neighbourhood interested prosumers, residents having rooftop solar based power plant, registered. The basic inventory needed is a computer interfaced with the smart net meters with a blockchain program running in background. The program gives each node a unique ID. During registration stage the fixed negotiating terms are also set in the form of smart contract.

The charging station having pre-arranged its vehicles and stack of batteries to get charged shall fix the load and shall get into a dynamic auction in a mobile based app; to get the cheapest power from the neighborhood rooftop solar plants. The lowest bidder gets to supply the power. The power transaction is recorded in a smart meter and the data is transferred to the computer. The data collected is validated using a decentralized mechanism. The purpose of the verification process is to achieve consensus on the content of the distributed ledger. Consensus-based verification is a decentralised (i.e. embedded on the blockchain itself) and automated process. The verified blocks combines with the already verified blocks to form a blockchain. The successful transactions payment can be done instantly by cryptocurrency, wallet money apart from bank transaction.

5 Conclusion

Peer-to-peer trading overhauls the system by opening up the market to allow households (prosumers) to buy and sell solar energy without a power company, trading their energy for more than the feed-in tariff while still undercutting power providers for buyers. This will fuel the growth of two environment friendly sector, EV and rooftop solar. And at the same time utilities shall be relieved of dealing with the slot of day when solar plants pump excess power to the grid. In India this will call for a second electricity reforms. The regulators will have to free the market and give the power to the people.

References

1. Press Information Bureau, GoI, Cabinet, dt: 17.04.2017
2. Denholm P, O'Connell M, Brinkman G, Jorgenson J Overgeneration from solar energy in California: a field guide to the duck chart. National Renewable Energy Laboratory, 15013 Denver West Parkway, Golden, CO 80401
3. Boivin JY (1995) Demand side management—the role of the power utility. Pattern Recogn 28 (10):1493–1497
4. Forum of regulators. Govt. of India, Study on impact of electric vehicles on grid
5. PWC global power and utilities. Bloackchain—an opportunity for energy producers and consumers

UEIC (Utility Excavation Information Center), Mobile App & Common Patrolling

Robin Giri and Mangesh Jarode

Abstract There are around 15 no. of utilities & authorities in Mumbai area having their cables and utilities underground on Mumbai Roads. Whenever authorities or utilities have maintenance, repair, new installation or new project work they excavate the roads. Generally underground utilities gets damage during excavation if machines are used for excavation or care is not taken during manual excavation. This was costed around crore of rupees for Tata Power to repair the cables damaged during excavation. Approx. Similar cost incurred to other utilities. Approximate 50% of damages were due to non-prior information about excavation from authority or utilities. So followings solutions are proposed, (1) This is proposed to have one utility excavation information center in each city of India. This information center will operate through app, phone and one common patrolling team among utilities. The mobile app will be in mobiles of all authorities & utility representatives. There will be clear govt. guidelines to all utilities and authorities to give prior intimation on this mobile app or to UIC by phone otherwise excavation will be considered illegal. If any utility or authority will go for excavation then he will share the details of excavation & location on Utility information app. This information will be immediately sent to concerned persons of all utilities related to that area as per data fed in the app. Everyone can see running excavation details on map in mobile app. Concerned persons will reach at site prior to excavation and will take safeguarding measures to avoid damages. If there is network issue then excavating party can call to common utility information center phone no. and can give details of excavation which will be immediately shared to all utilities or authorities immediately through message. One point contact will reduce burdens on utilities as they will have not to talk to 10 different persons in which few could be missed. (2) This is proposed that there will be only common agency for patrolling for illegal excavations and common agency for safeguarding in the city among utilities. Common patrolling will be monitored on GPS and all uninformed excavation details will be captured through mobile app. As currently 10 utilities are having 10 separate

R. Giri (✉) · M. Jarode
The Tata Power Company Limited, Mumbai, India
e-mail: robin.giri@tatapower.com

M. Jarode
e-mail: mangesh.jarode@tatapower.com

R. K. Pillai et al. (eds.), *ISGW 2018 Compendium of Technical Papers*,
Lecture Notes in Electrical Engineering 580,
https://doi.org/10.1007/978-981-32-9119-5_18

patrolling activities costing average 2 crore to each utility and total average 20 crore. So 18 crore of utilities will be saved. Similarly there are separate safeguarding agencies which do same job but separately so that cost also Rs. 1 core each agency and total 10 crore. This cost will also reduce to 50% if common safeguarding will be used.

Keywords Sharing of resources by multiple utilties in a city · Management of multiple utilities in a city

1 Introduction

Underground utilities like power, telecommunication, gas, water line, sewerage line gets damaged during excavation by various authorities and utilities on roads for various new development infrastructural work or repair and maintenance work. These damages results into safety risk to nearby persons, interruption in utility service and high cost for repairing the damaged utility. For one 11 kV high voltage power cable damage repair it costs around Rs. 1.5 Lacs. So it costs around Rs. 1 crore annually if 70 damages happens to 11 kV Cables of a power utility. Thus increasing damages to underground utilities during excavation activities has become major problem in present days.

Other problem is that each utility have different patrolling and safeguarding activities for avoiding damages during excavation. Thus each utility spends crore of rupees individually for patrolling and safeguarding activities while they are patrolling and safeguarding at common roads.

It has become very difficult to avoid damages to utilities as current system is working in unorganized way. There are many reasons of damages but it is observed that 50% of damage cases are due to non-prior information about excavation to utilities. In some cases when excavation parties pre intimate about their excavation to other utilities then some of utility get missed for pre intimation as excavation party do not bother to inform all utilities or sometimes they forget. That is because there is no organized structure for sharing the information.

Secondly it becomes difficult and very costly to protect cables & utilities during machinery excavation as different utilities work with different resources in different way at same location to protect their utilities. This results in high cost on individual utility and damages due to delay in protection activities as each utility safeguard his lines separately.

In this paper futuristic model of Utility Excavation Information Center (**UEIC**) is proposed. This model is having three components as below.

1. Information Collection and Sharing.
2. Common patrolling for monitoring excavations on roads.
3. Common safeguarding activities during excavations.

First component is to have common utility excavation information sharing center which will operate through mobile app, call center or CCTV cameras. Excavation utilities or patrolman have to share details through mobile app or through call to **UEIC**. CCTV cameras installed throughout city will be attached to **UEIC** server so that round the clock monitoring could be possible. There will be Artificial Intelligence program in the server to monitor CCTV data which will pop pup those locations where excavation or odd activity is happening [1]. If any pop up for excavation or odd activity will arise then patrolman will get the message and then he will go there and share the information to **UEIC**. Then all information will be clubbed together in the server and sent to all concerned area representatives of all utilities.

Second and third component is to have common patrolling and safeguarding activities. By adopting common patrolling wastage of crore of rupees in multiple patrolling activity can be saved to all utilities. Similarly each utility spends crore of rupees individually on safeguarding activities that cost can also reduce up to 50% of current cost.

2 External Damages to Utilities

As we can see in our cities majority of utilities like power, gas, telecommunication, water line, sewerage line are laid underground. If we take example of Mumbai then there are following major underground utilities & authorities.

1. Tata Power
2. Reliance Infra
3. BEST (Power Utility)
4. Mahanagar Gas Ltd.
5. Reliance Jio
6. Airtel
7. Vodafone
8. Idea
9. Reliance communication
10. Tata Telecommunication
11. MTNL
12. Sewerage Line (MCGM & MBMC)
13. Water Line (MCGM & MC + BMC)
14. DTH Services
15. CCTV & Signal systems (MCGM & MBMC)
16. MCGM & MBMC for various (Roads, drainage and other works)
17. MMRDA & PWD (Bridges & Highways).

Total road length of Mumbai City is 1941 km [2]. Each utility has multiple lines on same roads so we can say that there is very huge network of utilities in thousands of kms on Mumbai roads. It is observed by Tata Power patrolling reports that daily minimum approximately 100 excavation happens on Mumbai roads during nine months of fair season from October to June. Majority of these excavations are for new infrastructural work by MCGM for roads, water line, drainage, sewerage bridges etc.

To make it easily understood that how external damage takes place let's take a case of new concrete road construction work which was started by MCGM at Mohan Gokhale Road on 4th Feb 2016. Tata Power cables were laid along that road. JCB operator and his supervisor was not aware about the cable or utilities laid under the road. MCGM or his vendor has also not pre informed to other utilities that they are taking excavation on particular road on particular day and time. MCGM road contractor suddenly started excavation work with JCB at 7 am in the morning. JCB bucket punctured Tata Power cable and other communication utilities also. Similar to above example utilities get damage during excavation work by other authorities and utilities. Such damages happen because excavation happen without pre information to other utilities due to which no utility representative can be present to guide excavation party about presence of his utility lines. There are also other reasons of damages to utilities but non prior information contribute in majority.

3 Effects of External Damages

In the above example external damage has costed approximately Rs. 1.5 lakh to Tata Power to repair the cable. Taking a case where minimum approximate 70 high voltage cable damages happen to a power utility which cost around Rs. 100 Lacs. Similarly huge cost occurs to other utilities also. So cost of damages repair is unnecessary financial loss to utilities. There is also revenue loss to utilities if interruption happens in utility services due to damages.

As damage to power cable also result into tripping of power supply consumers get shutdown till restoration. Some time when there is no back up or main and back up feeders get damage then consumer face complete shutdown sometimes up to 24 h. We can imagine that how difficult is that to live in a city like Mumbai without power for 24 h. There are critical consumers like hospitals, schools, colleges and airports which gets affected by such power failure. Patients who are in ICU can face emergency situations if backup is not available. In similar way damages to other utilities like Gas, Water Department, communication etc. creates difficulties to public.

Whenever damage happens to high voltage power cables it result into big flashover due to direct grounding of high voltage power to earth. There are incidences when nearby persons got injured due to such flashovers. In similar way damage to Gas line can injured to nearby persons or fire can take place in the area.

4 Current Practices to Avoid Damages and Difficulties

A. ***Current practices for excavation information collection and sharing.***

- ***Govt. authority website***—Mumbai authorities also give details of approved tenders of project works on their website like new road, sewerage, waterline & other infrastructure project works but these details are generally displayed couple of months before starting of actual excavations. There are only project work details available on the website and no detail of other daily excavations is provided.
- ***Pre-intimation letters***—Govt. authorities and utilities also share pre intimation of planned project jobs to each other through letters. These letters are also mainly related to project works and generally not for day to day maintenance and repair activities. There is no mandatory guidelines to utilities and authorities to pre intimate each other about their excavation activity so it depends on individual discretion to share the details.
- ***Contacts among utilities***—There are contacts among individual representatives which sometime share information on individual basis. Here also there is no mandatory guidelines to utilities to pre intimate each other so this is also depends on individual discretion that whether he will inform or not.
- ***WhatsApp Group***—There is common WhatsApp group of Mumbai utility forum and some other WhatsApp groups of area wise authorities where utilities share their excavation details. One of the screen shot of WhatsApp information sharing is shown below.

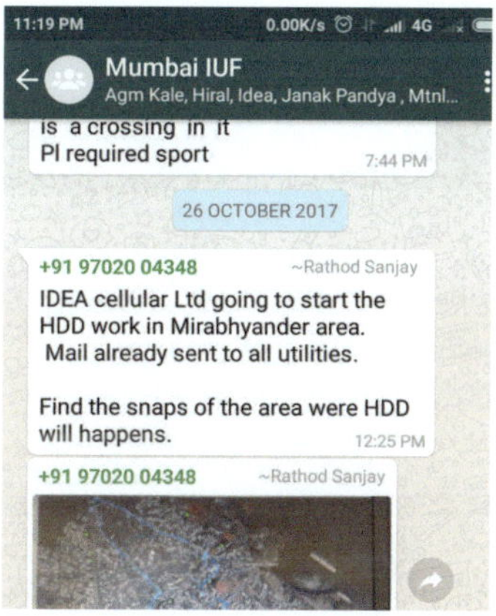

Difficulty in this is similar to above cases as there is no mandatory guidelines to utilities to pre-intimate each other so this also depends on individual discretion that whether he will inform or not.

- ***Individual patrolling activities***—Patrolling activity means covering all roads where utility lines are present with the help of bike or other methods. Patrolman covers their assigned roads on daily basis and reports if any excavation is observed on their utility routes. As per current practice Individual utility have individual patrolling activity to cover their network on roads. Following are the difficulties in this.

(1) In Patrolling activity patrolman cannot stay at one location all the time as he has to move to cover certain route in a day. So there are cases when excavation starts before he reach at excavation location or after he leave that location. So patrolling activity can capture only those excavation information which are already started and continue for at least two days. So organized pre-intimation system is required as back up to capture 100% information.
(2) There are around 10 utilities which monitor their utility routes on Mumbai roads with individual patrolling activities. Generally multiple utilities can be found on same roads in Mumbai as all utilities are basic needs of public residing in all over of Mumbai. Though there are multiple utility on same road but that road is being patrolled multiples times by different utilities. If there are 10 different utilities then 10 different patrolling teams are assigned to patrol their individual utility routes on roads

    (a) One of the consequence of individual utility patrolling activities is that sometimes few utilities get missed for pre intimation as working authority or utility do not bother to inform all utilities or they forget to inform due to large no. that is 10 no. of utilities.
    (b) Second consequence is burden of unessential cost on individual utilities by multiple patrolling activities on same routes. For an example if a utility spend Rs. 2 crores on their patrolling activity then it will cost Rs. 20 crore for 10 utilities. It is clear that same activity can be done by spending only Rs. 2 crore on single patrolling activity. So Rs. 18 crore are being wasted which can be saved with common patrolling activity.

B. ***Current practices for utility safeguarding activities***

Safeguarding activity is related to handling, supporting, shifting and protection work of utility pipes and cables by providing manpower for excavation and other works during critical excavations with JCB's or other machines. Similar to above patrolling activity there are 10 different utilities having 10 different safeguarding activities. Following are the consequence of having separate safeguarding activities.

1. One of the consequence of multiple safeguarding activity is that it creates delay in safeguarding work as only one utility can work at a time. This delay in safeguarding work further delays authorities work like roads developments etc. As authorities also have limited time period for completion of their projects so they do not give further time to safeguard the utilities and start doing their work in speed. This also results in damage to utilities. So due to multiple safeguarding activities each utility get limited time for their safeguarding work which further results into damages to utilities.
2. Second consequence is similar to separate patrolling activities which result into wastage of money on multiple safeguarding activities for same work. For an

example suppose there are multiple utilities at same site and if one utility comes up to safeguard its utility first then he has to depute his separate manpower for that. That utility will do his safeguarding work separately costing to him only. In similar way other utilities will come up one by one and do their safeguarding activities separately which will cost to them separately. Suppose cost of safeguarding activity for one utility was Rs. 1 Lakh then total cost incurred to all 10 utilities will be Rs. 10 lakh. So if there could be single agency for safeguarding of all 10 utilities then this could have resulted into saving of minimum of Rs. 5 lakh. Thus we can say that if an utility is spending Rs. 1 crore individually in its safeguarding activities then this will cost only Rs. 50 Lacs in case of common safeguarding activities.

5 Proposed Model of Utility Excavation Information Center UEIC with Common Patrolling and Safeguarding Division

Basic idea behind creation of Utility Excavation information center is to make one point utility damage protection system among utilities & authorities. This center will collect information from various utilities, authorities, CCTV cameras and common patrolmen and then share this information to all utilities and common safeguarding divisions of UEIC (Fig. 1).

Following are the guidelines and procedure to be followed by utilities, authorities and UEIC to ensure 100% efficiency of UEIC for avoiding damages.

1. There will be three components of UEIC as below (Fig. 1).
    1.1. Information Collection and Sharing (Fig. 2).
    1.2. Common patrolling for monitoring excavations on roads.
    1.3. Common safeguarding activities during excavations.
2. There should be govt. rule that each authority and utility must pre inform to UEIC before doing any excavation.
3. If any excavation is reported without pre information then that will be considered illegal and penalty should be levied on such cases.
4. Pre information should be given to UEIC three days before in case of planned activities and for emergency activities any time.

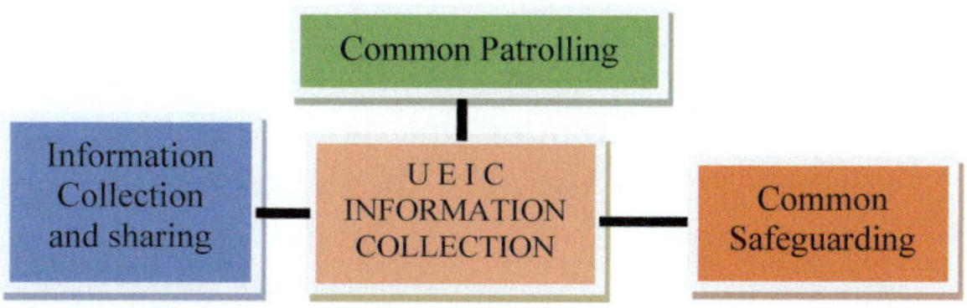

Fig. 1 System general block diagram

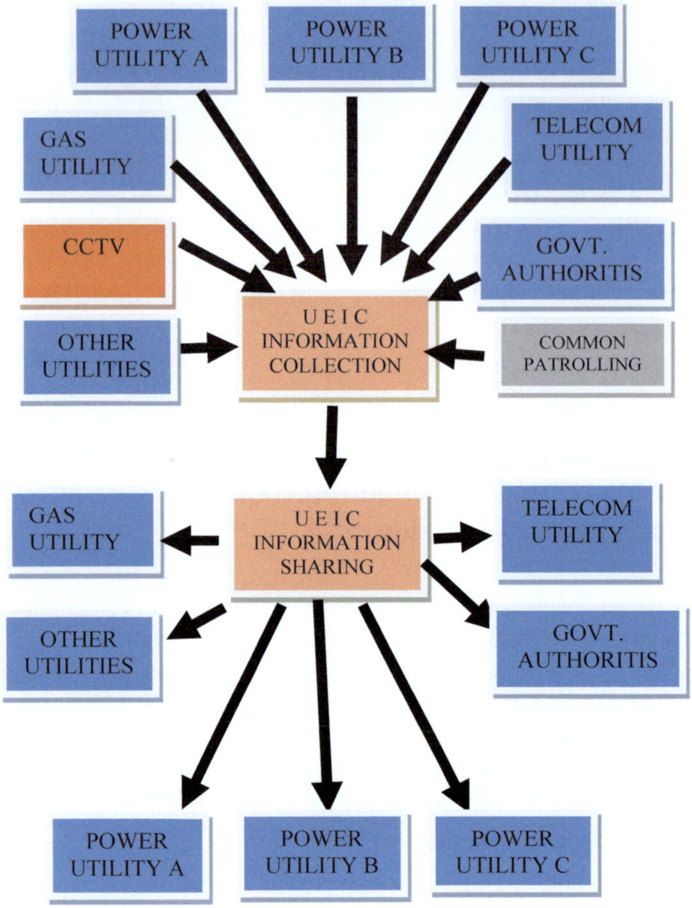

Fig. 2 Information collection and sharing

5. Excavation information can be shared to UEIC thorough mobile app or through calling to customer care (Fig. 3).
6. Excavation information should have
   - Name and details of contact person.
   - Starting date and time of excavation.
   - Location of excavation or Lat long value.
   - Work to be done.
   - Utility or authority details.
7. There are possibilities that few excavation can happen without sharing pre information to UEIC. For such cases there will be common patrolling teams in each area which will observe the excavation on roads and will share information to UEIC through mobile app or by calling (Fig. 4).

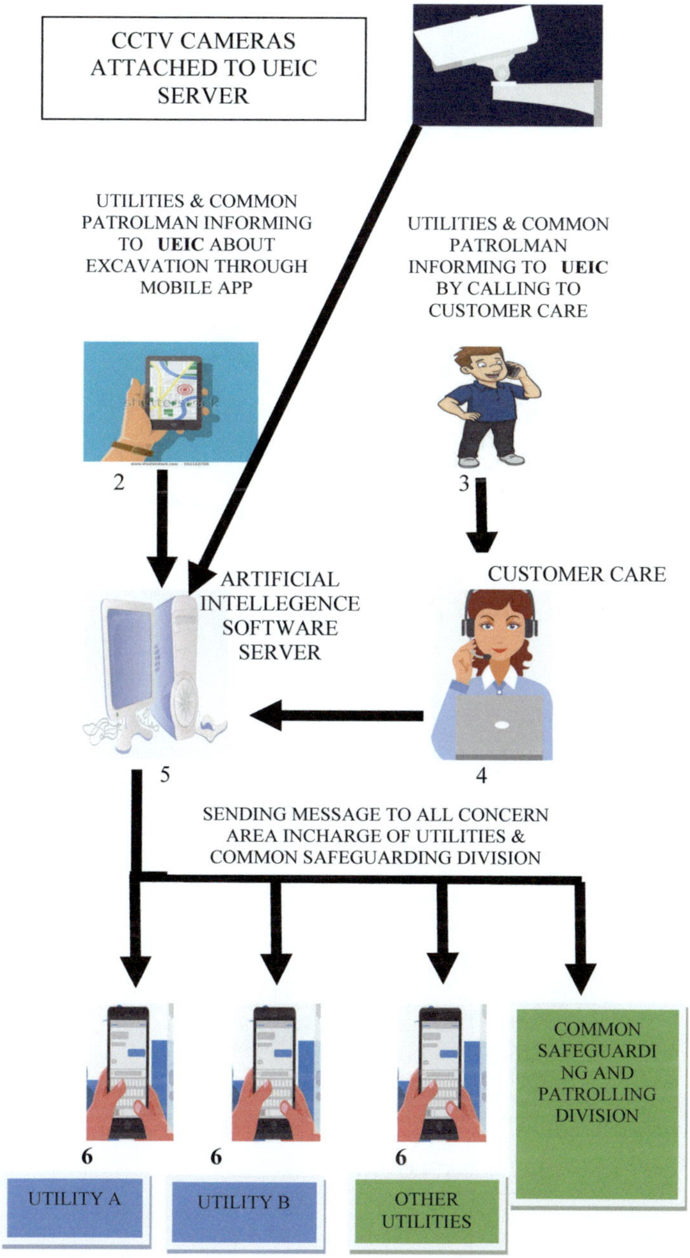

Fig. 3 Information sharing through mobile app and call center

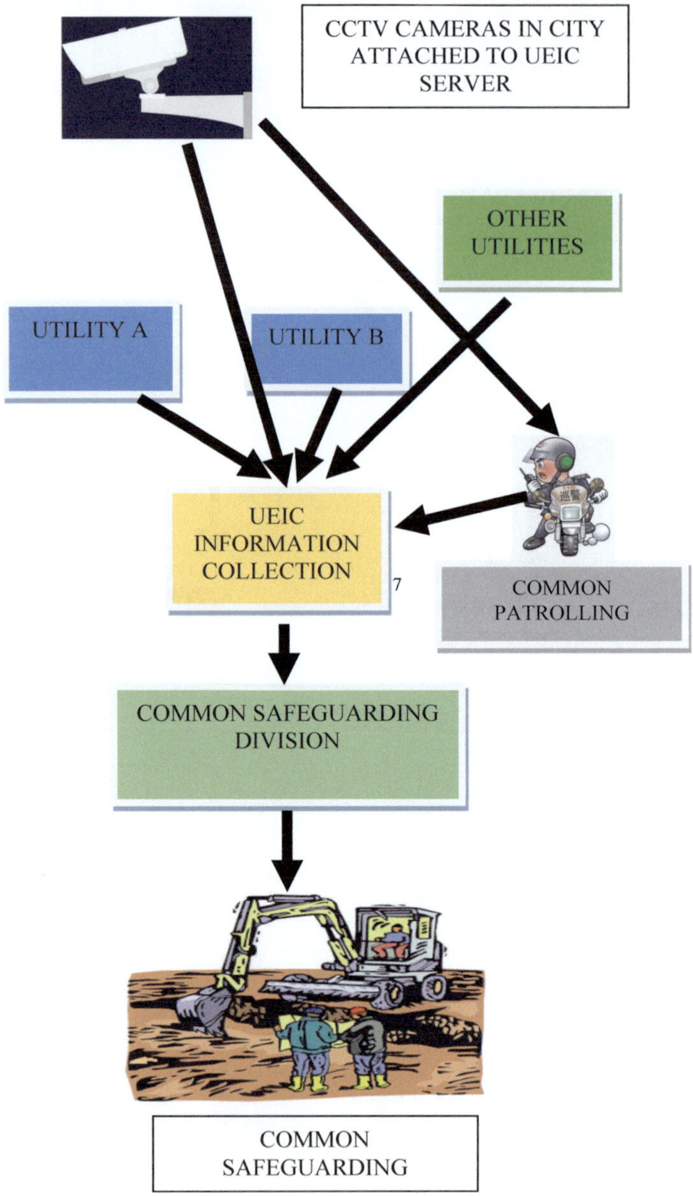

Fig. 4 Common safeguarding division

8. If information is given through mobile app then this will first go to the UEIC server and then artificially intelligence software will share that information to common safeguarding division and all concerned area managers of utilities related to that excavation location (Fig. 3).
9. If information is given by calling to customer care then customer care will enter all above details into UEIC mobile app and then that information will go to UEIC server. After that artificially intelligence software will share that information to common safeguarding division and all concerned area managers of utilities where excavation is to be happen (Fig. 3).
10. There are CCTV cameras installed by police throughout city for security reasons. These CCTV cameras will be attached to UEIC server with Artificial intelligence program. These cameras will be capable to pop pup if any excavation will be observed. Automatically message will be sent to patrolman by server to check the location (Figs. 3 and 4).
11. After getting excavation information utilities will share their lines route alignment related to these excavations to common safeguarding division and will also provide their representatives at site for better coordination about the safeguarding work (Fig. 4)
12. After that common safeguarding division will depute posting persons and safeguarding team at excavation sites. Safeguarding division will ensure that no damage should happen to any utility by providing utmost care and protection arrangements during excavation (Fig. 4).
13. Safeguarding division will arrange to shift or divert the cables on requirement of excavation utility or authorities (Fig. 4).
14. Utility must declare their road length with area wise in the starting of UEIC services. Individual utility will receive only those location information where their network exists.

6 Cost Sharing Among Utilities for UEIC, Common Patrolling & Common Safeguarding Activities

- ***Cost sharing for UEIC***—We will consider the case where UIEC will work only for information collection and sharing. We propose that cost sharing of utilities can be in proportion of their network road lengths. There are 10 no. of utilities having major network on Mumbai roads but just for easy understanding let's take an example of 3 utilities A, B & C with road length 1000 km, 500 km, 250 km, respectively. Suppose first year estimated cost of UEIC including one time setup and profit margin of service providing agency is Rs. 50 Lacs annually and from Second year it is Rs. 25 Lacs then proportion sharing of cost will be as below (Table 1).

Table 1 Cost sharing for UEIC

Utilities	Road length X	Proportion Y = X/ H * 100	First year cost proportion = Y % of 50 Lacs	Second year cost proportion = Y % of 25 Lacs
Utility A	1000	57	29	7
Utility B	500	29	14	4
Utility C	250	14	7	2
Total H	1750	100	50	13

We can say that if no. of utilities will be more then cost of services on individual utilities will reduce.

- ***Cost sharing for common patrolling***—*Here* we are taking the case where UEIC is also providing common patrolling services. Similar to above case cost sharing for common patrolling can be in proportions of road length of networks of individual utilities. Let's take above example of 3 utilities A, B & C with road length 1000 km, 500 km, 250 km, respectively. Suppose cost of patrolling of 1000 km road length in a city is Rs. 1.25 crore annually then proportion sharing of cost will be as below (Table 2).
- ***Cost sharing for common safeguarding***—Safeguarding activity depends on no. of excavations on roads. There can be locations where few utilities cannot have their networks and other utilities can have their networks. So safeguarding activity cost cannot be linked to total network road length in the city as in above cases but this can be linked to location wise network lengths. So cost incurred on safeguarding at any location should be shared in proportion of network length of individual utilities at that location only. Let's take above example of 3 utilities. Utility A is having four lines, utility B is having two lines and utility C is not having network at particular location where excavation is happening with JCB and utilities can be damaged. Suppose total cost of safeguarding of utilities for 15 m length during excavation is estimated to Rs. 2 Lacs then proportion sharing of cost to utilities will be as below (Table 3).

Table 2 Cost sharing for common patrolling

Utilities	Road length X (km)	Proportion % Y = X/H*100	Annual common patrolling Cost proportion = Y % of 125 Lacs
Utility A	1000	57	71
Utility B	500	29	36
Utility C	250	14	18
Total H	1750	100	125

Table 3 Cost sharing for common safeguarding

Utilities	Network length X (m)	Proportion % Y = X/H * 100	Common safeguarding cost proportion = Y % of 2 Lacs
Utility A	60	67	1.3
Utility B	30	33	0.7
Utility C	0	0	0.0
Total H	90	100	2.0

7 Benefits of UEIC Common Patrolling & Common Safeguarding

- ***Performance improvement***—There are approximately 50% utility damage cases which happens when authorities and utilities do not provide prior excavation information to other utilities. With the help of UEIC it is expected that 99% prior information about excavation will be received. So 50% damages can reduce to minimum 20% as all excavation information will be received with help of UEIC. Utility performance index is linked to uninterrupted availability of utility services. So if damages will reduce then tripping will reduce and this will improve utility performance index.
- ***Cost saving***—11 kV High Voltage Power cable damage repair cost is approximately Rs. 1.5 Lacs. Considering a case that a utility is having 70 damages in a year costing then it cost around Rs. 105 Lacs. If 30% of 50 damages that is 20 no. of damages will reduce with the help of UEIC then this will save around Rs. 30 Lacs annually. As per Table 1 first year cost of UEIC on Utility A is Rs. 29 Lacs and second year cost is Rs. 7 Lacs when there are only three utilities. If no. of participating utility will increase the individual utility cost will reduce. From above it is clearly understood that with the help of UEIC damages will reduced which will result in reduction of cost of repair of damages and increase in profit to company.
- ***Safety to society***—Damages to power and gas lines can create flash over and fire which can cause injury to nearby passing persons. Thus reduction in damages to power and gas utilities with the help of UEIC will improve safety to public.
- ***Improvement in utility services to public***—As damages will reduce tripping to essential utility services like power, gas, telecom, water will reduce thus utility service to public will *improve.*
- ***Revenue through uninterrupted service***—If damages to utilities will reduce then tripping in utility services like power, gas, telecom, water will reduce. Uninterrupted service will provide more revenue to utilities.
- ***Help to Govt. authorities***—Government municipal authorities will also be the member of UEIC so data of excavation activities will be available for them to track the excavation on roads coming under their judiciary. Police can also use that data to manage traffics and issue traffic permission to utilities.

Table 4 Annual profit of Common Patrolling

Utilities	Network road length X (km)	Annual cost of common patrolling in proportion of road length L	Annual cost of individual patrolling in Lacs M	Annual profit in Lacs N = M − L
Utility A	1000	71	125	54
Utility B	500	36	63	27
Utility C	250	18	31	13
Total H	1750	125	219	94

Table 5 Annual Profit of Common safeguarding

Utilities	Network length X (m)	Proportion % Y = X/ H * 100	Common safeguarding annual cost M	Individual safeguarding annual cost L	Annual profit in Lacs N = L − M
Utility A	60	67	1.3	2	0.7
Utility B	30	33	0.7	2	1.3
utility C	0	0	0.0	0	0
Total H	90	100	2.0	4.0	2.0

- ***Additional financial benefits of Common Patrolling***—Considering a general case Individual patrolling activities costs around Rs. 50 Lacs to Rs. 3 crore annually to different utilities. We can see Table 4 which is referred to table no. 2 example. Cost of Individual patrolling for utility A is Rs. 125 Lacs and in common patrolling it comes down to Rs. 71 Lacs. So there is Rs. 54 Lacs profit annually. Thus cost to individual utility will reduce if common patrolling activity will be adopted.
- ***Additional financial benefits of common safeguarding***—We can see in Table 5 which is referred to Table 3 example. We can see that cost of safeguarding to individual utility is Rs. 2 Lacs while in common it is costing to Rs. 0.7 and Rs. 1.3 Lacs to A and B utilities respectively. So total cost is reduce from Rs. 4 Lacs to Rs. 2 Lacs. Profit to utility A is Rs. 0.7 and Rs. 1.3 Lacs. Thus cost to individual utility will reduce if common safeguarding activity will be adopted.

8 Conclusion

Above model of UEIC through mobile app, Common Patrolling and common safeguarding states that damages to individual utilities during infrastructural or repair and maintenance work can be easily reduced by adopting an organized system of common activities. We have taken example in above tables where only three utilities have participated and we have got some profit value. On the other side if participating utilities will be more then profit value will be more. As we have seen above this model can help in multiway to not only utilities but to general public also in terms of financial, safety, revenue, uninterrupted services and other intangible benefits.

References

1. https://en.wikipedia.org/wiki/Artificial_intelligence_for_video_surveillance
2. https://timesofindia.indiatimes.com/city/mumbai/No-of-cars-up-34-in-6-years-Mumbai-roads-same-for-50-years/articleshow/17864082.cms

Artificial Intelligence (AI) Applications and Techniques in Smart Grid

Vinay Gupta

Abstract Smart Grid aims to provide grid stability, balance supply & demand, ensure reliability of power supply and enable greater integration of diverse generators & consumers. To enable these objectives, amongst others, use of Artificial Intelligence techniques plays a pivotal role. This paper gives a brief about the various applications of AI for successful implementation of Smart Grid. It also highlights the features of various AI techniques and their relevance in the emerging power eco-system.

Keywords Artificial intelligence · Smart grid · Grid balancing · Demand side management · Artificial neural network · Operations & maintenance · Self-healing grids · Fuzzy logic · Particle swarm optimisation · Convolutional neural networks · Evolutionary algorithms

1 Introduction

There are various ways to define the Smart Grid System. One of the way to define is—Smart Grid is an integrated system of varied types of generators, consumers, distribution elements & DISCOMs, which seamlessly balances the demand and supply to ensure reliable, 24×7 and high quality of power at the least cost, by utilising the communication, computing and control technologies. The ultimate Smart Grid vision is ANY POWER, ANY WHERE, ANY TIME at LEAST COST.

In order to achieve this vision, multifaceted actions by multiple agencies need to be undertaken, which includes design of grid compliant wind turbines & solar invertors, real time communication systems integrated in transmission & distribution systems, smart metering at consumer-end, robust forecasting & scheduling algorithms, near real time computational systems, self-healing & diagnostic networks and reliable cyber security enabled digital platforms. Thus Smart Grid

V. Gupta (✉)
Suzlon Energy Limited, Pune, India
e-mail: vinay.gupta@suzlon.com

R. K. Pillai et al. (eds.), *ISGW 2018 Compendium of Technical Papers*, Lecture Notes in Electrical Engineering 580,
https://doi.org/10.1007/978-981-32-9119-5_19

operations represent one of the most complex optimisation problem for data scientists with least cost and uninterruptible supply as the two main objective functions. Modern AI techniques can be effectively used in finding solutions to these Smart Grid computational challenges in real/near real time.

Artificial Intelligence techniques enables finding patterns in complex data structures and types from varied sources and protocols, which can be used for planning the nature of power sources to be marked for scheduling or planning the per unit cost for consumption by the consumers. Further, it enables us to adapt and continuously learn based on the patterns, which may be seasonal, local or event-based. Thus the unpredictability in climatic conditions (wind, solar, hydro energy sources are dependent on weather conditions) and consumer behaviour (varies based on social events, time of day, season, etc.) can be modelled with reasonable degree of accuracy and outputs fed to determine the suitable ramp rate, synthetic inertia or per unit cost, to enable the Smart Grid objective.

In order to execute the AI models, the key ingredient is availability of data viz. real time, near real time & historical. Rapid advances in Internet of Things (IoT) technologies in last few years have enable us to capture this required data from various elements, nodes, devices and users of Power System network and transfer them to store-compute-analyse processing units located at different levels of network.

2 Applications of AI in Smart Grid

Three main applications of AI in Smart Grid are being highlighted in this paper. These are

- Balancing the Grid
- Demand Side Management
- Operations & Maintenance of Smart Grid elements.

Details of these applications are covered in the subsequent sections of this paper.

3 Balancing the Grid

Effective grid operations requires stability of various parameters viz. voltage, frequency, harmonics & power quality. Real time and continuous matching of demand and supply is one of the methodology to achieve the required stability. Thus balancing the Grid at different nodes and different time of the day is essential and pre-requisite for Smart Grid.

However, with the recent changing scenario and requirements, the need for both generation & demand side management in near real time has become more

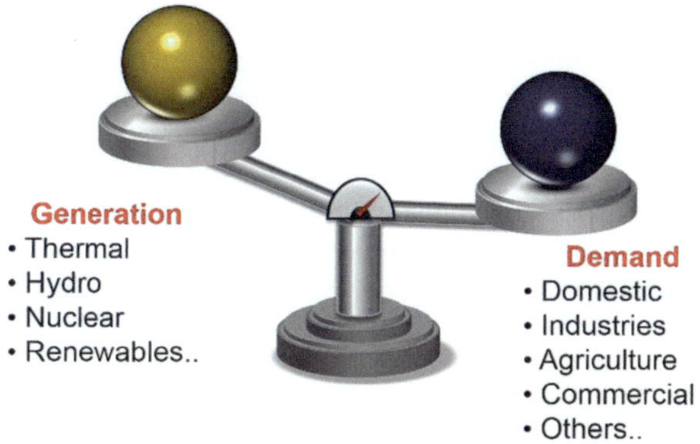

Fig. 1 Balancing the generating sources & consumer demand, a critical requirement of Smart Grid operations

prominent. This is primarily happening due to increase in the share of Renewable Energy in overall generating sources mix. The variability and unpredictability of wind and solar makes the task of distribution companies difficult & challenging. This aspect gets accentuated by the fact that government has granted renewable energy as ALWAYS-RUN status, thus no curtailment is permitted. Owing to this constraint, there exists numerous instances when the gap between demand and supply is positive or negative. To overcome this gap there is a need for robust forecasting and scheduling algorithms, in which human interface in minimal or zero. Seamless flow of forecasted data & real time data parameters with integration in scheduling algorithm and suitable alert mechanism (in case of any deviation from laid down rules) will enable the aim of balanced grid operations (Fig. 1).

Due to the increased thrust by Indian government on penetration of Solar energy, in the form of roof-top solar plants, a new breed of generator-cum-consumer profile has emerged, which are being termed as PROSUMERS. This new profile has bring both challenges & opportunities in Smart Grid operations. On the Generator side, they form part of the distributed generators thereby adding on the overall generation capacity, reducing the burden on transmission lines (and thus no AT&C losses), and harnessing natural energy. On the consumer side, they have the tariff cost advantages, reliable power supply and control over their consumption. Policies issued by State Government on Net Metering is enabling this initiatives, leading to increase in the number, scale and variety of this categories.

In last few years, the deployment of smart meters has enabled real-time measurements of consumption, and calculations of tariff plans based on Time of Day (TOD)—low tariff during lean hours of consumption, to enable demand side management. In addition, controlling (switching on/off) of air-conditioners is being done in the West Coast of the U.S as part of Critical Peak Pricing (CPP). However, these measures are likely to create additional peaks as and when the low tariff period begins or when the air-conditioners are switched on, after the CPP period is over [1].

In order to overcome the limitations of TOD based tariff plans and smoothen the peaks in demand, the concept of Real Time Pricing can introduced. In real time pricing the time slot is further reduced to 30 min or 60 min period and decided based on the demand and supply similar to stock prices. However, as in stock market suitable checks have to be introduced to avoid the sudden rush/fall in consumer behaviour. Therefore, for true grid balance, we need an intelligent systems that can accurately predict the generation and demand in short term, has the ability to responds to varying weather conditions and holiday surges/local function/events to accurately predict the changes. In addition, industries, which are heavy consumers of power, will have the flexibility to work out the best combination with respect to their workloads and tariff cost [1].

Thus from the computational point of view this will be an iterative process, where deep learning models once built will be learnt, re-learnt and continuously adapted to the changing behaviour and profile of the consumers. At this stage, it is also important to understand that the granularity of such consumers become an important parameter for consideration. If we take a larger data set, viz. city/large locality, we may not be able to predict the usage pattern correctly. To overcome this challenge, the data from individual smart meters will play an important role, as it will enable profiling the consumer behaviour at each individual metering unit level, which can then be aggregated to the sub-node and node level. The optimisation algorithm built on these datasets can be then used to determine the prices for different load profile, consumer and TOD.

Recent work using AI technologies has already begun to research the use of adaptive, learning and autonomous systems, which can determine the individual behaviour and patterns. The output so obtained are fed into the scheduling algorithms for Distribution Companies (DISCOMs) leading to balancing of demand and supply. Key applications of AI in balancing the grid are [1]

- Developing adaptive & self-learning models of consumer (different types, profiles & usage patterns) behaviour by combining deep learning techniques and big data computational power.
- Building simulation models and systems, which will determine the ramp rate, synthetic inertia and least cost price based on the varying atmospheric conditions and consumer behaviour.
- Determining the best combination and options for prosumers in terms of usage pattern, storage system capacity and least overall cost. This will greatly enhance the satisfaction and adoption of PROSUMERS model.

4 Demand Side Management

Demand Side Management entails greater participation of consumers in enabling the Smart Grid operations. Considering the varied nature of users, they can be broadly categorised in the following categories:

- Residential Users with Smart Meters installed.
- Industrial Users with Smart Meters and Building Management System installed.
- Residential/Industrial users with normal electricity meters, in which monthly usage patterns is only available, but have a suitable metering units at Node/Sub-Node level.
- Residential/Industrial users with Battery storage capacity, which can be supplied to the grid and they have signed up for net-metering tariffs (Prosumers).
- Users with Electric vehicles, and can actively participate in charging/discharging from the grid.

Based on the above categories, the Utility companies can work out a different model of engagement with them. The Residential/industrial users with battery storage capacity and EV users form the most beneficial category for participation as they can be asked to supply the power in case of any additional demand or vary their consumption as per the feedback from the grid.

The second set of categories, which are also useful for DSM are the users with Smart Meters installed. They can also participate actively as their load profile, usage patterns and price sensitivity can be analysed in greater details and accordingly planned. Further, the Smart Meters can be made more intelligent, where-in on sensing the additional load/demand at the grid level, they can automatically reduce the consumption at the residential/industrial level, based on the pre-specified rules/priority levels.

The users without the smart meters are most difficult to manage, however by installing suitable real time metering mechanism at the node level, this problem can be reduced. The models developed would be at an aggregate level, but with the availability of long term data and patterns, the accuracy can be greatly improved.

Thus considering the above approaches, the DISCOMs can incentivise the users based on their participation level in the grid stability. This will also encourage new users to move towards the better priced category. The users can accordingly share the data with the Grid operators and help them in achieving the mutually agreed objectives.

In this ecosystem, health of batteries play an important role in ensuring the reliability of power supply from the prosumers end. Thus the battery monitoring mechanism and charging system healthiness need to be checked at periodic intervals. Suitable prognostic models can thereafter be built for the same using the long term data, considering the large users of such category. Users, who maintain good health of batteries and spare capacity can also be incentivised based on the results of this model.

When it comes to Electric Vehicles, there may be a large number of passive users, who only utilise the charging stations. But the users, who make an additional effort to connect EV to the charging/discharging station needs to be incentivised. Further, it will also put an indirect pressure on various government/civil agencies to build such infrastructure as part of the greenfield/brownfield projects. Due to movement of these EVs and likely concentration of charging time during the certain period viz. pre-office & post-office timing [2], there is need analyse the demands imposed on the local distribution network, and the variability of renewable energy output.

Coordinating and integrating all these users of different hues and colours requires building a complex control and optimisation problem. IoT based edge analytics and AI Particle Swarm Optimisation techniques can prove to be effective in providing the solution for this.

Few applications of AI in the field of DSM are [1, 3]:

- Developing methodology for residential load consumption & management taking into account consumer's preferences, by simulating the household environment under different scenarios, with a prime objective of cost savings.
- Developing intelligence in the Smart Meters, wherein they can control the consumption/load based on the feedback received from the grid in an automated fashion, while ensuring key objectives/requirements are met.
- Developing suitable algorithms for EV integration based on their daily movement, shifting load profiles and identifying regions where the EVs can be used for supplying the grid and battery capacity.

5 Operations & Maintenance of Smart Grid Elements

Four main AI applications in Operations & Maintenance of Smart Grid are:

- Self-healing Grids
- Fault Detection System in Transmission Lines
- Image analysis of Transmission line elements
- Predictive Maintenance of Power Plants.

A. *Self-Healing Grids*. The three key factors of self-healing technologies are digitalization, decentralization and adaptability through artificial intelligence (AI), machine learning and expert systems.

Power utilities are now deploying self-healing concepts in densely populated metros such as Mumbai and New Delhi to reduce the time taken for power restoration to less than a minute (as against more than 15 min earlier). That, too, with minimal or no human intervention, thus reducing the impact of hurdles like geographical challenges [4].

Typically fault occurs in a system due to overloading of transmission lines, poor maintenance of earthing systems or non-operation of protection systems. In order to predict these failures/ensure reliability, the SCADA systems forms the backbone in providing the real time data or timely alerts and warnings. Detailed analysis of data parameters obtained from various IEDs, relays, metering system and transformers helps in predicting the failures. In last few years, the standardisation of communication protocols (IEC-61850 & IEC-60870-5) has enabled seamless data flow and integration across different OEMs.

Thus with the use of intelligent and autonomous systems at transmission and distribution level, the signals can be sent to reduce the generation from the sources, activate the protection circuit or isolate the system, till restored.

The computational challenge lies in handling large number of active elements in real time with complex interplay of protection and power circuits. Any minor error, or delay in operation can be catastrophic due to high power & voltage levels involved. AI techniques can play a significant role in designing computationally efficient algorithms that can predict voltage and phase information at different nodes in the (partially observable) distribution network, in real time.

B. *Fault Detection System in Transmission Lines*. In the traditional method, conventional distance relays are used to identify fault and tripping, considering the power swings. These methods results in the late detection and inaccurate results, and can have serious consequences for power system stability. To improve the performance, Artificial Neural Network (ANN) architecture can be used, which results in the earlier fault detection. Different fault like single line to ground fault and double line to ground faults can be detected using ANN. Seema Singh et al. [5] have suggested the use of feed forward neural network, which is trained with back propagation algorithm for detecting single line to ground and double line to ground for all the three phases. The ANN allows rapid learning based on the past behaviours and various Earth Fault/Over Current Relay trips in the transmission line networks.

C. *Image Analysis of Transmission Lines Elements*. D Kumar et al. [2] have proposed image processing technique that helps to find and diagnose faults in transmission lines. One such technique makes use of Digital image processing wavelet shrinkage function [6] for fault identification and diagnosis. In this, the images are captured using thermal imaging camera with suitable overlays of GPS coordinates (as provided by.kml file), which provides the detailed layout and condition of transmission lines. Thereafter, for automatic diagnosis and decision making process, a hybrid system is used for monitoring and diagnosis viz. combination of fuzzy logic and the neural network system, it is also called neuro-fuzzy system. This system model uses the IF-THEN fuzzy rule to find the membership function for input and output variable of the system. It claims to improves the operation process speed, achieve reliability of about 95–98%, with involvement of human factor to only 5%. UAV based transmission lines inspection has been found to be more effective than line inspection robots, manned aircrafts and optical satellite remote sensing. Based on the optical or thermal imaging camera used as payload, UAV line inspection gives the video output [7]. These video output enables us to identify the physical defects such as broken insulators, conductors with broken strands, abnormalities in vibration dampers and loose fasteners, In addition, the thermal contrast based images enable hot spot identification, which are caused due by loose connections, defective conductors/clamps, strained tubes, splicing sleeves and insulators. Intelligent deep learning technologies including Convolutional Neural Network (CNN) for defect detection from images enables us to analyse these videos in a near real time so as to identify, classify and record the defects with minimal human intervention. This greatly enhances the effectiveness of video imaging efforts & thereafter transmission line maintenance process.

Angelika WRONKOWICZ [6] have suggested the methodology to detect hot spots in the IR images of Transmission lines using various image processing and analysis methods. The suggested algorithm was tested on a set of IR images taken from vision inspections of 110 kV high voltage power transmission lines. The results turned out to be very effective in identification of likely defect areas.

D. *Predictive Maintenance of Power Plants*. Predictive Maintenance can play a great role in ensuring high availability of Power Plants and reducing operating costs by timely planning the repair and associated resources. However, it has been marred with large number of alerts, huge data from sensors, with only a fair degree of accuracy. AI techniques can take this initiative further by analysing and integrating various sensor parameters coupled with analysis of past failure patterns to identify the likely failures with greater accuracy and also provide suggestion to the site teams for repair/maintenance. Advanced analytics already demonstrate the benefit of intelligent maintenance. Wind Turbine generators and thermal power plant owners have been able to predict the timing of failures within few weeks to months in advance, with nearly 75% accuracy. Through these efforts of optimizing predictive maintenance, coupled with Prognostics analytics, Remaining Useful Life prediction, automating fault prediction, and increasing capital productivity through AI applications we can increase power plants availability by 4–5% and thereby generation earnings by 10–20%.

6 Artificial Intelligence Techniques for Smart Grid

The term "Artificial Intelligence" (AI) is used to describe various computational and analytical techniques, which are based on the study of living beings behaviours and decision making methodology. A lot of computational techniques have been inspired by animal kingdom such as Particle Swarm Optimisation, ANN, Fuzzy Logic, Genetic algorithms etc. Out of the many AI based computational techniques, seven techniques have been identified for Smart Grid related applications and enumerated in the succeeding paras.

A. *Complex Event Processing (CEP)*. Smart grid is a unique system, which involves an interplay of various nodes, elements, devices and metering devices by utilising the various data being generated at various levels. These data points are all along the spectrum viz. generation, transmission and consumption side. Analysis of these datasets involves real time computations and this challenge is being mainly addressed from the perspective of Complex Event Processing (CEP) techniques [8]. CEP, has the capability to combine, integrate data from various sources and derive meaningful patterns, which can be used for Smart Grid operations. One of the primary objective is to automate the processing and decision making thereby reducing the OODA (Observe-Orient-Decide-Act) loop. CEP techniques address event filtering to seek for relevant patterns. However, the selected events need to be semantically enriched in some way so that they can lead to an understanding of the ongoing situation. One such application could be to model the frequency variation, voltage dips and reactive power in order to anticipate their possible evolution and negative effects.

B. *Artificial Neural Networks (ANN).* Artificial Neural Networks (ANNs) simulates the functioning of biological neural systems. Thus it can be not only programmed but can also be trained by feeding large number of data parameters. The learning capability is enabled by feed forward and back propagation techniques. The patterns in large sets enables the model to learn on a continuous basis. These models can be effectively used in Power forecasting and Hybrid farm optimisation/control algorithms. Neural networks have demonstrated their capability to detect in real time the changing dynamics of power systems.

C. *Evolutionary Algorithms.* Evolutionary algorithms (EA), or genetic algorithms, has found be effectively used in optimisation problems in Power systems. Due to relatively low demand of computational resources, the Evolutionary Algorithms have become very popular. The work in [9] also resorts to a genetic algorithm approach for efficiency enhancement, demonstrating that those fuzzy controllers implemented by means of genetic algorithms obtained optimum results (both for entire and discrete time intervals).

D. *Fuzzy Logic.* The conventional control systems in buildings is improved by the introduction of artificial intelligence based intelligent system and Fuzzy logic controller. Multi Agent Systems (MAS) is a distributed computational intelligent technique, which is capable of making autonomous decision without human intervention [10]. Thereby, a smart home based MAS can be considered as a smart self-sustainable system. Fuzzy Logic is similar to our method of reasoning and decision making and thus has a very low overheads on computational techniques. The fuzzy rules are very critical task in the development of Fuzzy Logic Controller. Fuzzy rules are a series of linguistic statements (viz. IF_AND_THEN_AND_) that describe how the decision is made by the fuzzy controller. Fuzzy rules present a helpful connection between the inputs and the outputs of the system.

E. *Self-Organising Systems.* A self-organizing system (SOS) is characterised by lack of centralised control and yet self-organising & self-healing in nature. Adaptation, Robustness, Decentralised control and Scalability makes SOS very effective [10]. SOS is capable of handling complex systems like Smart Grid. It also provides methodology for mapping the interactions between various components/devices. This aspect is liked by Load Despatch companies. On the other side, the bossless structure of SOS, allows bottom-up processes, which keep the responsibility and decision making at customer level. Thus making them feel that they are in control. Rules for formulation are simple, but finding or predicting the correct rules is difficult in complex Smart Grid system. One of the application of SOS is to evolve an energy trader algorithm at consumer/prosumer level, where there is no initial knowledge about market rules. The trader rules are learned implicitly.

F. *Particle Swarm Optimisation (PSO).* This optimisation techniques makes use of group behaviours in animal kingdom such as birds, ants etc. Such group of animals which does not have any organizer in their group. The individuals makes the decision by continuously identifying the best condition and communicating with each other. These members will intimate it to the group then they will also go to that place. This process will be repeated until the best solution or a food source is found out [11]. Thus the PSO algorithm will be optimize throughout the system and it will find out the real and most suitable value in this. PSO can be effectively used to maximize the system energy production and meet the load demand with minimum cost and highest reliability [3].
G. *Convolutional Neural Networks (CNN).* Convolutional neural nets are a deep learning technology used to automatically classify the content of the image based on previously trained models. CNNs are comprised of three types of layers viz. convolutional layers, pooling layers and fully-connected layers. Convolution layer forms the backbone of the complete network, whereas the Pooling layers interspersed between the convolution layers carry out data size reduction and non-linear mapping. As regarding the Classification layer, it does not require a large amount of computation. These layers jointly forms the Convolutional Neural Networks. CNNs are being extensively used to analyse the images of Solar Panels, Wind Turbine blades, Transmission line elements and Sub Station equipment obtained from Drones/Spotter scope. The analysis leads to quick identification and classification of defects and thereafter making it available to the site teams for action, with minimal human intervention.

7 Conclusion

In this paper, we have highlighted the various applications of AI in Smart Grid System and brief on various AI techniques, which can be employed on these applications, to derive the meaningful output. The Smart Grid System thus has a great potential to embrace Artificial Intelligence in the coming years. At every step of the value chain, from power generation, transmission, distribution to end consumers, tremendous opportunities exists for distributed intelligence, advance analytical techniques and deep learning technologies exist that could help domestic/industrial users and electric utility companies to reduce the overall operational cost and enhance the satisfaction/end-users' experience.

Employing Artificial Intelligence techniques in a large number of use cases open a myriad of possibilities for this sector. It may lead to a world where power generation, distribution, and transmission operations are carried out in an automated fashion, where the grid is balanced independently or in an autonomous manner, where power trading decisions are made with lightning speed, and where end-users never don't have to worry about bills/power quality. Further research in the field of AI techniques and its applications in Smart Grid can reap great dividends for its smooth and successful implementation.

References

1. Ramchurn SD, Vytelingum P, Rogers A, Jennings NR (2012) Putting the 'Smarts' into the smart grid: a grand challenge for artificial intelligence. Commun ACM 55(4):86–97
2. Kumar D, Kumar A, Ansari MA (2016) Fault analysis in Transmission Line using Image Processing with Neuro-Fuzzy approach. Int J Electr Electron Commun 4(12):54–60
3. Pathak AK, Chatterji S, Narkhede MS (2012) Artificial intelligence based optimization algorithm for demand response management of residential load in smart grid. Int J Eng Innov Technol (IJEIT) 2(4):136–141
4. Sharma S (2017) Self healing grids to self healing factories. [Online]. Available: utilities are now deploying self- healing concepts in densely populated metros such as Mumbai and New Delhi to reduce the time taken for power restoration to less than a minute (as against more than 15 minutes earlier). That, too, with minimal or no human
5. Singh S, Mamatha KR, Shankar S (2014) Intelligent fault identification system for transmission lines using ANN. IOSR J Comput Eng (IOSR-JCE) 16(1):23–31
6. Wronkowicz A (2016) Approach to automated hot spot detection using image processing for thermographic inspections of power transmission lines. Diagnostyka 17(2)
7. Liu K. Hu, Inspection and Monitoring Technologies of Transmission Lines with Remote Sensing, Academic Press, New York
8. Santofimia-Romero MJ, del Toro-García X, López-López J-C (2011) Artificial intelligence techniques for smart grid applications. In: *CEPIS UPGRADE,* vo XII(4):41–48
9. Jahedi G, Ardehali M (2011) Genetic algorithm-based fuzzy-PID control methodologies for enhancement of energy efficiency of a dynamic energy system. Energy Convers Manage 42 (33), 2011
10. Elmenreich W (2014) AI techniques for smart grids. In: ISGT-ASIA 2014, May 2014
11. Tibbie Pon Symon VA (2017) Control and optimization of the smart grid by using particle swarm optimization (PSO): to reduce the distribution losses. Int J Eng Technol (IJET), 8 No 6, Dec 2016- Jan 2017

Vinay Gupta is presently working as Head Data Analytics & Business Excellence, Suzlon Global Services Limited. He is leading the Digital Transformation & Analytics Program of wind & solar farm operations. Prior to this, he was in Enercon India Limited for three years in the similar role. Earlier, he had served in Indian Army for 22 years and as a Colonel in Army, he had played a pivotal role in establishing the first Centre for Data Analytics and developing analytics driven Military Equipment Management system. Has rich & diverse experience of 26 years in the field of Data Analytics, Industrial IoT, and Electronics Engineering. He has participated as speaker/panellist in various international/national seminars/conferences on Big Data Analytics, Industrial IoT and Renewable energy. He is a member of IETE, PMI and IEEE associations. Also, a member of Working Group of LITD-27, related to Artificial Intelligence & Machine Learning. He is an M.Tech (Electronics & Telecom), PMP and Master Black Belt Six Sigma.

Degradation of Organic Pollutants in Drinking Water by Non-thermal Plasma

Nilakantan Ajay Krishnan, Jürgen Kolb, Rafig Azzam, Uwe Kaltenborn and R. Sarathi

Abstract Water is a scarce resource that is continuously threatened by pollution, growing population and strain on available reservoirs due to fast developing economies. A problem of considerable concern is posed by anthropogenic pollutants, foremost artificial organic compounds. Especially pharmaceutical residues with their intended stability are found in increasing concentrations in potable water in regions such as Chennai. Despite high standards and blanket coverage of water purification technologies for water, currently available methods most likely fail to remove many of the accumulating pharmaceuticals sufficiently. This shortcoming includes advanced oxidation technologies that are currently investigated, such as UV exposures, ozonation, hydrogen peroxide admixture and combinations of these approaches. The objective of this paper is the development of a novel and flexible technology for water purification. The immediate target is the removal of pharmaceutical residues in drinking water but other pollutants and also wastewaters will be addressed as well. Using non-thermal plasma, the closed process does not require chemical supplies or direct manual processes control. Therefore, the technology is

N. Ajay Krishnan (✉)
Maschinenfabrik Reinhausen GmbH, Regensburg, Germany
e-mail: a.nilakantan@reinhausen.com

J. Kolb
Research Program Decontamination, Leibniz Institute for Plasma Science and Technology, INP Greifswald, Greifswald, Germany
e-mail: juergen.kolb@inp-greifswald.de

R. Azzam
Institute for Geology and Hydrogeology, RWTH Aachen University, Aachen, Germany
e-mail: azzam@lih.rwith-aachen.de

U. Kaltenborn
Highvolt Prüftechnik GmbH, Dresden, Germany
e-mail: kaltenborn@highvolt.de

R. Sarathi
High Voltage Engg Division, Indian Institute of Technology, Madras, Chennai, India
e-mail: rsarathi@iitm.ac.in

R. K. Pillai et al. (eds.), *ISGW 2018 Compendium of Technical Papers*, Lecture Notes in Electrical Engineering 580,
https://doi.org/10.1007/978-981-32-9119-5_20

appropriate in conditions where it is difficult to maintain a supply infrastructure. The method is only relying on electrical energy and can be adopted according to pollution levels and also operated as batch process by a discontinued power supply. With this approach, methods based on electrical power only are intended as a corner stone of a smart water treatment management that can be implemented at different locations along the supply chain on different scale. Accordingly, it is specifically anticipated a suitable enabler of smart grids.

Keywords Pharmaceutical · Hydrogen peroxide admixture · Non-thermal plasma

1 Introduction

Water is an ever scarcer resource that is continuously and increasingly threatened by pollution. Increasing demands by growing populations in particular in India and Asia with simultaneously fast developing economies are putting a strain on available reservoirs. A problem of considerable concern is posed by anthropogenic pollutants, foremost artificial organic compounds. Especially pharmaceutical residues with their intended stability are found in increasing concentrations in potable water [1]. This has been shown in several studies for Europe [2, 3] and also for India [4, 5], with Chennai as one of the hot spots of the production of pharmaceuticals [6].

Despite high standards and blanket coverage of water purification technologies for water in Europe, currently available methods fail to remove many of the accumulating pharmaceuticals sufficiently. This shortcoming includes advanced oxidation technologies that are currently investigated, such as UV exposures, ozonation, hydrogen peroxide admixture and combinations of these approaches [7–11].

An exception is the application of non-thermal plasmas that are generated by short high voltage pulses either in water directly or close to water. The in situ generation of hydroxyl radicals by a plasma provides an efficient mechanism to break down even stable compounds, such as for example radio contrast agents [12]. In general, the complete oxidation and mineralization of organic compounds can be achieved. In addition, the plasma treatment simultaneously eliminates bacteria and any reactive species generated from water eventually recombine to water again.

A further inherent advantage is that no supplies, other than electrical power, are needed. Therefore, it is possible to scale the method for different volumes and for different locations within drinking water supply networks. Changing pollution levels and other water parameters can be readily addressed by adjusting electrical operating parameters, which advocates the method as a key component in a smart water treatment management. This is seen as a suitable enabler of smart grids, integrating the concept of rural electrification through microgrids. Water treatments could therefore be an essential component of the control and management of the electrical power available through smart grids. Using a batch process would enable the utilization of intermittent availability of renewable energy in local microgrids

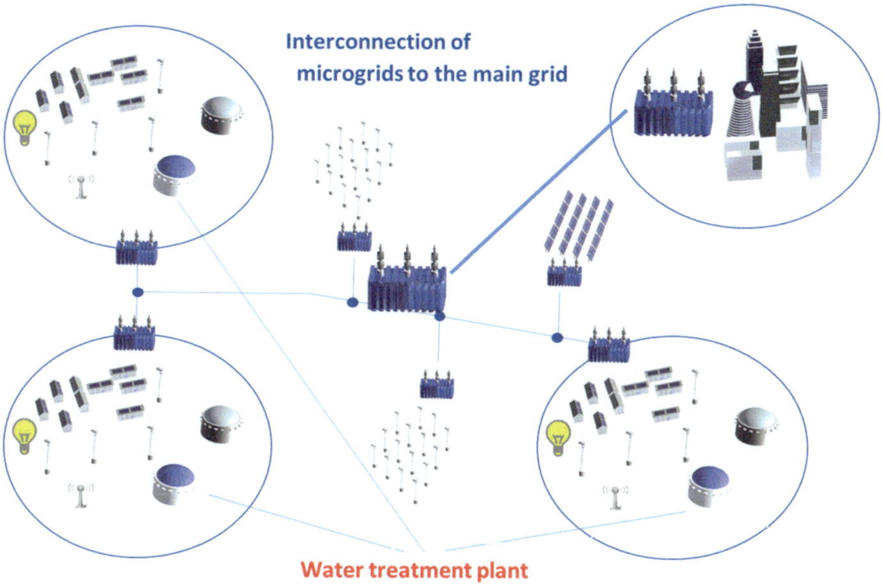

Fig. 1 Microgrids as enablers of smart water treatment

for water treatment. In a second step, connecting microgrids to larger units would enable higher amounts of renewable energy to be used for the water treatment process, even larger and continuous running treatment units can be used as shown in Fig. 1.

2 Technical Concept Process Description

The overall objective is to demonstrate the use of non-thermal plasmas for degradation of pharmaceutical residues and other organic compounds on problematic water matrices that are found in the drinking water supply in India.

A general concept for a purely electricity based water treatment system is described in Fig. 2.

In case of waste water, in a first phase suspended particles has to be removed. Here electro-coagulation is a suitable method.

In the process of electrocoagulation, the electric current from the electrode material produces metal ions. These metal ions react with water and produce hydroxides and increase the concentration of hydrogen ions. The metal hydroxides are coagulants that normally act as adsorbents. The applied current causes the corrosion of the anode which is associated with the formation of the coagulant. The flocculants can be removed in some cases due to the formation of hydrogen gas which stimulates the flotation process.

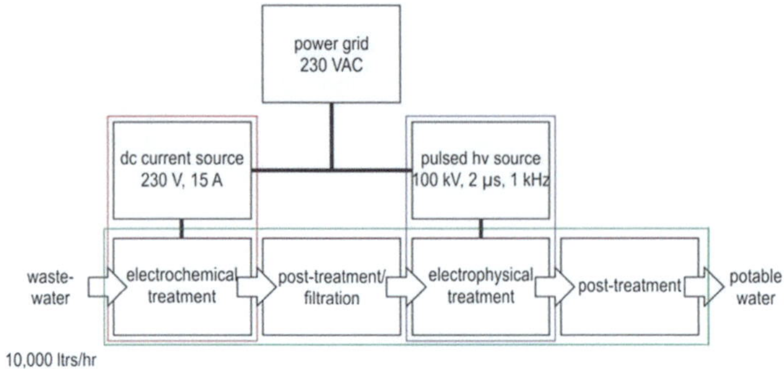

Fig. 2 Fully electrical driven water treatment process

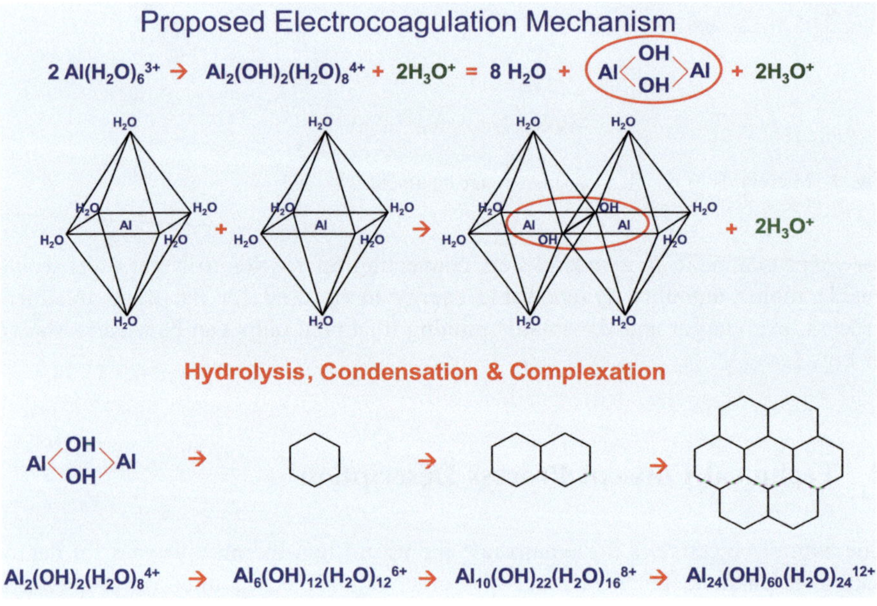

Fig. 3 Mechanism of electrocoagulation for aluminium metal

The metal ions tend to form metal hydroxides followed by a process of hydrolysis, condensation and complexation. This mechanism of adsorption of pollutants that have been destabilized by the above processes has been shown in Fig. 3.

Colloidal particles are formed in this process. These attract other particles in the suspension, which can be easily separated from the suspension. In a second phase, the organic pollutants have to be treated (Fig. 4).

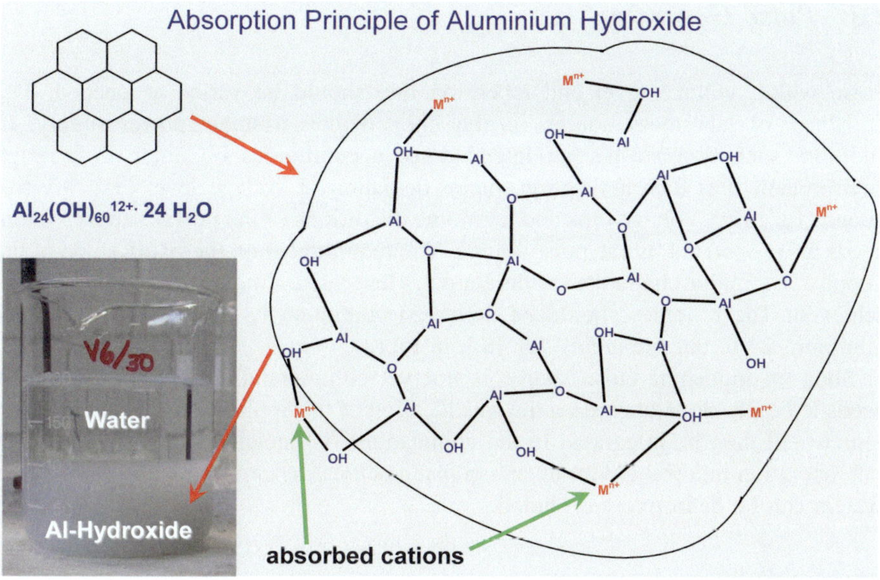

Fig. 4 Adsorption principle of metal hydroxide

2.1 Plasma System

For the demonstration unit, a treatment volume of 50–100 l/h is anticipated. The system will consist of coaxial modules with segmented ground and high voltage electrodes that will be operated in parallel by the application of 2 μs high voltage pulses of between 50 kV up to 80 kV. For the anticipated demonstrator showcase, a pulse generator will provide the pulses with a variable repetition rate between 10 and 100 Hz.

2.2 Evaluation of Sites and Conditions

The discharge of waste water is one of the major concerns of the pharmaceutical and textile industries. In India, only 25% of the waste water that is generated, is treated. This increases the scope of new technologies in the domestic wastewater treatment sector. A database will be created by measuring the composition of organic, inorganic pollutants present in the waste waters from different textile and pharmaceutical industries in India.

2.3 Pulse Generator

Pulse width, voltage level and repetition rate should be varied as needed. The drawback of laboratory sources is that they require a stable power supply. In addition, such sources are not intended for a continuous operation in a rough environment. For this application, major deviation in voltage level (±20%), frequency (±2 Hz), voltage dips and harmonic distortions of the power supply system needs to be operated by the pulse source. The requirement on the rising slope of the impulse is a major challenge for the output. Here a rise time of 400 ns needs to be achieved. These values should be delivered continuously over a time period of minimum 24 h, but preferably for 48 h or more.

Such an industrial pulse source is not yet commercially available. Hence, it needs to be developed based on the specification of the application. This ready built source will then be integrated in the application. A special requirement will be a safe operation in a wet environment, so that electrical hazard to the operators of the system can be definitively excluded.

2.4 Water Treatment System

A water treatment system will be developed taking into consideration the local requirements of water that will be treated. The pollution levels and the water flow depending on individual sources need to be addressed in particular. Different pre-treatment stages (filters) will be implemented if necessary. Different treatment methods, which bring water in contact with plasma, such as split up water flows or water sprays, will be investigated. The plasma module and the high voltage connection will need to be planned during the design stage.

2.5 Integration of Plasma System, Pulse Generator and Water Treatment System

Integration of pulse power unit, reactor with waste water circulation and purification setup will be carried out, before it is integrated with a pulse source, which is fully tested separately. A methodical experimental study will optimize output voltages and pulse frequencies. The parameter sets have to be optimized for the use of waste water from textile and pharmaceutical industry, domestic waste water by adopting hybrid model of using additionally an electro coagulation process to enhance the efficiency. Aspects, such as rates of oxidants produced by the plasma, fate of pollutants during treatment and eco-toxicity will be assessed before installation at test sites.

3 Input & System Requirements

The application of short voltage pulses (with durations of about 1 μs or less and amplitudes on the order of 100 kV) could generate non-thermal or cold plasmas for water treatment. These so called corona discharges consume only low electrical energy on the order of 1 J (=0.0000003 kWh) per discharge for a treatment volume of 0.25 l (i.e. 3–4 orders of magnitude lower compared to a spark discharges in the same volume) (Fig. 5).

The investigated pharmaceuticals [14] include:

1. Carbamazepine (antiepileptic)
2. Ibuprofen (analgesic)
3. Diclofenac (analgesic)—EU watchlist
4. 17α—Ethinylestradiol (hormone)—EU watchlist
5. Trimethoprim (antibiotic)
6. Diazepam (hypnotic)
7. Diatrizoate (X-ray contrast agent).

Plasma filaments react differently, as shown in Fig. 6, against organic chemical pollutants and microbiological contaminants simultaneously:

1. UV light destroys DNA [15] and hence the ability of cells to proliferate. Radicals can be generated when they are combined with catalysts such as TiO_2
2. Cells are torn apart by the shock waves [16]

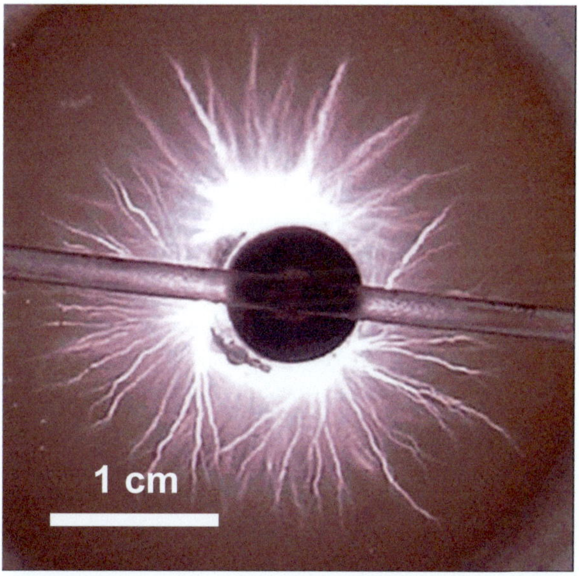

Fig. 5 Streamers in tap water around a tungsten wire [13]

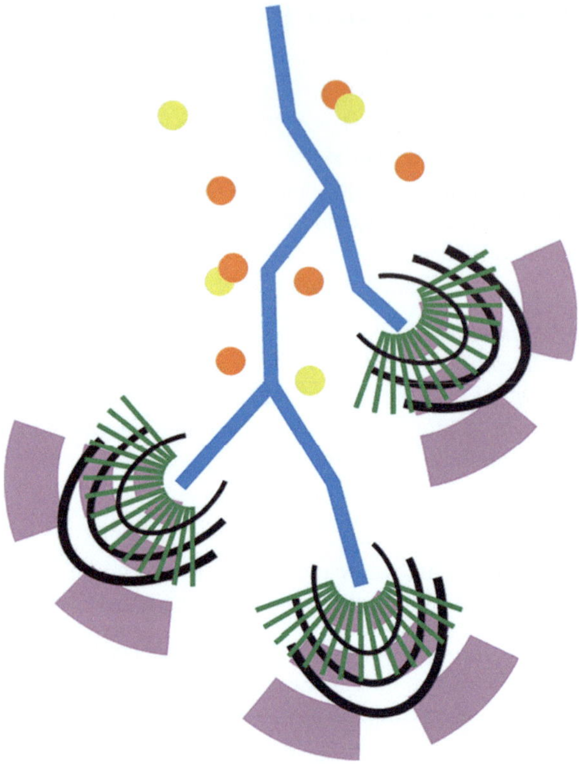

Fig. 6 Reaction pathways of streamer discharges. Radical Species OH*, O*, H_2O_2, O_3 …—orange; UV—violet; electric field—green; shock waves—black

3. Cell integrity is destroyed by high electric fields [17], which affects intracellular functions adversely.
4. Chemical reactive species such as hydroxyl radicals [18], provide a strong agent for oxidation of stable chemical compounds and resilient bacteria.

4 Test Setup

The pre-field test analysis was carried out in laboratories at INP Greifswald. It consists of a high voltage pulse generator, a pulsed corona discharge reactor, water infrastructures (pump, cooler and expansion reservoir) and measuring devices (probes, monitor, oscilloscopes), as shown in Fig. 7.

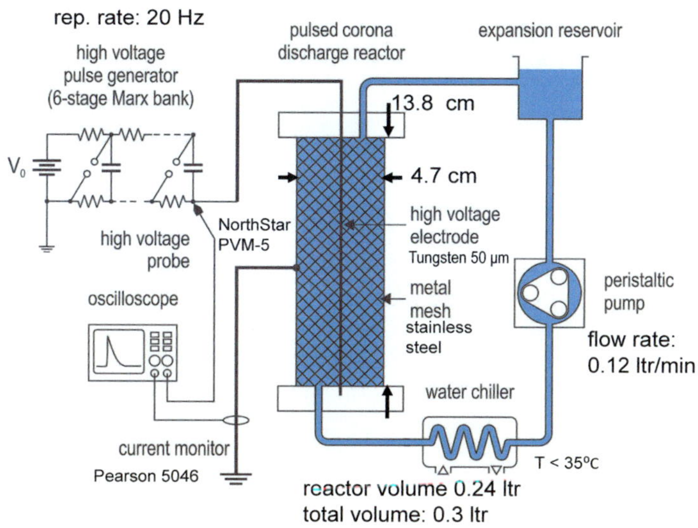

Fig. 7 Test setup for generation of pulsed corona discharges in water

5 Results

The corona discharges generated in water, does not cause increase in nitrite or nitrate concentrations, as shown in Fig. 8, or significant acidification. The other advantages include:

- High penetration depth, regardless of turbidity
- No chemicals or catalysts are necessary
- No pre- or post-processing other than filtration is required.
- Significantly more effective compared to chlorination or ozonation. However, combination with other treatment methods is also possible.

The water quality analysis conducted from water samples taken from the expansion reservoir showed that the pH values were not changing, conductivity increases by approx. 12 µS/cm and concentration of dissolved oxygen double (see Table 1).

Although there is a slight increase in nitrite and nitrate concentrations, these are well below the legal allowable limits of less than 0.5 mg/l for nitrite and 50 mg/l for nitrate concentrations as shown in Fig. 9.

Decomposition of organic compounds is primarily mediated by the formation of transient species such as hydroxyl radicals. This is also due to long-lived chemical reaction products such as ozone and hydrogen peroxide. If they are not consumed during the oxidation process, they eventually recombine to purify water.

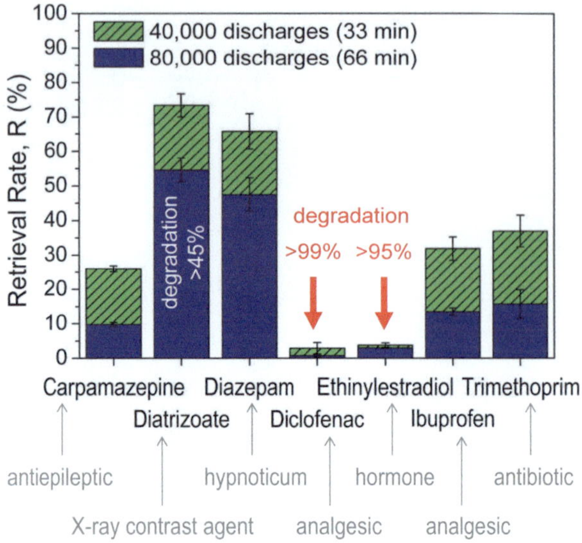

Fig. 8 Retrieval rates of degradation of pharmaceuticals

The success of the method depends primarily on the possibility to effectively generate discharges with short high voltage pulses of appropriate parameters (duration and amplitude) with respect to the treatment volume.

6 Conclusion

Pulsed corona discharges have been generated in water by the application of exponentially decaying high voltage pulses with a rise time of 20 ns and pulse width of 300 ns. It has been found that stable pharmaceuticals, even X-ray contrast agents, are effectively decomposed by pulsed corona discharges generated in water. An effective outcome of this process is that the water quality parameters such as pH-value, concentrations of nitrite/nitrates are not affected by the discharges. Furthermore, these could serve as a potential application in future for agriculture irrigation with direct high concentrations of nitrites/nitrates, which could serve as fertilizers.

Further development and research is required scaling the method to larger volumes, development of appropriate pulsed power supplies, and investigation of efficacies and efficiencies under real conditions for different levels of contamination and water quality.

Table 1 Water quality analysis

	pH		Conductivity in µS/cm		Dissolved oxygen in mg/l	
	Start	End	Start	End	Start	End
Carbamazepine	6.41 ± 0.06	6.38 ± 0.05	30.44 ± 0.54	42.33 ± 2.16	4.93 ± 0.15	10.11 ± 0.33
Diatrizoate	7.14 ± 0.30	6.63 ± 0.15	29.00 ± 0.00	44.29 ± 1.06	5.90 ± 0.41	11.01 ± 0.83
Diazepam	6.74 ± 0.21	6.26 ± 0.04	30.67 ± 0.82	43.33 ± 1.08	4.51 ± 0.04	9.44 ± 0.56
Diclofenac	6.50 ± 0.53	6.39 ± 027	28.67 ± 0.82	40.07 ± 3.24	3.57 ± 0.44	6.91 ± 0.50
Ethinylestradiol	7.00 ± 0.25	6.72 ± 0.22	30.33 ± 0.41	42.59 ± 1.21	5.35 ± 0.10	11.17 ± 0.23
Ibuprofen	6.44 ± 0.16	6.26 ± 0.01	30.01 ± 0.02	40.00 ± 1.63	4.28 ± 0.14	8.70 ± 0.65

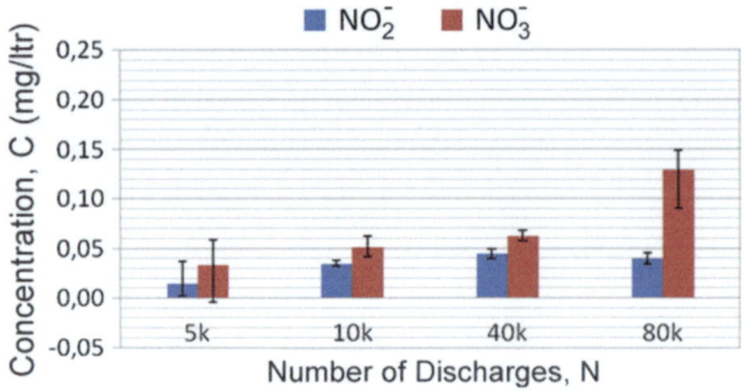

Fig. 9 Nitrite and nitrate concentrations—water quality analysis

References

1. E. Commission, Environment and Water: proposal to reduce water pollution risks, In: European Commission, European Commission—Press release
2. Hirsch R, Ternes T, Haberer K, Kratz K-L (1999) Occurrence of antibiotics in the aquatic environment. Sci Total Environ 225:109–118
3. Fatta-Kassinos D, Meric S, Nikolaou A (2011) Pharmaceutical residues in environmental waters and wastewater: current state of knowledge and future research. Anal Bioanal Chem 399:251
4. Balakrishna K, Rath A, Praveen Kumar R, Guruge KS, Subedi B (2017) A review of the occurence of pharmaceuticals and personal care products in Indian water bodies. Ecotoxicol Environ Safety 137:113
5. Satapathy SS, Ray SK (2014) Pharmaceutical residues in India: impact on aquatic environment. Curr Sci 107:1634
6. WHO Report 2012, www.who.int/water_sanitation_health/publications/2012/pharmaceuticals/en/
7. Hillenbrand Th et al. Texte 87/2014, Environmental research of the Federal Ministry of the Environment, Nature Conservation, Building and Nuclear Safety. www.umweltbundesamt.de
8. Giri R, Ozaki H, Ota S, Takanami R, Taniguchi S (2010) Degradation of common pharmaceuticals and personal care products in mixed solutions by advanced oxidation techniques. Int J Environ Sci Technol 7:251–260
9. Huber MM, Canonica S, Park G-Y, von Gunten U (2003) Oxidation of pharmaceuticals during ozonation and advanced oxidation processes. Environ Sci Technol 37:1016–1024
10. Poyatos J, Munio M, Almecija M, Torres J, Hontoria E, Osorio F (2010) Advanced oxidation processes for wastewater treatment: state of the art. Water Air Soil Pollut 205:187–204
11. Ikehata K, Jodeiri Naghashkar N, Gamal El-Din M (2006) Degradation of aqueous pharmaceuticals by ozonation and advanced oxidation processes: a review. Ozone Sci Eng 28:353–414
12. Banaschik R et al (2015) Water Res 84:127
13. Kolb JF, Joshi RP, Xiao S, Schoenbach KH (2008) Streamers in water and other dielectric liquids. J Phys D Appl Phys 41:234007
14. Umweltbundesamt, Arzneimittel in der Umwelt - vermeiden, reduzieren, überwachen, Hintergrund, April 2014
15. Lukes P et al (2008) Ultraviolet radiation from the pulsed corona discharge in water. Plasma Sources Sci Technol 17:024012
16. An W et al (2007) Underwater streamer propagation analyzed from detailed measurements of pressure release. J Appl Phys 101:053302
17. Sarkisov GS et al (2006) Observation of electric field enhancement in a water streamer using Kerr effect. J Appl Phys 99:083304
18. Malik MA et al (2001) Water purification by electrical discharges. Plasma Sources Sci Technol 10:82

Smart Water—An Automation of Existing Waste Water Filtration and Recycling System

Vrushali G. Nasre, Sudarshan Rao, Sushant Tupparwar, Disha Bhatt and Shriyash Deoghare

Abstract Clean and hygienic drinking water is a elementary requirement of human beings. Automation of water treatment plants plays a vibrant protagonist in the safe and consistent action in providing household usable water. This system emphases on an advanced, smart control & Checking system for water purification by using "IOT" And "microcontroller". This proposed system is an attempt to design a cost effective water filtration with smart control recycling system. In this system the bathroom & kitchen basin waste-water is reused for gardening, and toilet flushing. The basin water is passed through the "purifier system" Which consist of different purification techniques such as sedimentation, charcoal purifier, silica beads treatment is used so the water get purified close to neutral pH value. Water recycled is neither acidic nor alkali. The system also eliminates the undesirable color & odor of the water. For this "microcontroller system" is used to control flow of water and check turbidity of water. It is an automatic device that can provide safe, reasonable and readily available water for household usage. "IOT System" is connected to incoming water supply and check's the real time billing system and water usage. It also detects the "leakage of water". This system has "Hydro-power generator" which generates power by flow of water & stores the energy in a Lithium-Ion battery, so in case of power failure it can provide power to the recycling system for uninterruptable service.

Keywords Water · Recycling · IOT—Internet of things · Purification · Sedimentation · Microcontroller · Leakage of water · Hydro-power generator

---

V. G. Nasre (✉) · S. Rao · S. Tupparwar · D. Bhatt · S. Deoghare
Priyadarshini College of Engineering, Nagpur, Maharashtra, India
e-mail: vrushnasre@gmail.com

R. K. Pillai et al. (eds.), *ISGW 2018 Compendium of Technical Papers*,
Lecture Notes in Electrical Engineering 580,
https://doi.org/10.1007/978-981-32-9119-5_21

1 Introduction

Water-H_2O is one of the key and well known natural resource and is played vital role for entire living things on the earth planet. As more than 60% of human body is filed with water, therefore quality and cleanliness of water is very essential. Now a day, only 3% fresh water resources are available on earth. According to the survey, 124 Million liters/day of water is lost in leakage & illegal connections. Flush of a toilet uses 6.5 gallons of water daily. 70% of the world is composed of water, salt and sea. Concrete cities have caused reduction in ground water level. However, due to human consumption and irresponsibility, water pollution was inevitable thus limiting the supply of portable water. On the basis of the study, the aforementioned problems also occur in water resources.

The drinking water must be hygienic, uncontaminated, microorganisms free and germ-free before consuming it. All water treatment plants treat the raw water from Reservoirs like river, lake, dams or other underground resources and provide harmless and clean drinking water to mankind. But the same water is used in daily household uses like bathroom and toilet flushing, kitchen purpose, gardening, washing (cloths and utensils), cleaning vehicles, and other many more. And that water can be called it as waste water. Considering the rate of change of increasing population, supplying clean and hygienic water to each individual is a most essential challenge. Exponential growth in population, industrialization and urbanization leads to wastage and scarcity of water. Increasing growth rate of Population in India is indications to significant increases quantity of waste water, which makes it an urgent imperative to develop cost effective and strong modern technologies for waste water management and treatment.

Recycled water is waste water that has been purified so that it is suitable for a range of non-drinking purposes. Improper waste management causes various environmental issues such as soil erosion and contamination of fresh water bodies. Thus recycling and proper management of waste water is the need of the hour. The methodologies or techniques and the level of purification of water depend on the financial feasibility. Grey water is dumped in open area, which causes soil erosion and contamination of soil which makes land unsuitable for agriculture. Common assumption among the people is that recycled water is not appropriate for daily usage. Valves are controlled manually, so there is a chance of human error. Leakage and water theft detection is difficult.

Unconventional technologies for waste water treatment are required to eliminate pollution. Automation is essential to water treatment plant monitoring and controlling since it has various perceptible and immaterial benefits. In order to address these problems, the researchers decided to design a system that can recycle waste water from the home to provide clean water and serves for conserving water for the next generation. The proposed system describes a modern technology that can automatically recycle the waste water coming from the outlet of the home except flushed water of the toilet, which will be useful for gardening, toilet flushing.

2 Existing Methods

2.1 For Water Purification

One of the existing methods for water purification is the ultraviolet water treatment process. UV rays decontaminate the water without adding any impurities like unwanted color, scent, chemicals or taste and there are no residual any products. UV treatment is one of the fastest and very cost effective and environment friendly process. UV technology is easy method for this kind of application since it provides immediate results without any by-products they leave behind. The size of the UV equipment is very small so device is easy to handle. The major drawback of UV treatment is it completely destroys the residual ozone present in a water stream. This process is depends on the flow rate of water, absorption of ozone in water, feed water quality and the temperature of the water stream and surrounding temperature. Investment cost depends on the system configurations and technologies used [1].

Solar water pumps were used first time for water provision in off-grid rural areas. The technology has developed around many different system designs and in certain water pumps the consistency and maintenance requirements have enhanced over the basic pumps introduced to the market [2].

The microbiological community must be kept alive for the design of better and more effective filter. In an orthodox slow sand filter, oxygen is absorbed by organisms through dissolved oxygen in the water [2].

Another method for water filtration is reverse osmosis, which is the key method for water filtration in the European countries. However, commonly due to high energy cost, these technologies are not realistic application for poor developing countries. Due to limited alternative water sources, mostly large-scale reuse water schemes are in Israel, South Africa, and arid areas of USA. In many developing countries like The Philippines, has a major sanitation issue for both their urban and rural area citizens. Specifically for countryside citizens, according to 2008 Survey 17% of this countryside citizens still had no access to better-quality sanitation, with 14% estimated to be practicing open defecation. The Ferro cement biogas septic tank was installed as sanitation technology for individual families or small clusters of families [2]. All of the mentioned studies are useful and carefully examined by the researchers. Wastewater management will be of great help in conservation of water especially in the near future. Many researches are focus on environmental concerns because of the possible portable water shortage in the near future. The above mentioned literature is shown in Table 1.

2.2 For Water Automation

P. R. Panditrao et al. proposed a system to provide quality water filtration and cost effective recycling of water. Author demonstrate the new water filtration methods for domestic purposes can be combined together & automated using Programmable

Table 1 Existing methods for water purification

Parameters existing system for water purification	UV technology	Reverse osmosis	Sanitation technology
Installation expenditure	High	High	High
Maintenance expenditure	High	High	High
Area requirements	Very high	High	High
Power consumption	Very high	Very high	Very high
Purpose	Drinking	Drinking	Drinking

Table 2 Existing methods for automation

Parameters existing system for water automation	Raspberry pi	μc based	PLC
Installation expenditure	High	Less	High
Maintenance expenditure	High	Less	High
Area requirements	High	Very less	High
Power consumption	Very high	Less	High
Purpose	Drinking	Drinking	Domestic purpose

Logic Control to create a consistent & cost-effective system to get hygienic useable water. Mostly in India waste water or cleaning water from industries is dispose into natural fresh water resources such as rivers and lakes. When these natural resources saturate by industrial waste and several chemical impurities, the quality of water will automatically get reduce and downstream water are no more longer usable deprived of costly purification treatments. As per the reports of UNICEF industrial water productivity of India is very less and similarly very poor water treatment capacity use related to Indian population [1].

Rhowel Dellosa provides fully automatic filtered drinking water vending machine, which offering a touch less nozzle that can be used without a switch. Furthermore, accuracy is obtained by using Microcontroller Based Wastewater system. It is an automated vending machine that can provide a hygienic, reasonable and readily available drinking water [2].

Raspberry pi based automation is proposed by Sonali S. Lagu and et al. Automation plays a vital role in the hygienic, safe and consistent working of a water purification plant for providing pure drinking water. The efficiency of Plant is increased by using Raspberry Pi processor. Use of raspberry pi makes system more economic [3]. Table 2 shows the comparison of existing methods for automation of water purification system.

3 Overview of Proposed Smart Water System

In this system the bathroom basin, kitchen basin water is reused for gardening and toilet flush system. The basin water is passed through the "purifier system" which is used to purify the water for gardening as well as flush system. In this design the charcoal purifier, silica beads, alum, gravels sedimentation used so the water get purified as normal pH, value as it doesn't contain any acidic or alkali as well as removes odor, for this "Microcontroller system" is used to control flow of water and check turbidity of water. Sedimentation processes are good replacements for elimination the toxic composites from wastewater. Furthermore, by the use of the microcontroller accuracy and efficiency is increased. It is an automated system that can provide hygienic, affordable and readily available water for household usage. The system also gets rid of the unwanted color and odor of the water.

As well as "IOT system" is to be connected for incoming water supply and to check the real-time billing system. It also detects the "leakage of water" and monitors the usage of water. By using IOT the system can be controlled from any part of the world. IOT assure that the cloud is safe enough from all the external attacks and threats so that the customer does not affect a loss of data or data theft. Block diagram of smart water system is as shown in Fig. 1.

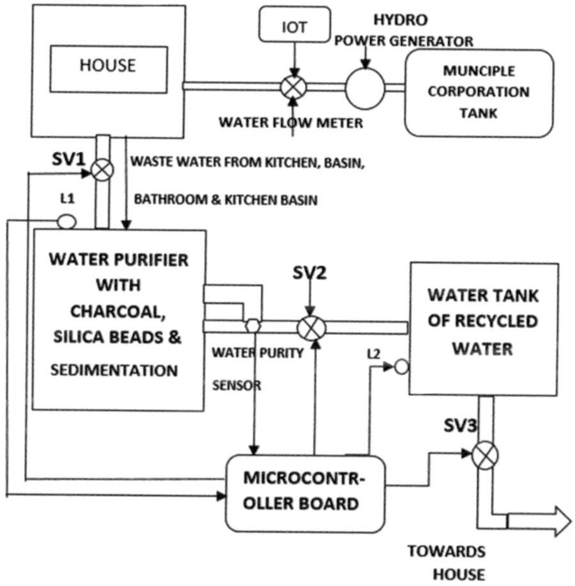

Fig. 1 Block diagram of smart water system

4 Water Purification

Polluted water is a major problem in front of today's world, since the high rate of population growth supplying pure water to each and every individual is a main challenge. The key idea behind the project is to make use of waste water by using simple filtration processes; combining it with the advance and cost effective technology to automate the whole water filtration plant to cut down the huge cost and man power required [1].

4.1 Block Diagram Waste Water Purification System

4.1.1 Proposed Model and Working

The waste water from houses is collected and subjected to filtration process. The filtration process involves the following steps:

(1) Cloth filtration
(2) Primary filtration using charcoal, silica, alum, gravels.
(3) Charcoal filtration and sedimentation
(4) Turbidity check
(5) Re-filtration (If required).

(As shown in Fig. 2), the water is passed through solenoid valve. Solenoid valve is an electromechanical operated valve. The valve used in this system is a normally-closed, direct-acting valve. The waste water from home passed through the solenoid valves, the solenoid valves are placed in system to control the flow of water between various stages.

In this purification process the water is then passed through cloth filter, the cloth filter reduces the pathogen count of almost 99%. Inexpensive cotton cloth or saree cloth, folded 4 to 8 times provides a filter of approximately 20 mm mesh size.

This is small enough to remove all zoo-plankton, phytoplankton and all v. cholera. Once the water undergoes to cloth filtration it needs to be disinfectant and further purified. Un-filter water is then passed over charcoal carbon bed. The carbon filtering is a method of filtering that uses the bed of activated carbon or charcoal carbon bed to remove impurities using chemical absorption process; it is most effective at removing organic compound like chlorine, sediment, volatile, taste and odor from water where as they are not effective at removing dissolved inorganic compounds and mineral salt. The water is than passed through Silica sand bed, Silica sand is used to clean and purify water. To capture suspended solids in water the natural silica filtration grade sand is used which has a sub-angular to rounded shape, making them an ideal filtration media. Next the water is passed through Aluminum sulphate when added to water reacts with the bicarbonate alkalinities present in water and forms a gelatinous precipitated. This floc attracts other

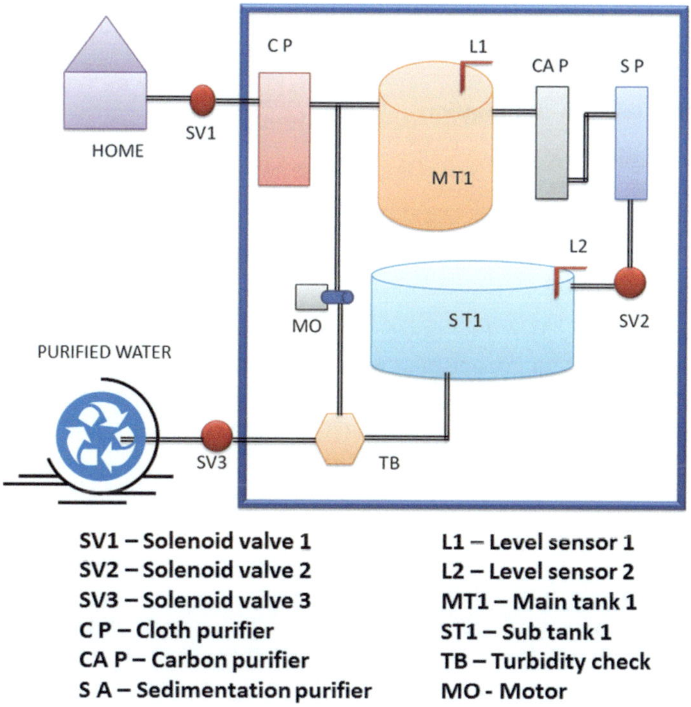

Fig. 2 Block diagram of waste water purification system

suspended materials and fine particles in water and settles down at the bottom surface. Next the layer of gravels are used as purification layer, gravel filter removes bacteria and other particles from water using simple filter and accessible technology. The water passes through the sand leaving suspended particle passing through the layer. The outcome should be clear water in carbon filtration, activated carbon works through a process called absorption whereby, pollutant molecules in the water to be treated are trapped inside the pore structure of the carbon substrate. Then water goes through sedimentation filter, Sedimentation is a process allowing suspended particles in water to settle down by allowing the water to stay still under the effect of gravity. Sedimentation is often use as a primary stage reducing the content of suspended solid as well as pollutant embedded in the suspended solid (as shown in flowchart). The purified water is then collected into the overhead tank; this purified water then goes through turbidity check using litmus paper. After going through the turbidity check if the water is found pure enough then it is supplied to the houses. Else, the water is sent to the primary tank for re-filtration. This process is repeated unless and until the water is satisfactorily pure.

4.1.2 Role of Microcontroller in the Automation of the Smart Water System

Components used in automation

1. Microcontroller 89C51
2. Level sensors—2
3. Water Flow sensor
4. ULN 2003 Solenoid driver
5. Solenoid valves—3.

(As shown in Fig. 3) A low power, high-performance Microcontroller is used, which is CMOS 8-bit microcontroller with 8 kb of flash programmable and erasable ROM (Read Only Memory). The main use of microcontroller is to control the flow of water and send the required information to the IOT.

Microcontroller is used to interface with solenoid driver ULN2003 which control the ON and OFF of the solenoid valves installed at various stages of purification and transmission of water.

Level sensor—The level sensor senses the water level at primary and secondary tank. Once the water in secondary tank is filled at its highest level it will give the indication through the level sensor to the microcontroller and solenoid valve 2 will be closed. The primary tank also contain level sensor, as the primary tank is full it will sense and will close the solenoid valve 1.

Water flow sensor—Water flow sensor are placed to detect the theft and leakage of the pipe at initial level of incoming water to house and give the information directly on the IOT. The water flow sensor will check the water pressure at the input and the output valve and show it on the IOT. If there is any leakage or

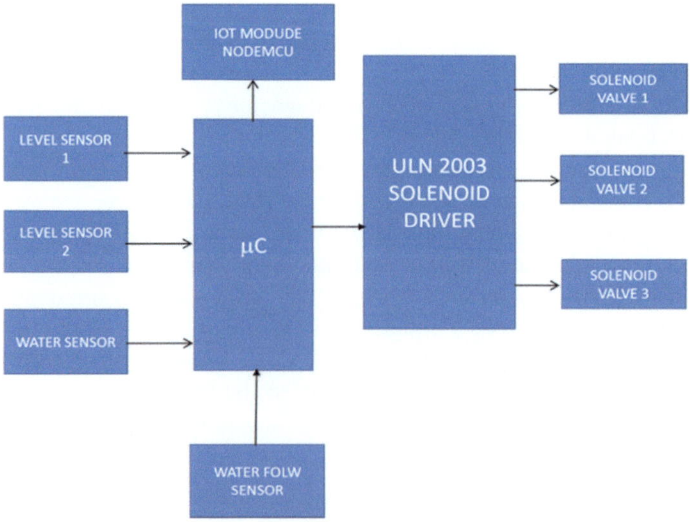

Fig. 3 Automation using microcontroller

theft then there will be back flow of the water and flow is calculated by determining the frequency of the pulse. To measure the flow rate following mathematical equation is used.

$$Flow\_Rate = Flow\_of\_water \times Common\_Factor$$

ULN 2003 Solenoid driver—ULN2003 Solenoid driver is used to interface all the 3 solenoid valves with the microcontroller 89C51. ULN203 is a high voltage, low power and high current Darlington array. It is a useful for driving a wide range of loads including solenoids, relay DC motors etc.

Solenoid valve—It is used to control water flow between various stages. There are 3 solenoid valve SV1, SV2, SV3.

SV1 is connected in front of flow meter which is used to ON and OFF of the process if the primary tank and Secondary tank is filled. SV2 is connected between primary and secondary tank. Once the secondary tank is filled up the level sensor will give the indication and SV2 will be closed. SV3 is connected at the end of the recycling system. Each and every solenoid valve are directly controlled by IOT module through microcontroller.

Water sensor—The water sensors is used to check that the water is present in the primary tank and secondary tank. It also checks the turbidity level of water by passing light through the water and checking the transparency of the water through photo detector (as shown in Fig. 3).

4.1.3 Flowchart for Sequence of Operation

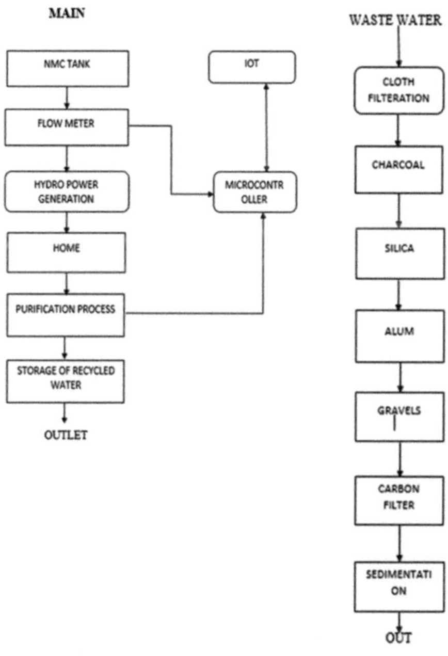

5 Role of IOT in Theft Detection and Online Billing

5.1 Back End

1. *Consist of a database that contains the water flow from the flow meter as shown in* Fig. 4.
2. *Values from all the components (level sensors, solenoid valves, flow meter) are monitored and controlled.*
3. *The values from flow meter in the form of water pressure is collected and monitored.*

5.2 Front End

Since it is monitored and a trigger analysis (as shown in Fig. 5a–c). Alters are set to the concerned authorities in the form of SMS and mail. The mail is sent to the authorities in case of water pressure goes below threshold level set in accordance with house. If water pressure goes down the threshold level then system senses water leakage or theft and concerned authorities are alarmed (as shown in Fig. 6).

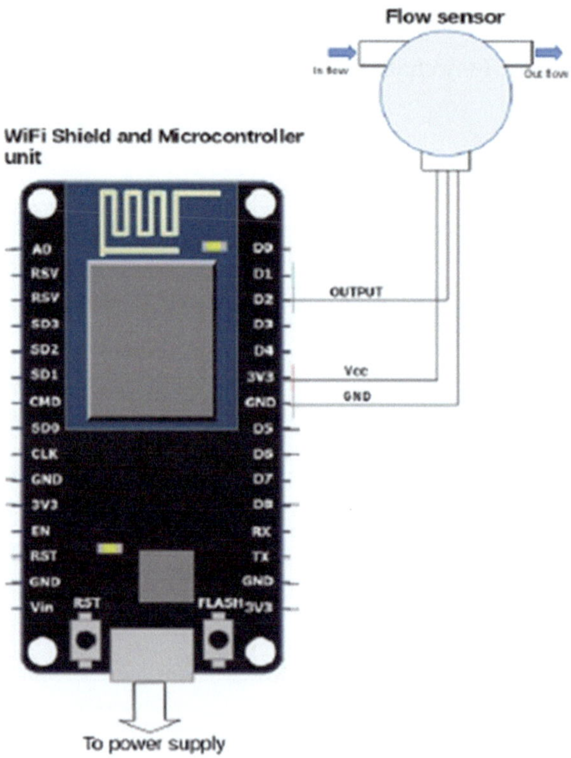

Fig. 4 Interfacing of node MCU with flow sensor

(a)

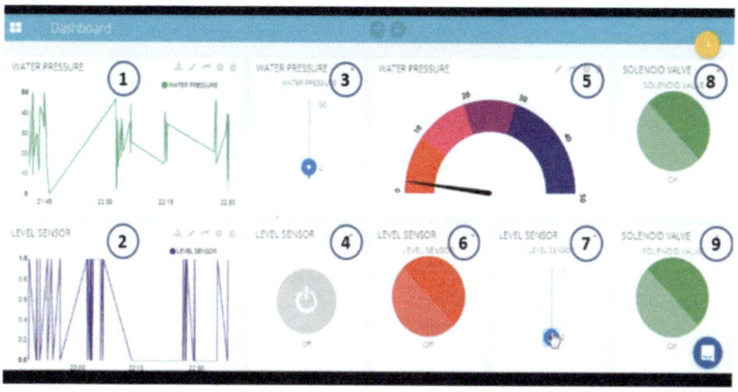

(b)

Window No.	Particular	Working
1	Water Pressure	Timeline
2	Level sensor	Timeline
3	Water Pressure	Water pressure controller (0-50)
4	Level sensor	Switch ON/OFF
5	Water Pressure	Water pressure meter (0-50)
6	Level sensor 1	ON/OFF showing full/empty(Red/Green)
7	Level sensor 2	0-OFF and 1-ON
8	Solenoid Valve 1	ON/OFF-[Green/Red]
9	Solenoid Valve 2	ON/OFF-[Green/Red]

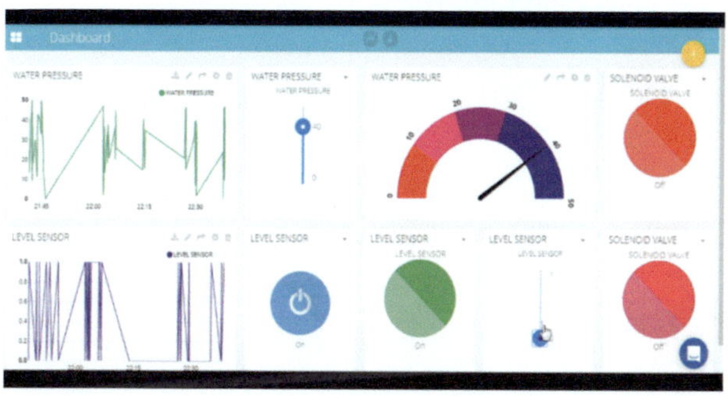

(c)

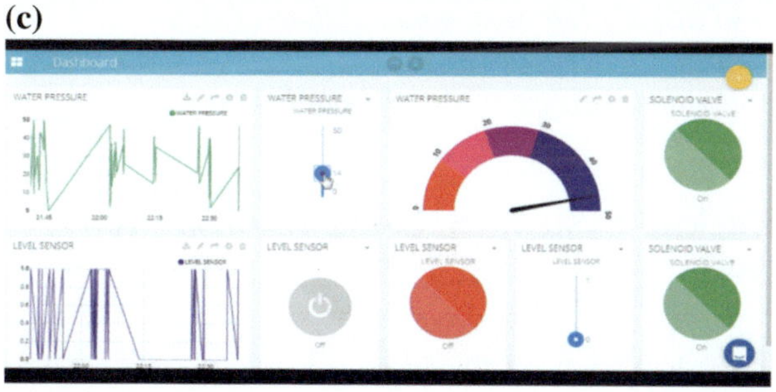

Fig. 5 **a** Initial stage of IOT monitor system. **b** IOT monitor system, Stage-1. **C** IOT monitor system, Stage-2

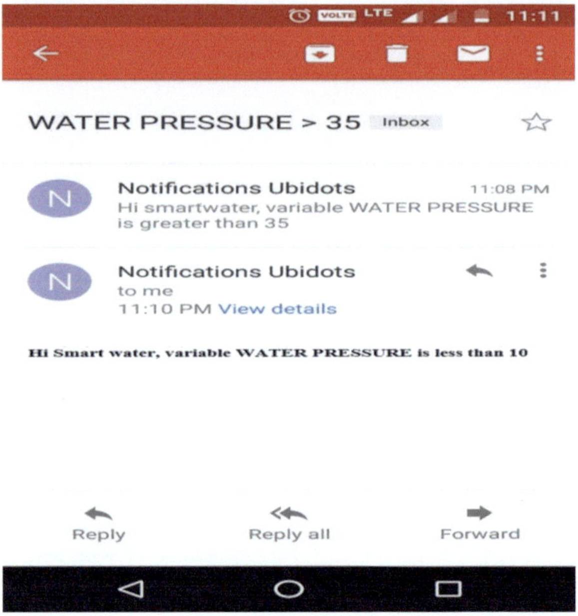

Fig. 6 Notification through e-mail

It monitors sensor reading and escalates the situations by altering to the authority and the municipal corporation [4, 5].

5.3 Email

Theft or leakage detection is as shown in Fig. 6 *i.e. notification through e-mail.*

5.4 IOT Purpose

4. *IOT is used to display real time value of flow of water.*
5. *Used for controlling solenoid valves.*
6. *Used to display the pressure of water from flow meter.*

6 Power Backup System by Using Hydropower Generation

Transformation of the energy of flowing water into electro mechanical energy is done by hydraulic turbine and a hydroelectric generator is used to converts this mechanical energy into electricity. The operation of a generator is based on the Faraday principles as shown in Fig. 7.

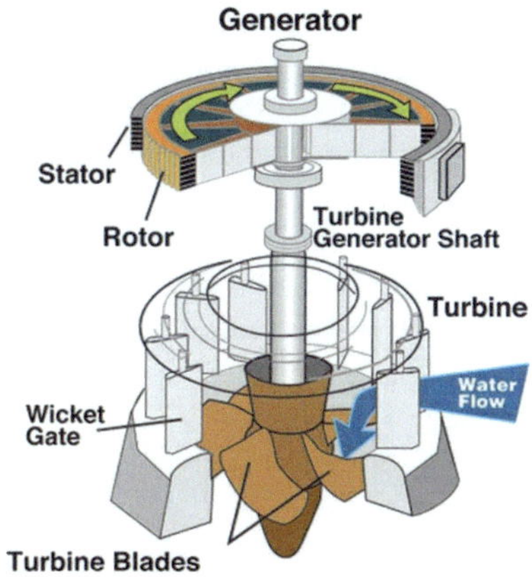

Fig. 7 Hydro power generator

The water strikes the turbine blades and turns the turbine, which is connected to a generator by a shaft. As the turbine blades turn, the rotor inside the generator also turns and electric current is produced as magnets rotate inside the fixed-coil generator to produce alternating current (AC).

The hydro power generator is connected between the municipal corporation tank and house provides power in terms of electricity by converting water pressure into electricity. Through this hydro power generator the amount of electricity produced is enough to power the system in case of absence of power or load shedding, so this system can work seamlessly and uninterruptedly. The power of system can be varied according to the water pressure on the turbine blades[6, 7].

As a working fluid, water in a hydropower system is not consumed, it is thus available for other uses. Diagram explanation (as shown in Fig. 7).Calculation of Power using Hydropower

$$Power = Head \times Flow \times Gravity$$

where;

The flow is the volume of water which can be captured and re-directed to turn the turbine generator. The head is the distance the water will fall on its way to the generator.

Where Power is measured in Watts, Head in meters, Flow in litres per second, and Acceleration due to Gravity in metres per Second Square. The acceleration due to gravity is approximately 9.81 m/s^2.

7 Conclusion

Use of IOT and microcontroller for system automation is a unique and advance technology. Computerized automation provides the hygienic and reliable operation of a smart water system in providing recycled water for gardening and flushing. Use of Microcontroller simplifies the automation process and increases the overall efficiency of the system. Use of microcontroller enables to reduce the price of the system. In this system, the purification level/process is limited to some extent but depending on financial feasibility and requirement the purification level can be extended to Ultraviolet (UV) or even Reverse Osmosis (RO) purification so that it can be used for drinking purpose also. In this system IOT is extensively used for theft detection and online billing. This system has "Hydro-power generator" which generates power by flow of water & stores the energy in a Lithium-Ion battery, so in case of power failure it can provide power to the recycling system for uninterrupted service.

References

1. Panditrao PR, Bhalerao CS, Bhor SB, Kad RS (2014) Cost effective automated water filtration and recycling system (an ISO 3297: 2007 Certified organization). 3(1)
2. Dellosa R, Hernandez JE, Mangay R, Robles R (2014) Microcontroller based wastewater management system. Int J Sci Technol Res 3(10)
3. Lagu SS, Deshmukh SB (2015) Raspberry Pi for automation of water treatment plant. In: International conference on cloud computing communication control and automation, Issue 10, Sept 2015
4. Bodhe AR, Singh R, Bawa A (2016) An internet of things solution for sustainable domestic water consumption. In: International conference on computational system and information system for sustainable solution
5. Sanamdikarand ST, Harne KR (2012) Advanced method for sewage water treatment. Int J Adv Technol Civil Eng 1(2). ISSN: 2231 –5721
6. Chakradhara Rao Ch, Ramana AV (2016) Data security in cloud computing. Int J Curr Trends Eng Res (IJCTER) 2(4):84–92
7. Yuan M (2017) Getting to know NodeMCU and its DEVKIT board, 7 Aug 2017 (first published 12 June 2017)

Intelligent Water Management System: Smart Approach Towards Sustainability (Smart Water)

Mohit Saluja

Abstract Whenever we address India, the Initial thought one can come up with is the population of this country which is going to surpass the China's population in 2024 and is projected to touch 1.5 Billion; said by UN Official. It is anticipated that India will be the highest populous country along with the fastest growing economy in the world but despite of economic opportunity, India's infrastructure is very crumbling. India's immense population makes it terribly vulnerable once it involves water shortage and scarceness. Concerning 330 million individuals within the country currently suffer from regular water shortage problems. India's economy is basically captivated with its agriculture. Water shortage and drought not solely impact on the agricultural districts but also have a calamitous effect on inflation and economic progress. With a mission to make India Smart, it requires Smart Minds who actively participate in governance and Promote Smart Building Management Systems through Integration, Information and Innovation. Building the foundation of Smart Water is the Need of the Hour and Even the Purpose of this Paper. Analyzing Water with Data Base Management System and Internet of Things to Measure Operations and Financial Controls Intelligently, Zero Percent Wastage, Improve Revenue and Efficiency by means of Advance Metering Infrastructure, Automatic Distribution, Leakage Management, Disaster and Emergency Preparedness, Pressure Regulation and Smart Grid Analytics.

Keywords Advance metering infrastructure · Automatic distribution · Leakage management · Disaster and emergency preparedness · Pressure regulation · Smart grid analytics

---

M. Saluja (✉)
Project Management Group, CBRE South Asia Pvt. Ltd, Gurgaon, Haryana, India
e-mail: mohit.saluja@cbre.com

R. K. Pillai et al. (eds.), *ISGW 2018 Compendium of Technical Papers*,
Lecture Notes in Electrical Engineering 580,
https://doi.org/10.1007/978-981-32-9119-5_22

1 Introduction

Water is coming gradually to an alarming situation on the planet. Earth is loaded with water yet 99% of water is in the ocean which is salty consequently can't be utilized. Others sources are secured ice sheet in ice frame consequently just 1% of water is accessible in groundwater and surface waters for human survival. Changes in the natural condition like low precipitation, atmosphere changes make water scarcer. Water assets like waterway lakes are contracting gradually. Water management has turned out to be significant issues in numerous nations. Notwithstanding to drink purposes. More than 800 million individuals don't have access to clean water for drinking. UN in recently has pronounced water and sanitation as fundamental human rights. More than 2 billion of individuals don't have proper sanitation facilities. It is also analyzed that almost 2 billion individuals till next rot will live in territories having completely shortage. With populace shut everything down 12 billion till 2050. This will expand the demand of water and other needs such as food by half at that point.

India is a land of agriculture. Here 70% of water is being used for the horticultural reason. It was normal that water demand will ascend by half and one can't be disregarding the way that India is as of now under high-feeling of anxiety. Just in India and, China 2.7 billion individuals live in water shortage. In India adequate water assets are accessible yet water is unevenly dispersed. There is the situation in which just a single tap is introduced for 100s of slum population and one tap for one house. Water Infrastructure exhibit in numerous urban communities was very maturing. This maturing pipeline experiences disintegration which eventually causing water spillages. General mindfulness in individuals with respect to water saving was additionally not found. Causing water waster consequently brings about additional water usages. Countries like India physical and business misfortune are the significant zone to work out. That can be lessened.

2 Need of Innovation in Water Management

There is obviously an incremental speculation required to convey a smart water metering framework. Notwithstanding the meter modules, forthright cost items include communications infrastructure, data management applications and a scope of extra innovation to help the substantial volume of information, middleware and informing programming that might be expected to handle communications between applications and alerts to and from field devices. This cost remains as a hindrance to smart water meter endeavors, as proved by the reality cost recuperation and degree of profitability (46%) and in advance, utility costs (42%) were considered by North American utility chiefs to be the signature requirement in actualizing such undertakings. Per unit metering costs differ extraordinarily and rely upon venture scale, prior foundation, and sort of innovation sent. While the wide range of stages restrain the legitimacy of general assessments, expenses of activities attempted in North America in the course of the most recent decade have tended to fall amongst US$150 and US$300 per introduced module (counting establishment, interchanges framework and application costs). Demonstrative unit costs for a brilliant meter contrasted with a simple ('imbecilic') meter are in the request of USD30 and USD60, individually. As far as payback periods or degrees of profitability, of the few assumes that have been openly cited, payback periods in North American and Australian cases have had a tendency to be in the scope of 3–15 years.

Asia is home to very nearly 60% of the total populace yet has just 36% of water resources. In Asia, around 29 million cubic meters of water have lost which cost around 9 billion $ every year. As per the report by World Water Development Report 480 million of people in Asia alone will face water scarcity. According to Asian Development Bank, it is expected that water demand will be double as compared to present till 2025.

Coordinating a Smart Water Network is an incredible route for utilities to give new life to speculations they have effectively made. Most Water utilities have established their fundamental frameworks in the two resources and innovation and have moved their concentration to utilize these consumptions in the facilities they have set up.

The introduction of Smart Water Networks is the utilization of data innovation to upgrade a utility's ability. The initial step for a water utility to push toward actualizing a Smart Water Network is to build up its needs to improve effectiveness. At that point, assess the data accessible and distinguish which bits of information are missing and could be accomplished by incorporating existing IT foundation. With this data, the utility will be prepared to design new speculations to fill in the innovation gaps.

The three mainstays of Smart Water Networks:

- Information: making full utilization of all information delivered by a water utility
- Integration: using current IT frameworks to expand past ventures
- Innovation: having the adaptability to address future difficulties.

A Smart Water Network isn't just an individual framework that streamlines a system's efficiencies yet rather a method for connecting together various frameworks inside a system to share information crosswise over stages. Considering a significant number of the regular difficulties confronting utilities, including leak management, control compliances and regulations and client benefit, utilities can enhance execution by coordinating frameworks in a way that pinpoints particular issue zones.

By tending to the gaps in the accessible data, a utility can precisely set objectives, design speculations, and unravel some of its greatest difficulties.

3 Real-Time Data Integration

There are fundamental levels of Smart Water Network reconciliation that will give utilities relatively immediate advantages. One particular ability that can rapidly pay profits is changing real-time data into significant data for the quick decision making in zones of the utility outside of the control room. By moving data out of the control room, utility administrators can remain mindful of what is happening in the field constantly and react rapidly and fittingly when an issue emerges.

Monitoring continuous information enhances support techniques in light of the fact that the framework is consequently creating data as event occur. For instance, when there is a defective bit of gear or when a pipe releases, utility chiefs can dispatch groups promptly to the correct spot of the incident with particular data for that kind of hardware.

Utilizing a propelled GIS framework to send groups to the correct spot of the incident additionally cuts labor costs. Utilities can significantly diminish the time between when an incident happens and when the issue is settled, in this way lessening the hazard and the cost related to that occasion.

Notwithstanding enhancing reaction times, connecting real- time information from the field has likewise demonstrated basic for building the exact water powered models that utilities use to think about what ought to occur in the field with what is really happening. By performing on the web reenactments, utility administrators have an intense apparatus that enables them to build up an exact benchmark to measure their system's operational proficiency (Fig. 1).

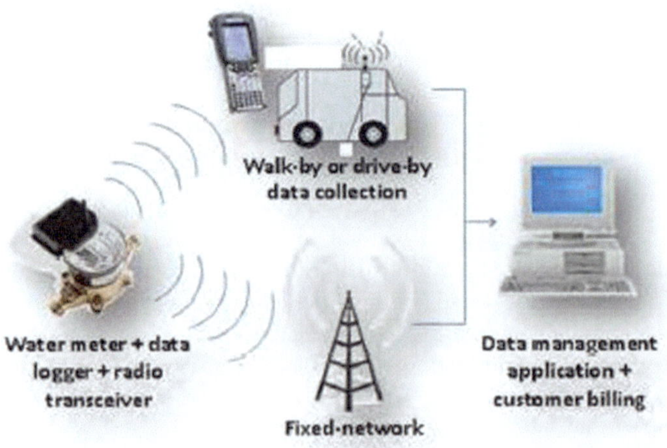

Fig. 1 Smart metering process

4 Enhancing Customer Satisfaction

Having an adaptable framework enables utilities to precisely set future destinations. Bringing in information from client data frameworks into the SCADA framework enables utilities to pick up a more exact perspective of generation versus utilization.

This enables utilities to build up a water balance and put forward advanced metering infrastructure (AMI) objectives in view of key performance indicators (KPI). Additionally, by connecting water driven model simulation with the client data framework, utilities can speak with clients all the more precisely about when—and to what extent—their administration will be hindered for maintenance and support exercises.

5 Precisely Forecasting Demand

Utilities depend on authentic information to make a demand–supply curve so they can legitimately adjust their production to guarantee they are working at peak efficiency. However, by utilizing an AMI framework, that procedure can be computerized so pumping regimens can be adjusted to more precisely fill the demand-supply gap. AMI still isn't as generally utilized as a part of the water business as it is in the energy sector, in spite of the fact that its capacity to streamline demand curve to set more precise edges will without a doubt be one of the principle bits of the baffle that will enable utilities to upgrade their creation and operations later on.

While utilities are monitoring water demand they additionally should be aware of the energy they are expending all through their own framework. By utilizing this System to track the water demand, utilities can program pumps to address the issues for particular circumstances of the day, saving both water and energy.

Water utilities around the world are faced with the challenges of giving a high level of services for clients while working in a way that is proficient and beneficial for the association. Numerous utilities have discovered that implementing artificial intelligence and advanced IT arrangements is a powerful method for striking the harmony between meeting the quality norms clients anticipate and augmenting assets will gather an exceptional ROI.

6 Smart Water Meters

Smart water metering refers to a framework that measures water utilization or reflection and conveys that data in a robotized design for observing and charging purposes. Smart meters contrast from traditional meters in that they measure utilization in more prominent detail also, transmit that data back to the service provider without the requirement for manual readings.

Smart metering systems can be designed from multiple points of view, and when extensively characterized, the term incorporates both Automated Meter Reading (AMR) and Advanced Metering Infrastructure (AMI) frameworks. AMR alludes to any framework that permits computerized gathering of meter peruses (for the most part by radio transmission), without the requirement for physical assessment. AMI is utilized to depict a framework that includes two-path correspondence with a water meter. That is, water utilization data is transmitted to utilities, while utilities can thus issue charges to water meters to attempt particular capacities. Throughout the last decade, most smart water meter arrangements around the globe have been AMR frameworks, notwithstanding due to their extra usefulness, the industry is beginning to move towards AMI and 'shrewd lattice' arrangements. Notwithstanding of the arrangement, all keen metering frameworks comprise of three primary components: (a) estimation; (b) correspondence; (c) programming application.

Smart water metering is encountering solid development all through the industrialized world with the annual growth projections shifting in the vicinity of 15 and 20% until 2017. By 2015 smart meters are relied upon to represent half of the worldwide water meter showcase, with a market size of around USD900 million. Up to this point smart water metering ventures have been embraced primarily in Europe and North America.

Activities change extraordinarily in the measure from little provincial towns with less than 1000 associations with substantial urban areas, for example, New York and Mumbai which serve up to 1 million associations. Over the past decade Boston, District of Columbia, Cincinnati, Philadelphia, Atlanta, Chicago, and Detroit have all embraced major AMR ventures. Substantial tasks are too at present in progress in New York, Kuwait, Malta, and Toronto. While brilliant water meter organizations have been amassed in created districts, late tasks in Mumbai what's more, Dar es Salaam show creating nations may figure all the more conspicuously in the coming years. Essentially, keen metering organizations are starting to tailor items to developing business sector needs, for example, Elster's keen standpipe arrangement.

Smart water metering offers a scope of advantages when contrasted with traditional water metering. These include:

- Faster and more effective meter reading
- Theft and leak detection
- Greater billing accuracy
- Enabling an adaptable duty structure
- Increased read recurrence, bringing about enhanced obligation gathering
- Ability to remotely monitor resources use.

7 Data Analytics from Questionnaire (Created at QuestionPro)

A questionnaire had been developed and floated on various platform of social media consisting 20+ questions on the various usage of water to analyze where do it get mostly waste. And it is highly seen and promoted by the professionals, students, colleagues, friends and family members. It is the data recovered and analyzed nearly from 500 participants and highly appreciated that they took time to fill this questionnaire and concerned about the water and its conservation. I got really amazed to observe the findings from the results. It made me think twice that where is the lag when people are already aware about the causes and actions.

Questions are given below:

1. What is the sort of Housing Ownership? (Fig. 2)
2. What is the Age of your Dwelling? (Fig. 3)
3. For Personal Hygiene, which one is Majorly Used? (Fig. 4)
4. To what extent does each shower last?
   The result of this question is shocking that 60% of individual answered 20–30 min. Only 12% have answered under 10 min which is nearly justified. But this is what I have experienced that people waste more water while bathing or taking showers.
5. What is the bathing and washing recurrence? (Fig. 5)
6. What is the recurrence does family unit go clothing in seven days?
   Majorly answered once in a three day. Only 27% of individuals do it every day.
7. What is the Major Source of family drinking water?
   97% is Supply/RO Water and 3% Bottle/Cans.

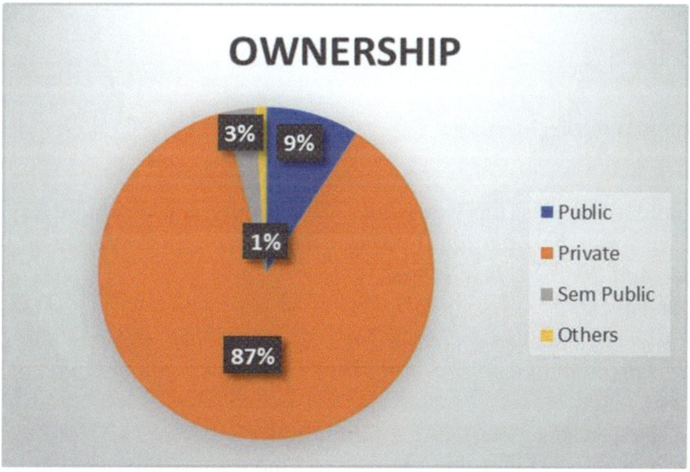

Fig. 2 Graph showing housing ownership

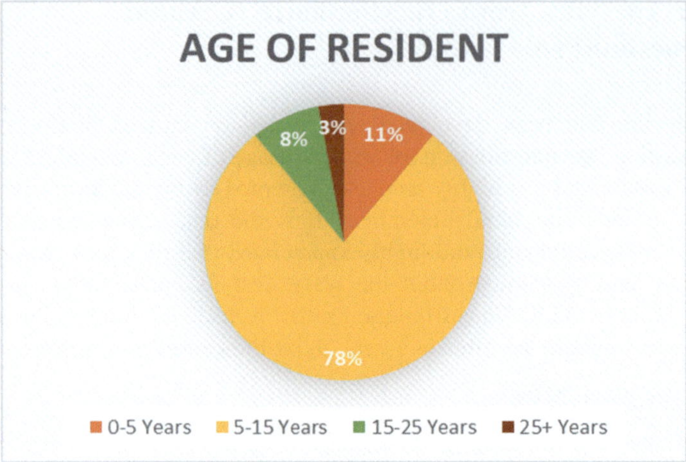

Fig. 3 Graph showing age of resident

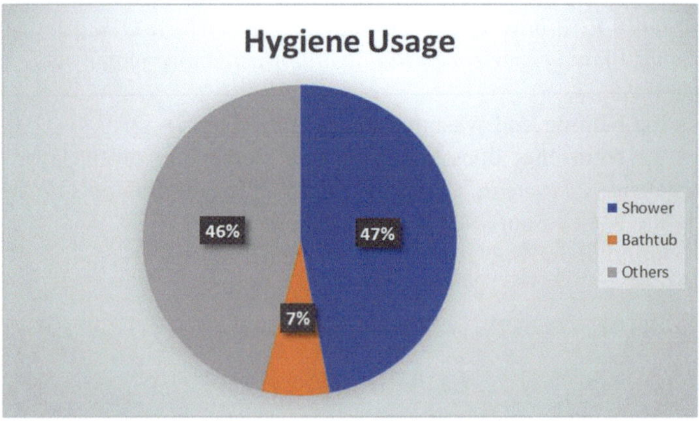

Fig. 4 Graph showing hygiene usage

8. What amount of drinking water does your family unit expend every day?
   Assuming 4 members per family, Major answers are in the range of 25–40 L in Summer and 20–30 L in Winters.
9. Do your taps spill?
   31% answered that they suffer leakages and it is also observed that the water supply from municipal corporations faced leakages problems and caused water shortages.
10. Do you have water meters in the family unit?
    More than 50% said they don't have water meters. So, the readings are directly observed and thus bills generated sent to customers by water corporations.

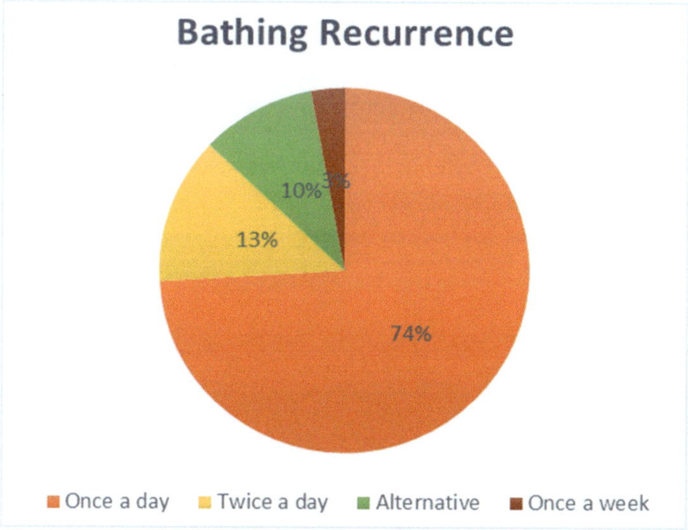

Fig. 5 Graph showing bathing recurrence

Customers don't have their real time data how much they are using and what do it costs?

11. How frequently is the water meter perused?
Few people answered this question that they do check meter readings once a month. Nobody cares about the cost related to water because as compared to all the resources, it is very cheap but the what cost it would be occurring in the future that nobody is thinking.
12. What are the Major Water Using Appliance? (Fig. 6)

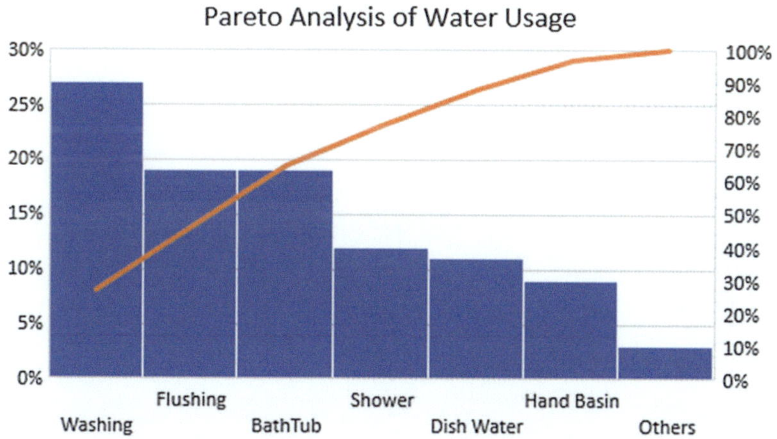

Fig. 6 Pareto analysis of water appliance usage

13. How is the flow water supply benefit? (Fig. 7)
14. What is the normal month to month cost of water in the family unit?
    All answers are in the range of Rs. 1000–1500.
15. To what extent did the most recent water shortage last? (Fig. 8)
16. Is there any water saving measures polished in the family unit?
    This result surprised me and also gives a large scope of business opportunities to this study. Maybe it has not reached to all the audience or only reached to the targeted audiences but nobody has any kind of technology for water saving purposes. Even I am one of them.

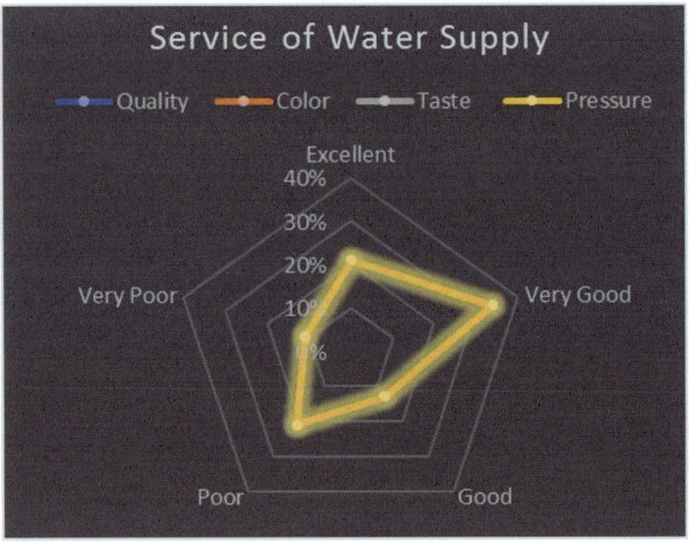

Fig. 7 Radar chart of water supply services

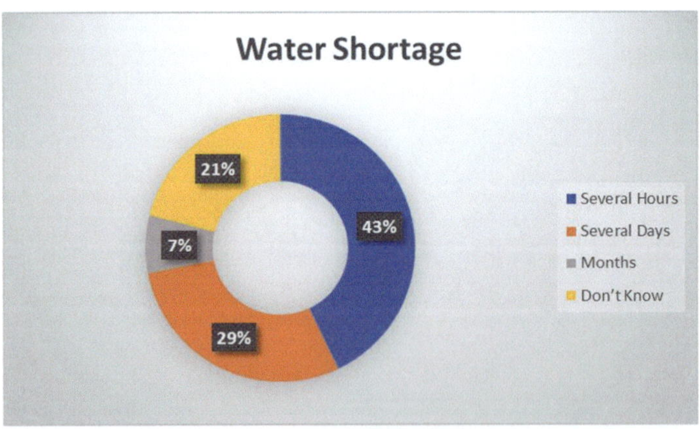

Fig. 8 Pie chart of water shortage in household

17. Have you at any point seen the purposeful publicity identified with Water Conservation?
71% answered that they have experienced and rest never cared about it.
18. Do you know What is Smart Water?
As expected, nobody knows the concept of Smart Water before this Questionnaire. I am sure many of them are aware now after researching about this terminology. So here also 99% answered that they are not aware of this technique.
19. On the off chance that the Govt. offers sponsorships to family units to enhance the current water framework. Would you take an interest?
But everybody has answered that they will support the govt. initiatives if they offer sponsorships to families.
20. Have you heard of Artificial Intelligence, IOT, DBMS?
61% answered that they are known to these terminologies, 39% don't.
21. Do you support the innovation diminishing your wastage and use of water?
Again, 100% have claimed that they will support the innovations which could reduce the wastages and usage of water.

8 Major Findings from This Data

- Individuals use 90% of their water in all other uses and 10% for drinking, which could be reverse because water is not meant for anything else but drinking.
- Washing and Flushing are the major cause of water wastage or usage.
- People are aware of the fact that water is a basic need to sustain and conserving it is the need of the hour but they really don't care about the cause because its cheap, easily available to them or anything else.
- Any individuals can't carry any estimation of water shortage, wastage, usage or can't figure out the leakage detection and what will be the cost projected regarding their current pace of usage.
- Business opportunities for Smart Water is humongous.
- Govt. Initiatives (and conference and events like ISGW) creating awareness among very large pool of audiences.

Smart Water Distribution Using Low Cost Communication Protocol

Pankaj Mohan Gupta, Moreshwar Salpekar and Pravesh Kumar Tejan

Abstract Water is one of the most critical natural resource and is essential for human survival. However, water resource management and distribution even today are not in the best shape. As per recent WHO reports it is estimated that almost 32 billion cubic meter of treated water is wasted during transportation from one place to another through the public water distribution systems. This paper intends to propose a smart solution for public water distribution and management system with IOT ecosystem. This solution will enable water supply management efficiently. The proposed system covers key aspects of the water distribution system i.e. reservoir management, leak management, water quality monitoring. On the supply and distribution side, the proposed system will evaluate and monitor parameters such as pressure, pH values and metallic and biological impurities in water. This would be done by data analytics (DA) server, machine learning (ML) for indicators like peak usage, low water supply etc.

Keywords Leakage · Quality · Sensors · Wireless sensor networks · Internet of things

P. M. Gupta (✉) · M. Salpekar · P. K. Tejan
Aerospace, Defence and Legacy, ST Microelectronics (I) Pvt. Ltd., Greater Noida, Uttar Pradesh, India
e-mail: pankaj-mohan.gupta@st.com

M. Salpekar
e-mail: moreshwar.salpekar@st.com

P. K. Tejan
e-mail: pravesh.tejan@st.com

R. K. Pillai et al. (eds.), *ISGW 2018 Compendium of Technical Papers*, Lecture Notes in Electrical Engineering 580,
https://doi.org/10.1007/978-981-32-9119-5_23

1 Introduction

Water is becoming scarce and valuable resource as population and hence water consumption rate is rising. Drinkable water meeting right standards should be available to all and should not be wasted, yet some of it is lost to leakages, some because it does not remain potable due to addition of impurities like dirt, chemicals, sewage due to seepage into pipes. As most of local water distribution system is underground, the problems are detected at a very late stage. It is often difficult to locate the source of impurity addition quickly because the pipes are underground inside cities and town. Sometimes water pressure also decreases between source and its final endpoint due to various reasons. This implies a need to measure the purity and pressure from end to end along the distribution path (i.e. pipes). With advent of IoT [1, 2], it is possible to monitor the system efficiently. A system that uses IoT is proposed to collect data in real time to correctly identify the problem points in distribution so as to take quick action.

1.1 Internet of Things

As per ITU-T recommendation [3]:

a. Internet of Things (IoT) is a global infrastructure for the information society, enabling advanced services by interconnecting (physical and virtual) "things" based on existing and evolving interoperable information and communication technologies. Through the exploitation of identification, data capture, processing and communication capabilities, the IoT makes full use of "things" to offer services to all kinds of applications, whilst ensuring that security and privacy requirements are fulfilled.
b. With respect to IoT, "Thing" is an object of the physical world (physical things) or the information world (virtual things), which is capable of being identified and integrated into communication networks.

1.2 How Will IoT Help?

IoT helps by constantly checking the state of system with respect to pressure, impurities, etc. that would otherwise require manual monitoring and control. It aids the water authority by taking up the aforesaid tasks leaving them to focus on other important tasks. IoT does not intend to replace need of water department personnel, it just aids them in monitoring and making decisions.

2 The Proposed System

2.1 Goals of the Proposed System

The proposed system intends to fulfil the following goals:

1. Automating pressure regulation in distribution network.
2. Detect leakage in water distribution system and its automated information seeding in water authority complaint redressal system.
3. Detect water impurities at various strategic positions in water distribution network.
4. Establish a personalised alert system which provides timely alerts with respect to leakage or impurity in water distribution network and its impact on localities.

2.2 The Design

The high level architecture of IoT enabled smart water distribution system has following major subsystems connected via LPWAN such as LoRa:

- Sensor Subsystem.
- Communication Backbone based on LPWAN (LoRa).
- Cloud Ecosystem augmented by Data analytics.
- Information Subsystem.
- Action Subsystem.

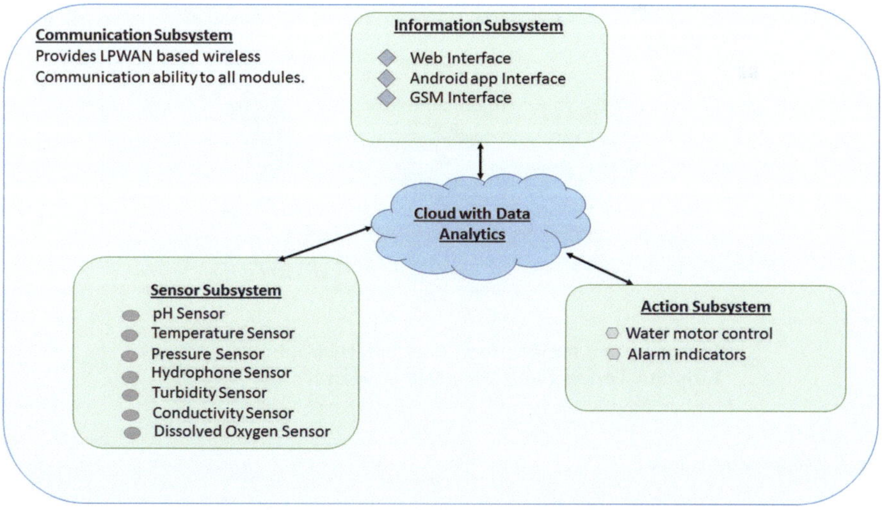

Fig. 1 Smart water distribution

Figure 1 shows how above are knit together.

Each of subsystems is described further in following sections.

2.2.1 Sensor Subsystem

Sensor subsystem consists of smart nodes—a microcontroller [4] based devices having sensors and LPWAN communication module capability. Each node

- Has capability to acquire sensor signals and send it further to cloud using LPWAN.
- Can have one or more sensors as mentioned in the sensor diagram (Fig. 2).

These Smart Nodes are used for periodic or on demand sensing of pressure level and other quality parameters viz. pH value, concentration, turbidity, dissolved oxygen, temperature, conductivity and fitted with a wireless communication module. The geological location of smart sensor node will determine which group of sensors will be part of a smart node and which smart nodes constitute a group.

The main task of sensor subsystem is to send real time or pseudo real-time data to Cloud Ecosystem for analysis.

2.2.2 Communication Backbone Based on LPWAN [5]

As described in section subsystem, smart sensors are placed in pipe network at strategic locations and they need to be connected to a network so that data can be

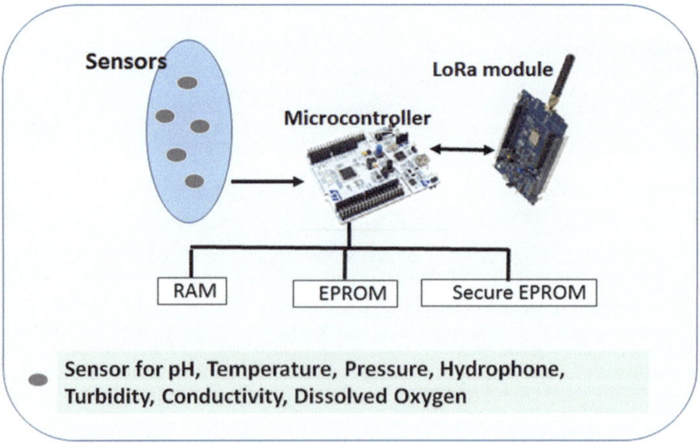

Fig. 2 Smart sensor node

assimilated at central location. For communication, the proposed system uses Low Power Wide Area Network (LPWAN) because of its low power and long range capabilities. LoRaWAN [5] is proposed as the system is intended for wireless battery operated "things" in a regional, national or global network. LoRaWAN targets key requirements of Internet of Things such as secure bi-directional communication, mobility and localization services. The main highlight of LoRaWAN is its unlicensed spectrum and is based on LoRa. LoRa also provides a steady RF coverage in rural. LoRa and LoRaWAN have high link budget greater than any other standardized communication technology.

LoRaWAN has three components End-nodes, Gateways and Network Server.

End-nodes: These are low-power nodes viz. a microcontroller which receive data from smart sensors. End-node will reconcile the data at real-time and will provide it to the LoRa gateway.

Gateways: Data is received at Gateway via a designated wireless channel. Gateways are located strategically based on distribution network size, geolocation of Sensor nodes. The gateway forward packets coming from end-nodes to a network server over an IP backhaul interface for a bigger throughput. The gateways serve as a link layer relay and forward the packet received from the end-nodes to the network server after adding information regarding the reception quality.

Network Server: Network server is responsible for detecting duplicate packets, choosing the appropriate gateway for sending a reply, consequently for sending back packets to the end-nodes (Fig. 3).

These smart sensors will send real-time data about leakage and quality parameters to network server (Cloud Ecosystem).

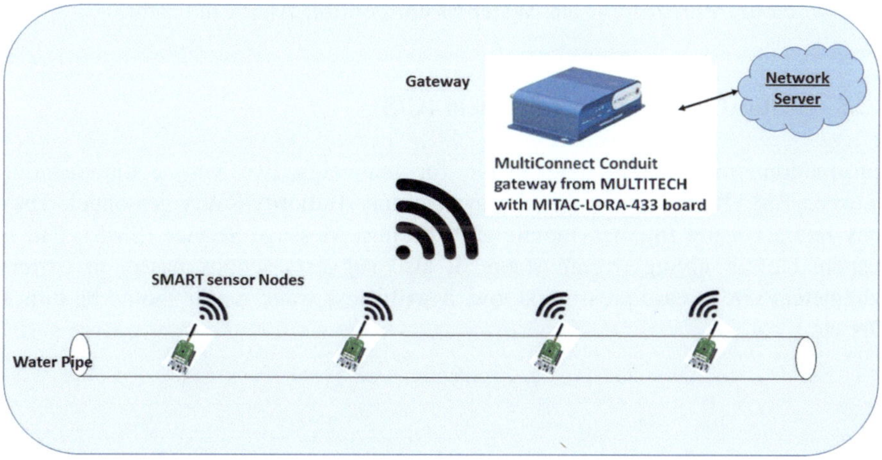

Fig. 3 Smart node communication backbone

2.2.3 Cloud Ecosystem with Data Analytics Capability

The Cloud is a Network Server in terms of LoRaWAN in this system which receives data from sensor subsystem via gateways. It is well equipped with Data Analytics capabilities.

The data analytics cloud, using real-time data from sensors, gives precision information in terms of quality of water, possibility of leakage in pipes in near future etc. Also it has ability to generate alerts in time to inform concerned authorities and RWA about any unforeseen situation like leakage or quality deterioration. Based on the inputs and applied algorithms, Cloud comes up with precision points and decisive suggestion viz. when to start/stop water supply. The Cloud can control Action subsystem in two ways i.e. in automated mode or authority assisted mode.

In the authority assisted mode, the Cloud can request authority to approve action while in automatic mode, it can perform suggested actions proposed by Data Analytics.

2.2.4 Action Subsystem

Action subsystem consists of network of connected smart devices which perform certain tasks based on remote commands sent wirelessly in LoRaWAN communication medium. It is responsible for managing all the actions to be performed based on the decisions taken by Data Analytics done on cloud OR instructed by water department personnel via path between Information Subsystem and Cloud.

Action subsystem is based on the precision values generated by Data Analytics (DA) in cloud. Primarily, it has Water Motor control, Alarm indicators.

2.2.5 Critical Information Subsystem (CIS)

Information subsystem is responsible for managing two way communication between SMART water distribution system and Authority/RWA personnel. They may interact using Internet, mobile app or Short Message Service (SMS). CIS is responsible for giving overall status. It also suggests actions based on current parameters viz. in case pressure is low, it will show water motor should be turned ON etc.

3 Use Cases

3.1 Leakage Detection

A large volume of the water is lost in transit from treatment plants to consumers in most water-distribution systems. Typically it is 20–30% of the water produced, whereas in some systems especially older ones water loss is huge up to 50%. This water loss can be because of several causes' viz. leakage occurring in different parts of water distribution system such as transmission pipes, distribution pipes, service connections, joints, valves. Usual causes for leaks are corrosion, material defects, faulty installation, excessive water pressure, water hammer, ground movement due to drought or freezing, and excessive loads and vibration from road traffic. Leaks waste both money and a precious natural resource, and they create a public health risk. It takes considerable amount of time to detect leakage. With Smart Water distribution system, leakage can be detected fast with the correct location so that proper actions can be taken for its remedy. Smart water distribution use following as sensor and the information hence gathered is given to cloud using LPWAN.

3.1.1 Acoustic Sensors Based on Accelerometers

Leak noise localisers and leak noise correlators are listening devices which use accelerometers to sense leak-induced sounds or vibrations caused by water escaping from a pressurized pipe. Leak noise correlators are more sensitive than localisers, as they are able to automatically pinpoint leak locations based on parameters like pipe size, distance and pipe material. The range of the sensors is 250 m.

3.1.2 Acoustic Sensors Based on Hydrophone Sensors

It is also possible to determine leak positions by analysing data from hydrophone sensors that capture sound waves in water. Hydrophones are more sensitive than accelerometers and are effective in detecting leaks in larger pipes. These sensors are best installed in 100 mm or 150 mm air valves located not more than 750 m.

3.1.3 High Rate Pressure Sensors

High Rate Sampling Pressure Sensors are used to detect patterns of pressure transients which occur during a pipe burst. These sensors have a much wider detection range of 1.5 km, but produce lower resolution results in a range of a few hundred meters, compared to tens of metres for acoustic devices. Moreover, there is a need to distinguish pressure transients caused by leaks against those caused by network operations or regular draw-offs.

3.1.4 Virtual District Metering Areas

Virtual District Metering Areas utilize flow meters to monitor inflows and outflows to sub-sections of the water network. Combined with automated meter readings from customers premises, mass balance calculations can be used to derive the actual water loss in an area. Operators can then focus their leak survey efforts on areas where high water losses are observed. Statistical analysis of pressure and flow Statistical analysis of flows and pressures can be carried out to reveal deviations that may indicate a new leak. However, water flow fluctuations can be attributed to a variety of reasons such as seasonal variations and public holidays amongst other causes. Thus, statistical analysis also needs to take other datasets into account to improve its accuracy.

The block diagram given below shows how leakage detection is envisaged. It should be noted that Data Analytics (DA) sub-system residing in the cloud is heart of system and uses:

1. Previous dataset for modelling using Machine Learning algorithm as also improving its learning (reinforcement learning) using new data.
2. Real-time data from sensors (Fig. 4).

This system, by way of real time monitoring and DA from cloud, will not only give precise leakage point but also it will help understanding the predictive analysis for leakage prone areas.

3.2 Quality Detection

Impact on water quality can be seen when water is distributed through piped networks due to the processes such as pipe material release, biofilm formation and detachment, accumulation and resuspension of loose deposits. This changes in

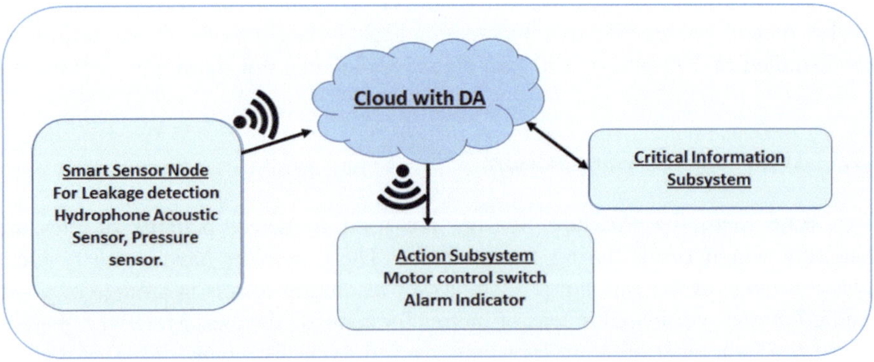

Fig. 4 IoT based leakage detection

water quality may cause physiochemical and microbiological de-stabilization of pipe material biofilms and loose deposits in the distribution system which have been established over decades and may dock components that cause health issues.

3.2.1 Quality Detection Sensors

There are sensors that directly measure specific water quality parameters such as pH, turbidity and conductivity. There are also sensors which aim to detect any deviations in generic properties of water (such as optics) in order to cover a broader spectrum of contaminants. Finally, there are biosensors which monitor the behaviour of living organisms in the water to assess the toxicity of water samples (Fig. 5).

Smart sensors [6] are located throughout the distribution network at strategic locations for measuring quality parameters viz. pH, turbidity, conductance etc. These sensors relay the information about quality parameters to Cloud via gateway. With this information overall information about water quality in whole area can be seen at real-time.

DA sub-system residing in the cloud uses:

1. Previous dataset of quality parameters in area for modelling using ML algorithm.
2. Current real-time values from quality sensors.

Together with real-time data, data analytics helps in giving realistic value of water quality as per standard. The information hence generated can be used for planning distribution network cleaning OR re-establishment. It helps in saving water as lot of water wasted in degraded quality water while RO process in individual households.

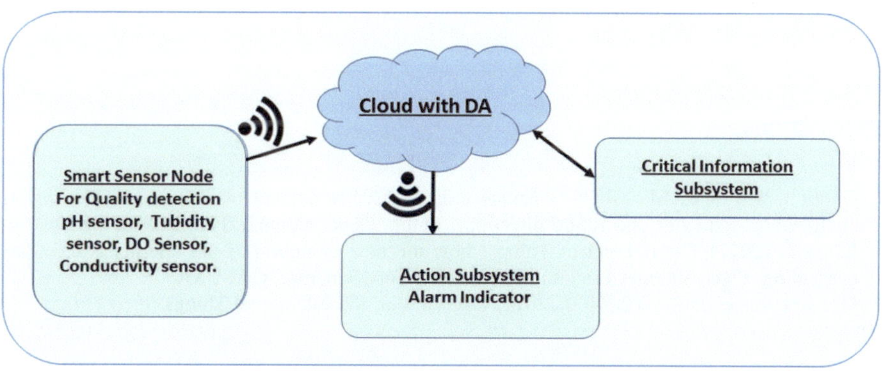

Fig. 5 IoT based water quality management

3.3 Anamoly Control on User Side

The data receiving module at centre sends data to local SCADA system which can aggregate and analyse data for anomaly detection. The aggregated data can then be passed to higher level system. For example, all intelligent sensors send data to connected billing centre which further sends data to SCADA system for analysis. The analysed report and aggregated data are then passed to control centre for further monitoring. Anomaly detection implies that the analysis is showing that the incoming data is above threshold in one or more parameters. This detection is done using SCADA system. For example, SCADA analysis may observe that one of intelligent sensors has detected current pressure is lower than previously reported pressure. This analysis may be acted upon by local system or passed on to higher level e.g. control centre for further action (i.e. if only one sensor is reporting lower pressure, sensor needs to be checked but if a series of sensors report lower pressure then problem is likely before first sensor and action is needed at the point where first sensor is present).

4 Conclusion

Developing countries has been looking forward to technologies to decrease water wastage and improve quantity and quality of water distributed. IoT, Cloud and Data Analytics can positively influence modern water distribution system with latest tools to give precise assisted support to take intelligent actions. The paper discussed a system design that monitors the water distribution smartly with help of intelligent sensors and decentralized monitoring centres. This will help in minimizing water wastage and improve water quality. The system may be augmented using data analytics to detect and predict anomalies and free up personnel for more important tasks like improving system and data analytics models.

References

1. Gubbi J, Buyya R, Marusic S, Palaniswamia M (2013) Internet of Things (IoT): a vision, architectural elements, and future directions. Future Gener Comput Syst 29:1645–1660
2. Evans D (2011) The Internet of Things: how the next evolution of the internet is changing everything. Cisco Internet Business Solutions Group, San Jose, CA, USA
3. ITU Recommendation Y2060 06/2014, Overview of the Internet of Things

4. IoT Connectivity Made Easier STM32 MCUs and LoRa: STMicroelectronics Pvt Ltd. http://www.st.com/content/ccc/resource/sales_and_marketing/presentation/product_presentation/group0/b5/72/c6/ec/c8/e3/4a/8c/IoTaugmented_stm32-lrwan/files/IoTaugmented_stm32-lrwan.pdf/jcr:content/translations/en.IoTaugmented_stm32-lrwan.pdf
5. LoRa Alliance (2015) White paper: a technical overview of Lora and Lorawan. The LoRa Alliance, San Ramon, CA, USA
6. Gutiérrez J, Villa-Medina JF, Nieto-Garibay A, Porta-Gándara MÁ (2013) Automated irrigation system using a wireless sensor network and GPRS module. IEEE Trans Instrum Measur 63(1):166–176
7. Allen M, Preis A, Iobal M, Srirangarajan S, Lim HB, Girod L, Whittle AJ (2011) Real-time in-network distribution system monitoring to improve operational efficiency. Am Water Works Assoc J 103(7):63
8. Hamilton S, Charalambous B (2013) Leak detection: technology and implementation (1st edn). IWA Publishing, London, UK; Loh CJ (2016) Smart devices trial extended to 3,200 households in Yuhua
9. Kim Y, Evans R, Iversen W (2008) Remote sensing and control of an irrigation system using a distributed wireless sensor network. IEEE Trans Instrum Measur 57(7):1379–1387

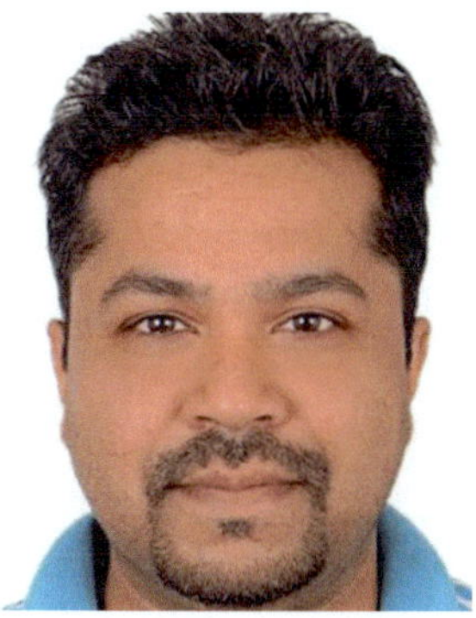

Pankaj Mohan Gupta possess expertise in system design and integration and capable to deliver quality products. Having 18 years of experience in Embedded Systems. He has worked on various domains viz. DVD players/recorders, STB decoders/encoders for sat and IP, IoT, realizing complex use cases requirements coming from around the world. He has keen understanding on Machine Learning and Cloud computing and good exposure on python and R. Pankaj has knowledge on STB security protocols and has presented in various conferences. He is a life member and invited speaker of The Indian Science Congress Association, ISCA. He presented a talk on AI Hardware in Delhi University.

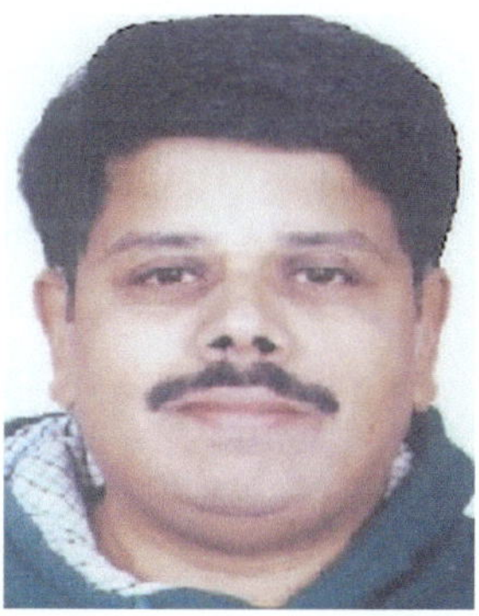

Moreshwar Salpekar has eighteen years of experience in software architecture, design, and implementation for Embedded Systems based on microcontrollers and microprocessors. He has worked on set-top boxes (STB), voice and data encryption/decryption projects. He is currently involved in recording and playback on STBs where security plays important role. He has done courses on machine learning. He has published and presented papers in 5 conferences including one published in ISGW 2017. His areas of interest include IoT, Data Analytics and Embedded Security.

Pravesh Kumar Tejan is a result driven PMP certified Embedded Systems Architect with 15 successful year in digital media products (STB/DVD) from ideation and product definition through design, development and release. A STB Cryptography and cyber security Expert, responsible for delivering software development solutions kit (SDK) for worldwide deployment. Deep architectural understanding of digital video technology, embedded systems, run time environments. Devoted to IoT R&D and Design for smart city, Cloud based data analytics using Machine Learning. Further presented 8 papers in events like ISGW-2017/ITC-2017. He is a life member and invited speaker of The Indian Science Congress Association, ISCA. He presented a talk on AI Hardware in Delhi University

Smart Micro-Cities: Decentralization of Access and Decision Making

Sravan Appana

Abstract A smart city is interconnected and data driven. Indian cities in particular are way denser than western ones, and heavily reliant on retrofitting. Urban population in India will double by 2040. An average smart tier-2 city is projected to have at-least 1000 surveillance cameras, 200 environmental sensors, 300 Wi-Fi hotspots generating more than a million gigabytes data per day. Data generated across the city should be processed to make any meaningful contribution. Big data analytics tools are essential but the insights may not be relevant when data is generalized across the city. Data decentralization for smart cities involves breaking up city administration into more manageable chunks. One way to achieve this is with smart micro-cities. Individual micro-cities will have the necessary set of tools to make localized decisions. The idea that all the data flows to a central data centre where the central authority takes all the necessary actions is short sighted if not insufficient. Decision making needs be decentralized with relevant data accessible by people on ground. Smart microcities offer a unique balanced approach towards decentralization. Smart micro cities embody the spirit and identity of the larger city and at the same time functions closer to the citizen. Individual micro city will look to optimize the local resources, engages citizens and reduces the number of external interactions needed. The implications of decentralization in each city component may be different. City leaders cannot take a 'one-size-fits-all' approach towards implementing smart solutions and need to be proactive in identifying the right approach for the needs of the city.

S. Appana (✉)
Fluentgrid Ltd., Hyderabad, India
e-mail: sravan.a@fluentgrid.com

R. K. Pillai et al. (eds.), *ISGW 2018 Compendium of Technical Papers*, Lecture Notes in Electrical Engineering 580,
https://doi.org/10.1007/978-981-32-9119-5_24

Keywords City · Smart and sustainable city · Micro city · Smart micro city · Data decentralization · Distributed network operations · Data generation · Data processing · Data normalization · Decision making · Central data repository · Edge computing · Distributed technologies · Technology disruption · Citizen engagement and ownership

1 Introduction

A city is not defined by the land or the boundaries, but the people. A large number of actors taking independent actions and interactions. A complex network with individuals playing a crucial role. Different sets of actors with different roles and responsibilities. One big application of game theory.

A smart city is usually defined by its infrastructure, quality of life to its citizens and a clean and sustainable environment. Few of the basic necessities for a city to be termed smart. A true smart city, however, is one where bulk of the aspirations and motivations of citizens, city leaders, and corporations are aligned.

Indian cities are unique in their challenges. They are much denser. None of them are planned. They are driven by need and growth of particular sectors. Acquiring large areas of land for planning is difficult. Illegitimate constructions and encroachments are common. ULBs are not capable and fast enough to cater all the growing needs and demands. Citizen engagement is minimal.

Much of the current efforts in smart transformation are towards digitization of assets and resources, bringing all the components of the city together to synchronize and optimize. This low hanging fruit is absolutely the need of the hour and will provide maximum results with minimum efforts.

The next phase of smart transformation, however, will be defined by citizen engagement, decentralization of power and democratization of decision making. We will go through other domains where this concept is being implemented. The benefits of a true distributed economy and society is best illustrated through following example from nature.

Ant colonies are one of the great examples of a naturally occurring decentralized system. In this system, individual biological agents act upon local information to collectively create a global behaviour. They achieve collective global goals without any central control by doing local tasks and reacting to change in conditions. Although this distributed system is a great end state to be desired, we can practically aim for a middle ground of decentralization.

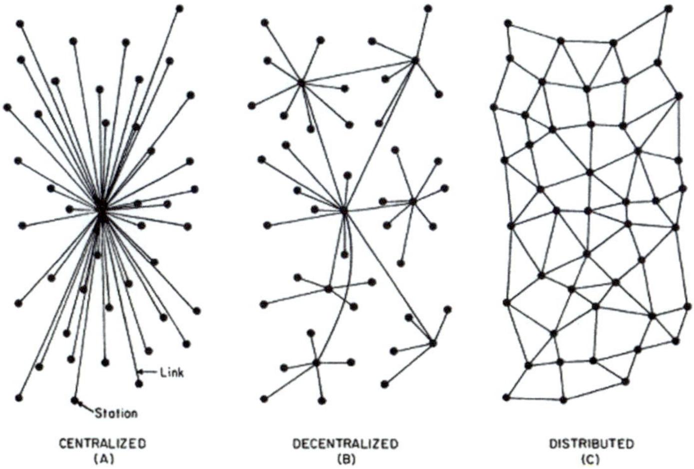

With all its clearly evident benefits, few problems in centralized network do not get enough attention:

- Storage
- Risk of abuse, manipulation, security
- Bandwidth
- Processing and Response time
- Lack of trust and participation.

2 The Eternal Tug of War: Centralization Versus Decentralization

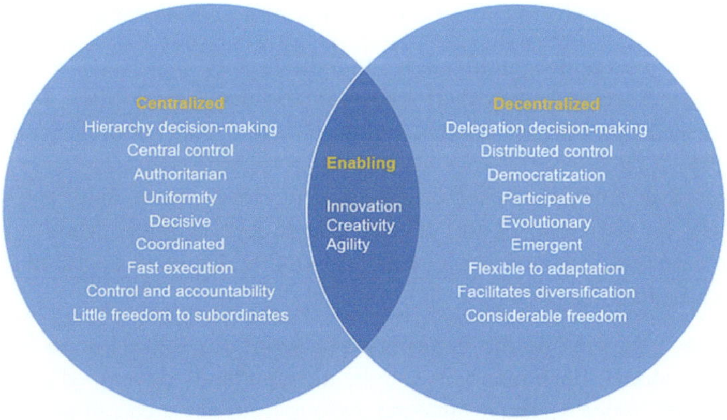

3 How Smart Micro Cities Can Help

Smart micro cities embody the spirit and identity of the larger city and at the same time functions closer to the citizen. Individual micro city will look to optimize the local resources, engages citizens and reduces the number of external interactions needed.

In the initial phase, each micro city may be envisioned to constitute a command center which brings together all the components relevant to the region. Even though bulk of the ICT infrastructure is centralized, the microcity command center serves as a relay for bulk of the information that needs to be transferred to central data repository before getting processed. Decentralized command centers at the micro level can be implemented through operational restructuring. Existing command center application can be reused and redeployed by configuring rules and roles specific to the microcity.

The central city command center thus receives fully/partially processed information and events generated at micro level. Predefined configurations will determine the level of information to be forwarded to the central city command center. Thus the level of centralization can be configured according to the size and needs of the city. The decentralized command centre architecture thus helps in

1. Reducing the time lag while responding to incidents,
2. Space constraints to accommodate various department representatives at one place,
3. Eliminates loss of focus owing to the handling of trivial and redundant issues at centralized level.

3.1 Microcities Hub and Spoke Models

The following two examples illustrate a couple of ways decentralization can be achieved without compromising on the benefits of a centralized system.

1. Component based decentralization: Individual city components can have a central task force and individual spokes distributed across cities. Main advantages include flexibility in customization, effective utilization of city resources. Refer Fig. 1.
2. Area based decentralization: Decentralized microcity command centers can be replicated on the basis of the central one except in scale. Main advantages include ease of design and implementation, higher degree of coordination across city components. Refer Fig. 2.

Microcities also provide the flexibility and scope for following an 'area based development' model. Each microcity can be independently developed of each other in a step-by-step process. The results and lessons from one area can be the basis for

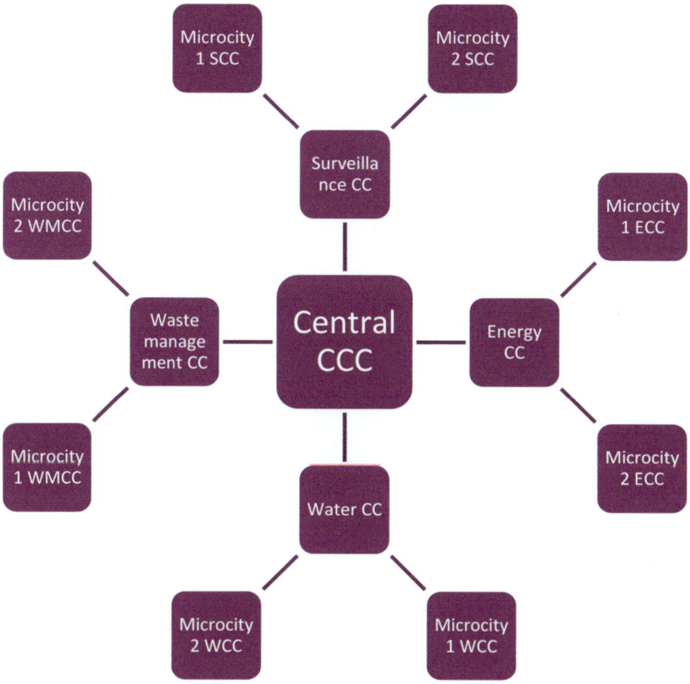

Fig. 1 Component based decentralization

progressive development. They also act as a failsafe and enable data continuity in case of failures at central level.

Although a command centre embodies the heart of a smart digital city, it has to be supplemented with the latest technological advancements, proactive city leadership and citizen participation to yield the greatest benefits and a sustainable future. Some of the emerging technologies that go along well with the distributed model are as follows:

- **Edge computing**: One of the key technologies gaining traction. It provides key benefits of low latency, near real time and parallel workloads. The increase in demand for more operational intelligence and a greater focus on the needs of the community will drive the need of edge computing.
- **Swarm intelligence**: Nano bots that mimic our ant colony example by acting locally and independently to achieve global goals. With the tremendous progress in AI, come machines that mimic humans but also take inspiration from other natural phenomenon.

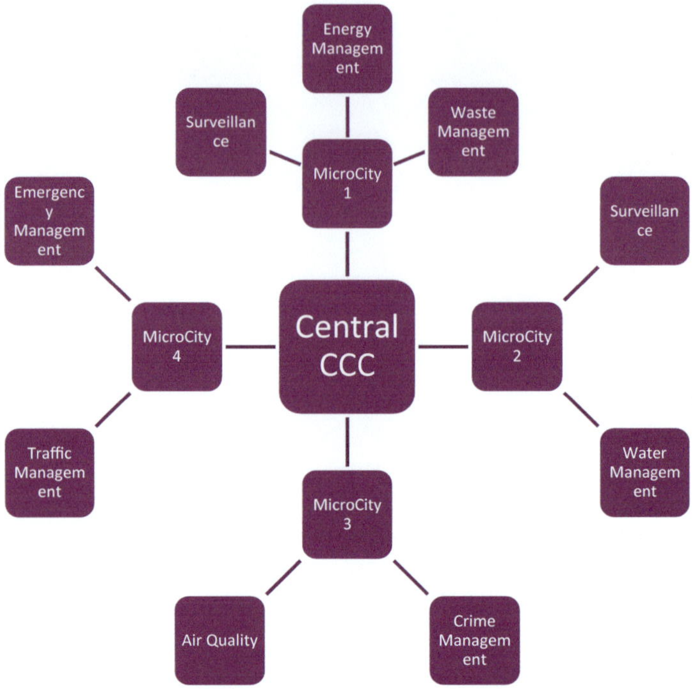

Fig. 2 Area based decentralization

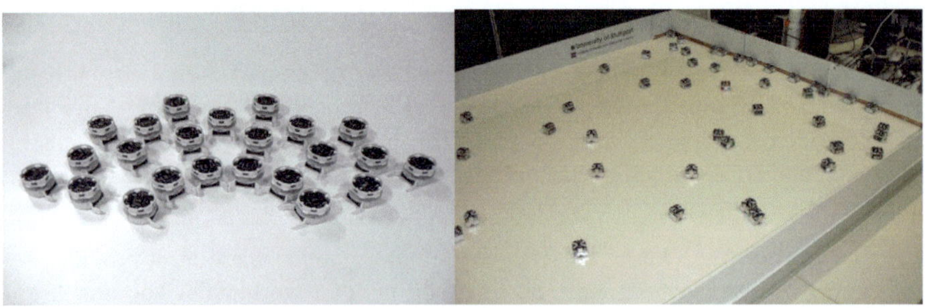

- **Blockchain**: Blockchain technology is set to disrupt multiple industries through eliminating central bodies of trust and distributing consensus mechanism. This may seem counter-intuitive for government bodies to give up control and intermediaries, but this technology is getting recognition and adoption across the world. It provides tremendous benefits to all parties by reducing intermediary costs and hack-proofing transactions in a data abundant world.

4 Implications in the City Components

Surveillance: One of the most discussed potential threats due to digital transformation is centralized surveillance. When the self-defence weapon becomes too powerful, it is at risk of being abused. Decentralized surveillance offers a balanced approach by not allowing concentrated data power houses and more efficient event identification.

Disaster response: A decentralized system allows for faster response during emergencies. The individual spokes can also act as relief and communication centres. They also reduce the risk of connectivity discontinuation and failure.

Energy Management: Decentralized energy generation and consumption systems are on their way to disrupt the existing power holding companies. Microgrids form the building blocks of the energy cloud shifting from the centralized energy production and distribution model to a highly distributed, dynamic and networked grid.

City Planning: City planning is one area where a centralized data visualization makes more sense. Components across the city can planned on top of each other without redundancy and duplication in efforts. Having said that, microcities can help in gathering and prioritizing the needs of people better.

Education: We saw tremendous leap forward in crowd sourced educational content generation. Even academic credentials can be verified and recognized worldwide without any paperwork. This will eliminate the fraudulent claims of un-earned educational credits.

Banking: Banking industry is among the first to be set for disruption by decentralization through Blockchain. Distributed public ledger allows greater control, trust and security to citizens than any bank could ensure.

Business Enablement: Disintermediation and digitization allows ease of setting up a new business. It provides greater transparency and connects the businesses to its customers.

Transportation and Mobility: Imagining a future with fully autonomous cars, trucks and public transport. Central ULBs or private entities such as Uber and Google cannot be given the entire control of such a network.

Governance: Grievance redressal is expedited.

Machine learning and AI are developing rapidly and the companies which are at the right position at the right time will end up controlling bulk of the world's data. Facebook and Google already have a head-start in the data oligopoly. National governments are not the best at securing citizen data.

Funding: Crowdfunding industry attempted to disintermediate capital by allowing individual investors the ability to directly pledge ideas and entrepreneurs. This allows individual investors to have more direct control on upcoming projects but also provides a share in the returns much earlier than an IPO. Crowd funding an infrastructure or public services project will be easily facilitated through these platforms.

Voting: One of the crucial components where there cannot be sole reliance on a single central authority. Even though the existing system of independent regulatory body separate from the government works, there still exists great risk of corruption, abuse across the world. Authentication of voter identity, secure records of vote tracking and trusted tallies are crucial. Blockchain technology provides a decentralized mechanism to handle these issues.

5 Conclusion

City leaders, today, have the right technological tools and intent to bring the smart transformation. There is no single ideal model for a smart city nor there a perfect end destination. City leaders, planners and regulators should consider:

1. Developing clear, actionable plans to realize the city potential building on existing strengths of the city.
2. Promoting flexibility and adaptability in designing the system architecture and innovation. Striking the right balance between the distribution of accountability and ownership between the central and local points determines the outcome.
3. Directing short term pilot projects and experiments towards long term city needs. Focusing on creating strong feedback loops as a base and continuous iterations.
4. Creating a collaborative ecosystem: Along with distributed architecture and data analysis, distribution of responsibility and ownership brings the most out of any city. Citizen engagement and participation are thus most crucial.

References

1. Acxiom, Corelogic, Datalogix, eBureau, ID Analytics, Intelius, PeekYou, Rapleaf, and Recorded Future
2. Goodall B (1987) The Penguin dictionary of human geography. Penguin, London
3. Kuper A, Kuper J (eds) (1996) The social science encyclopedia, 2nd edn. Routledge, London
4. Gordon D (2010) Ant encounters: interaction networks and colony behavior. Princeton U Press, Princeton, NJ
5. https://www.ftc.gov/system/files/documents/reports/data-brokers-call-transparency-accountability-report-federal-trade-commission-may-2014/140527databrokerreport.pdf
6. https://medium.com/@solyarisoftware/centralized-or-distribute-chatbots-2615fbc88290
7. https://chatbotsmagazine.com/encryption-ai-and-the-myth-of-incompatibility-9afca1ca115
8. https://channels.theinnovationenterprise.com/articles/158-7-uses-for-analytics-in-smart-cities
9. https://venturebeat.com/2017/10/08/can-blockchain-decentralize-the-internet/
10. http://smartcities.gov.in/upload/uploadfiles/files/What%20is%20Smart%20City.pdf
11. http://futurethinkers.org/industries-blockchain-disrupt/
12. https://techcrunch.com/2016/10/09/a-decentralized-web-would-give-power-back-to-the-people-online/

13. https://medium.com/world-bug-repository/zerosearch-a-google-for-the-decentralized-web-b940ee66d9f3
14. https://www.buhlebenkosiconsulting.com/single-post/2017/12/14/Is-decentralization-the-future
15. https://www.ncbi.nlm.nih.gov/pmc/articles/PMC5516840/
16. http://sandhill.com/article/is-data-decentralization-the-new-trend/
17. https://www.linkedin.com/pulse/edge-computing-key-smart-cities-internet-things-iot-robyn-goldstone/
18. https://www.navigantresearch.com/wp-assets/secure/WP-ECLOUD2-16+Navigant+Research.pdf

Mobile Electricity Restoration Application-Outage Management System (MERA-OMS)

Alka Malik, Pushpendra Chaudhary, Samant Nagpaul and Santosh K. Sarangi

Abstract Tata Power-DDL is in the business of electricity distribution over North and North West Delhi serving approximately 14 Lakhs consumers. New consumers are added each day, as a result the responsibility for handling outages and restoration of power in respective area are increasing day by day, in a scenario like this, for improving reliability, service levels and most importantly, increasing customer satisfaction there is need for technology adoption. One such technology is Outage Management System (OMS), it supports Tata Power-DDL's network management and restoration process, providing dispatchers with the situational intelligence required to quickly assess critical inputs and variables to determine the best course of action to be taken. It helps optimize outage restoration by efficiently predicting the source and location of outages, based on multiple inputs such as customer call data and SCADA alarms. But in situation of disaster and heavy call rush this immobile technology itself is not enough for the functioning of a Utility, such events need operating technology that will perform when it is needed most. This is when comes the need for a mobile technology that will help the field crew handle the call rush by on the fly determination of the best crew to confirm outage and perform repair work in the field and same to be updated in real-time. This led to introduction of MERA-OMS (Mobile Electricity Restoration Application), which has helped in achieving productivity improvements through the reduction of voice dispatch time, route optimization, elimination of idle times and task optimization.

Keywords OMS · MERA-OMS · TATA POWER-DDL · SCADA · Tata Power-DDL

A. Malik (✉) · P. Chaudhary · S. Nagpaul · S. K. Sarangi
IT-Mobility and OT-Solutions, Tata Power-DDL, New Delhi, India
e-mail: alka.malik@tatapower-ddl.com

P. Chaudhary
e-mail: pushpendra.chaudhary@tatapower-ddl.com

S. Nagpaul
e-mail: samant.nagpaul@tatapower-ddl.com

S. K. Sarangi
e-mail: sk.sarangi@tatapower-ddl.com

R. K. Pillai et al. (eds.), *ISGW 2018 Compendium of Technical Papers*,
Lecture Notes in Electrical Engineering 580,
https://doi.org/10.1007/978-981-32-9119-5_25

1 Introduction

Tata Power Delhi Distribution Ltd. (TATA POWER-DDL), a 51:49 Joint Venture of Tata Power Company Limited (Tata Power) and the Government of Delhi, was formed on July 1, 2002, as an outcome of the Electricity Reforms Process in Delhi. TATA POWER-DDL is authorized to distribute electricity in North & North West part of Delhi having about 1.4 Million registered consumers. TATA POWER-DDL is in a regulated business and governed by Indian Electricity Act and regulations issued by DERC (Delhi Electricity Regulation Commission). In a short span of 15 years, TATA POWER-DDL reduced Aggregated Technical & Commercial (AT&C) losses from 53% to less than 8.8%, show casing one of the few success stories of PPP model in power distribution.

In 2006, TATA POWER-DDL implemented SAP ERP modules (PS, PM, MM, HCM, FICO and BI) to strengthen operational efficiency. In 2011, to improve commercial efficiency, it implemented SAP ISU modules (CS, DM, FICA, Billing, UCES and CRM) and discontinued thirteen home grown commercial applications. In 2013, it implemented state of art SAP-BCM (Business Communication Manager) application for its call center operation. SAP BCM handles approx. 3.5 lacs calls per month. Organization has established itself as a front runner in implementing industry best practices.

To excel in operational efficiency the company has implemented high-tech automated Operational Technologies such as Geographical Information System (GIS), Outage Management System (OMS) and Distribution Management System (DMS), Supervisory Control and Data Acquisition (SCADA) which are the cornerstone of the company's distribution automation project.

Over the last few years, Team TATA POWER-DDL has steered the company to greater heights through business process re-engineering and IT & OT adoption. This has helped in making TATA POWER-DDL a reference point in the Power Distribution sector globally. And hence its strategy is to continually improve its operation effectiveness through the use of updated technologies. Various TATA POWER-DDL organizations responsible for Distribution Operations, Sub transmission Maintenance etc. are challenged with meeting various business and operational performance goals. To end this, organization has developed MERA-OMS bringing the OMS from zonal dispatch centers to on-site in hands of field crew.

2 History

GE's Power On Restore provides end-to-end outage management and dispatch capabilities that are an integral part of a utility's day-to-day outage management and storm restoration processes [1].

Out of the various features provided out of the box, there are quite a few which we have to develop to suit our landscape and environment. They are namely,

- Various Outage restoration workflow & Dispatch specific to Tata Power-DDL domain.
- Decentralized remote dispatch (PORD)
- Reporting and analytics.

In reporting Tata Power-DDL has used SAP BO (Business Objects) and for analysis HANA is used, which completed our system and the organization didn't have to rely on the provided components.

3 MERA-OMS

MERA-OMS works with Tata Power-DDL's OMS to handle different types of planned and unplanned outages. The objective of the mobile application is to reduce operational cost by eliminating/reducing duplicate efforts as the data or information provided by the field crew is directly updated in the source system on real-time basis, thus ensuring more effective process management due to time efficient and accurate update of information in the system. The outages as soon as created in OMS, based on input from various source system namely SCADA, DMS, and SAP BCM etc. are reflected in MERA-OMS thus removing the need for a middleman TO (Telephone Operator) for communication of the same from zonal OMS dispatch system to field crew. The field crew based on the available information in the PORD component in OMS stands for "Power-On Remote Dispatch" which was developed by GE keeping in mind United States Power Discoms/Utilities, primarily to be used by control room staff at home when they can't come to control room/ office in cyclone/tornado season. Here PORD is used to implement OMS in a decentralized manner, where control room dispatches outage information to zones (local decentralized) offices who then close them in system after resolving them in field. This PORD is later on extended to mobility platform (MERA-OMS) maintaining the integrity of the Specifications.

All these features together contribute to the GE's OMS used in Tata Power-DDL domain (Fig. 1) the figure above is a snapshot of the control window [1].

created order and displayed in MERA-OMS proceed with the execution of the work. Based on several order lifecycle the status is communicated to the back-end OMS dispatch system. And as soon as the work is completed on-site same is updated to back-end system by field crew. This ensures:

- Seamless communication of all outage and investigation orders as they are managed by OMS dispatch centers into the field.

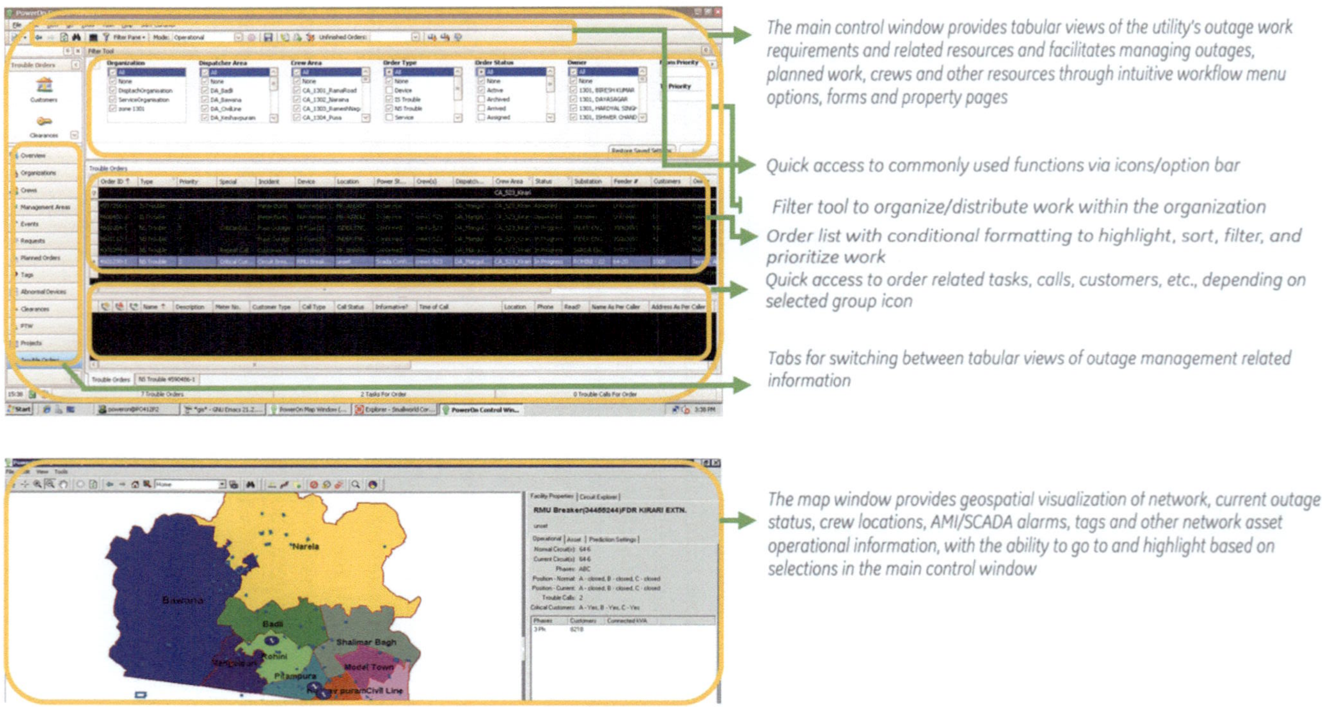

Fig. 1 OMS control window and map window view (Tata Power-DDL region)

- Dynamically update the predicted outage location, crew and workflow status, estimated restoration times and other key operational data from either OMS PORD (PowerOn Remote Dispatch).
- Electronically receive and process critical information from field crews regarding the outage confirmation, location (move-up or move-down), restoration time, cause codes, damage and repair data and information for the immediate generation of follow-up orders.

3.1 Features

Below mentioned are features incorporated in the mobile application as per availability in the OMS PORD.

(1) *Authentication & User Interface*: Role based authentication to the user allowing access to domain specific data only after two-phase authentication process. The user can login using Tata Power-DDL domain ID once it is ensured the person is Tata Power-DDL employee based on internal crew to zone mapping the user is provided access to the complaints of its respective area. The application is provided with Bilingual (English/Hindi) interface to facilitate the skilled/semi-skilled user for easy understating.

(2) *Shutdown Creation*: The field crew can take both HT reference/LT shutdown using the application on-site, also the Permit to work (PTW) can be issued corresponding to a particular No current complaint thus saving the time and overhead of contacting the zonal OMS dispatch office and Power Control room.

(3) *Complaint Transfer*: Provision to transfer Low Voltage complaints from ZSO (Zonal Safety/Shift Officer) to Linemen and High Voltage complaints from Lineman to ZSO (Fig. 8). Also the option to transfer complaint from one zone to other zone in case of boundary-case complaints.

(4) *Follow-Up Order*: Creation of follow-up for Meter Burnt, Service Line damage & Tyco Box damage as notification in SAP using the application corresponding to a particular consumer complaint.

(5) *ETR and Status Update*: Ability to increase & decrease the ETR (estimated time to restore) for a particular order is provided on-site and the order lifecycle status if assigned to field crew, in-progress, completed etc. are communicated to the back-end office

(6) *Geo-location Capture*: Capturing of user location for convenient tracking along with imparting responsibility and also at the time of complaint closing, Image capture with geo-tagging to analyze current situation.

(7) *Merging of Complaints*: If more than one complaint belong to same area or device then, all the complaints can be merged from the mobile application itself and new complaint order is created for execution.

(8) *Predicted Device Change*: The crew and confirm the predicted device from on-site, also the crew can request a change of predicted device in case the device predicted by system and on-site are not same (Fig. 10).
(9) *Emergency Alert pop-up in case of Disaster*: In case of any disaster and failure disaster alerts are send to the field crew based on individual zone or district or PAN-Tata Power-DDL level (Fig. 12).

Apart from the salient features stated above some more functionalities are also added to the application for its user-friendly experience. A notification pop-up (Fig. 11) is flashed on screen with the number of pending orders in crew's zone, whether the application is logged-in or not.

3.2 Architecture

Figure 2, states the current scenario of integration of MERA-OMS and related systems in Tata Power-DDL's domain. Now the complaints are simultaneously transferred to the zonal PORD and mobile devices of the field crew based on the prescribed zones. This has removed the gap between the entire communication between the zone fields and zonal offices, thus adding to cycle time reduction, quick complaint resolution, communication of the same to the consumers, skill enhancement of the crew along with efficient tracking.

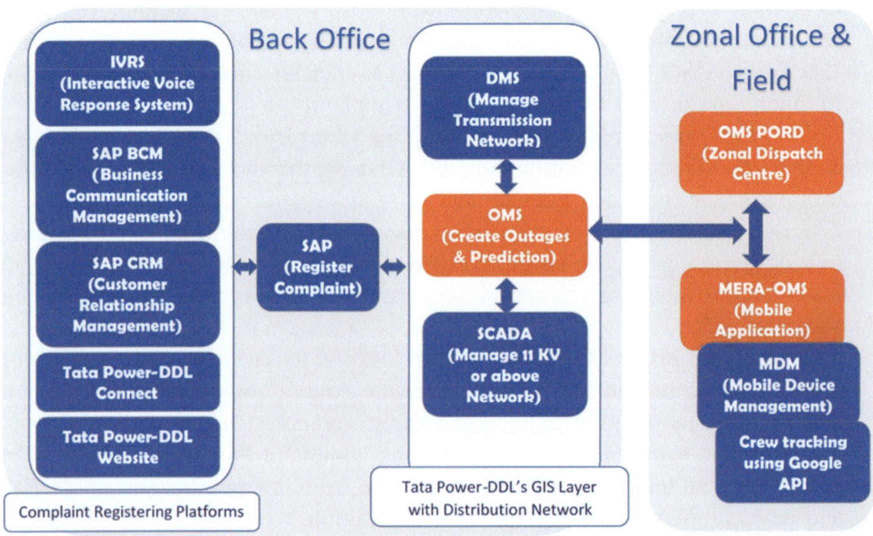

Fig. 2 Architecture of MERA-OMS landscape in Tata Power-DDL

3.3 Layout

Following are the mobile application UI (User Interface) screens.

Figure 3 is the welcome screen of the application where both the authentication process are executed altogether before the user is directed to the safety checklist screen. Figure 4, is an audio checklist, the crew member need to check all the safety measures else will not be directed to the Order List screen, where all the complaints pertaining to a zone are displayed.

Figure 5, is the details displayed to the User for a respective complaint and the details of the caller or consumer for the particular NCC (No current complaint).

The HT-reference and LT shutdown (Fig. 6) can be applied by selecting the hierarchy of devices till HT-fuse. Zone Transfer and merging of order (Fig. 7) is made easy thus eliminating call to the back-end office (Fig. 8).

Follow-up for Meter burnt cases can be directly requested by closing of the complaint from mobile application on-site. A notification is SAP corresponding to the CA number is created. Also the ETR (Estimate time to restore) can be updated from device for timely intimation of the same to the consumers (Figs. 9 and 10).

Figure 11 displays the SURAKSHA form, to capture the safety violation on-site (if any) and the display of complaint in Hindi language for ease of use of application (Fig. 12).

Fig. 3 Login and password change screen

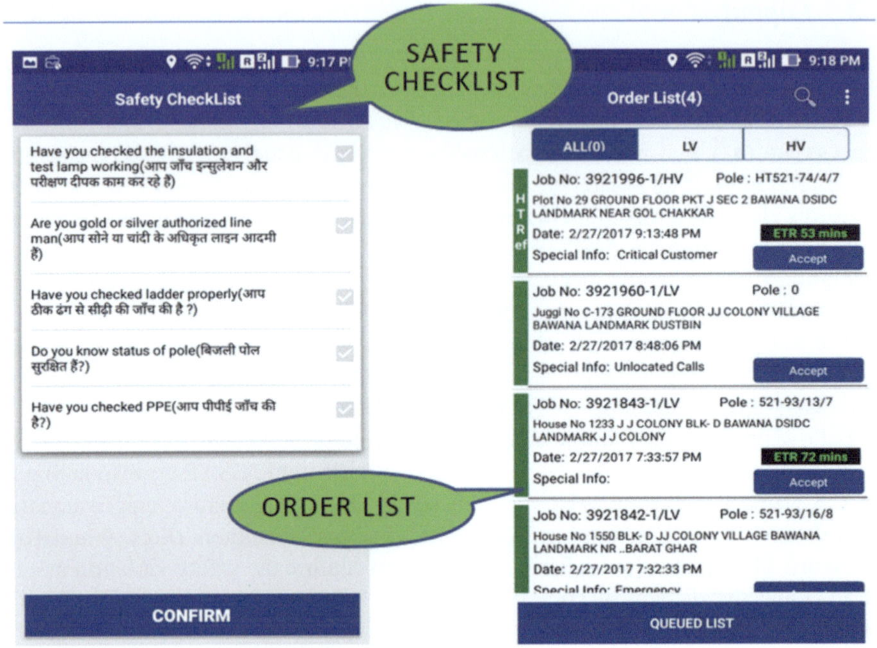

Fig. 4 Safety checklist and order list screen

3.4 *Other Integrated Systems*

Apart from integration to OMS system MERA-OMS is supported by other system and technology for seamless functioning.

(1) *MDM (Mobile Device Management)*: The MobileIron MDM platform enables to secure and manage modern operating systems of mobile devices. It incorporates identity, context, and privacy enforcement to set the appropriate level of access to enterprise data and services. The MobileIron platform secures data-at-rest and data-in-motion as it moves between the corporate network, devices, and various on-premises and cloud storage repositories. With MobileIron, IT can secure corporate information wherever it lives while preserving employee privacy [2].

(2) *Crew tracking with Google Map API*: The application send continuous location update of the field crew which can be tracked and analysed using a tracking portal integrated with google map api (Fig. 13).

(3) *Exchange Server*: For secured Tata Power-DDL user login authentication.

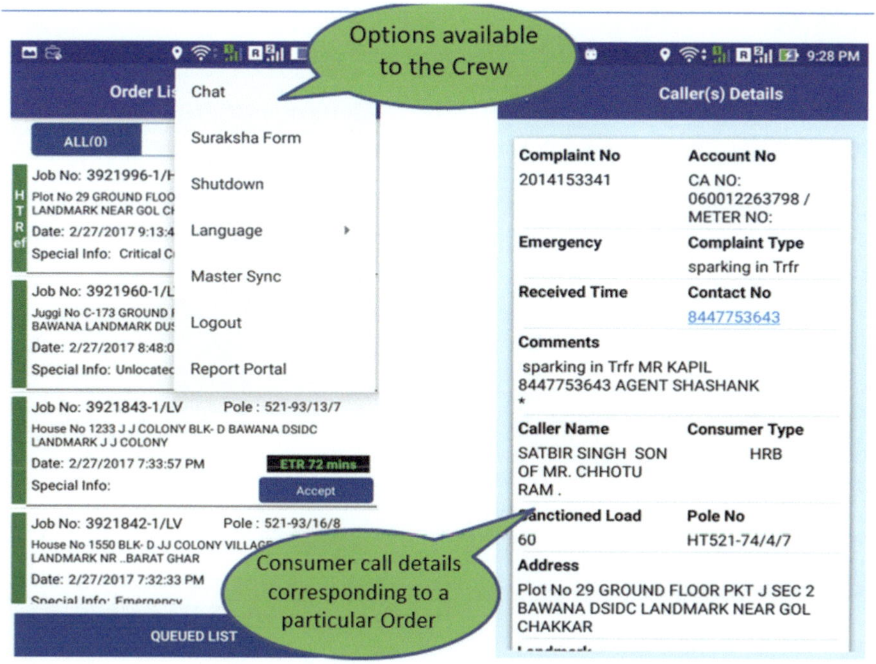

Fig. 5 Menu bar options and caller details

4 Comparison

The adoption of MERA-OMS mobile application have led to both qualitative and quantitative benefits. Some of the quality improvement advantages are stated below:

- Transferring functionality like HT reference, LT PTW from back office to field crew which helps in getting statutory compliance.
- Skill enhancement of field crews in technology transition as the partly automated process of complaint closure is now fully automated using MERA-OMS mobile application.
- The field crew can be easily tracked using the Geo-location captured on-site in coordination with Google location API.
- Any ‘Suraksha’ incident can be captured and reported using the application thus contributing to more safe work environment.
- Reduction of carbon footprints, no need of print-outs of the complaints as all the required information in available through mobile application.
- Multiple visits avoided between zonal offices & on-site location all the messages can be communicated through the application.

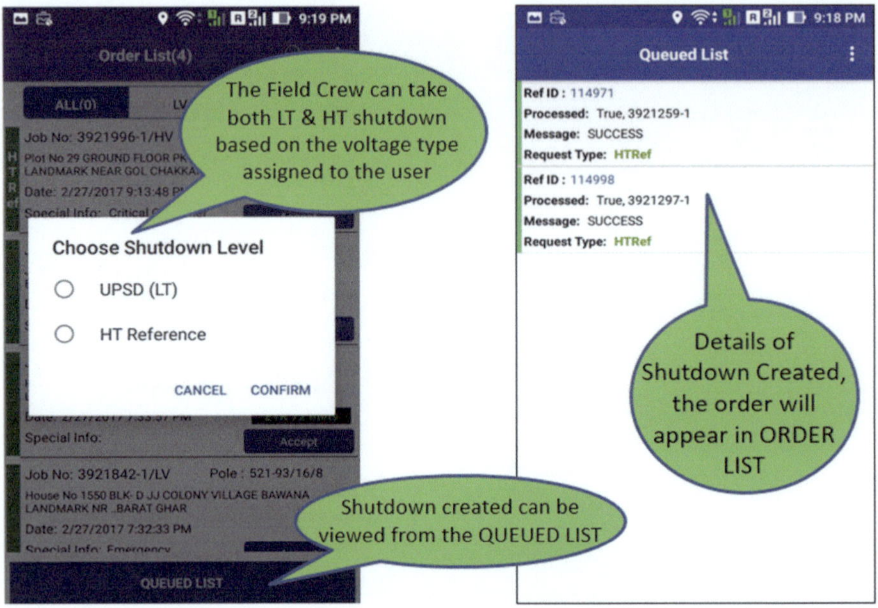

Fig. 6 Shutdown and queue list screen

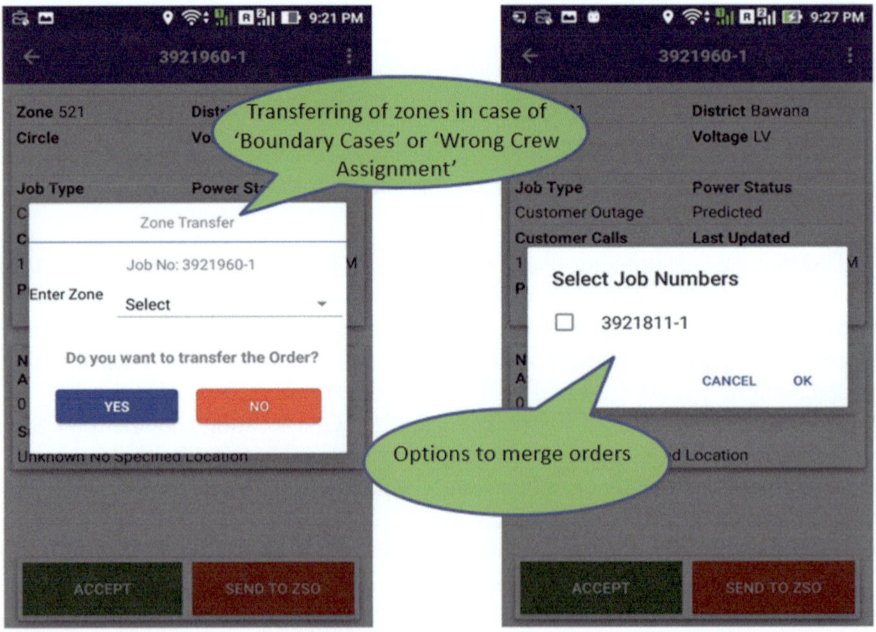

Fig. 7 Zone transfer and merge order screen

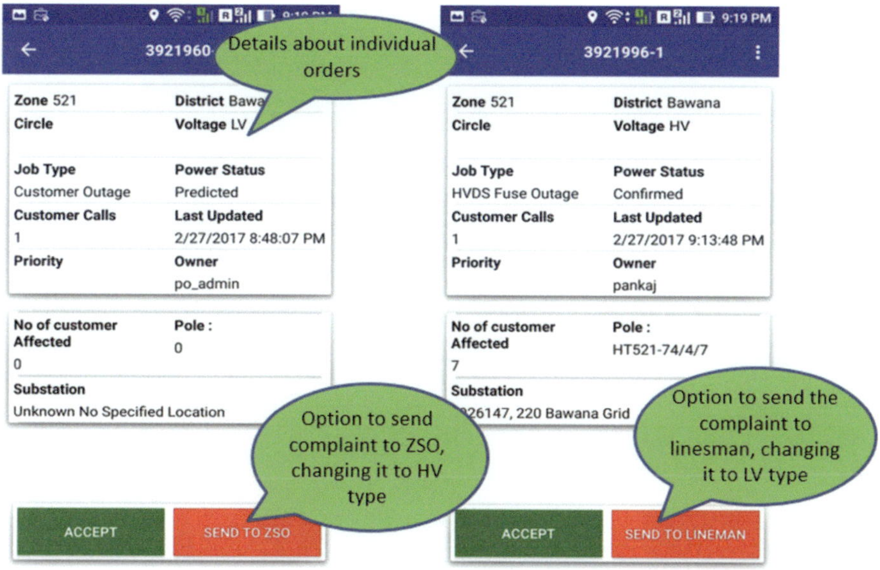

Fig. 8 Send to ZSO and lineman screen

Fig. 9 Order close, ETR update and follow-up screen

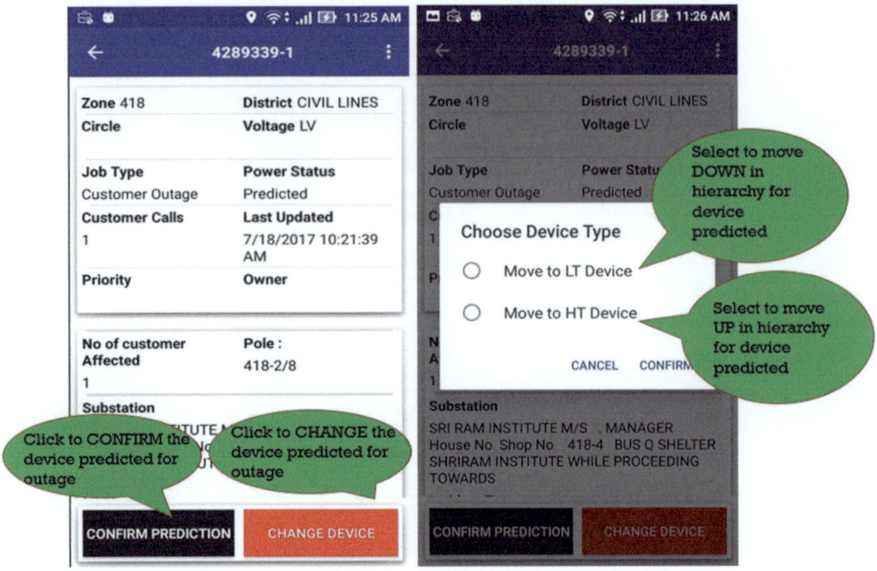

Fig. 10 Prediction change and confirm screen

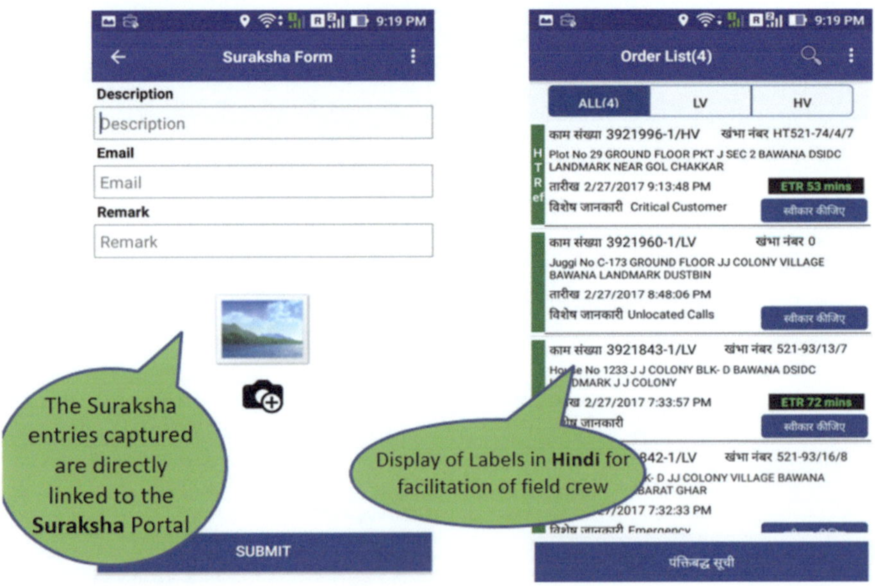

Fig. 11 Suraksha form and hindi UI screen

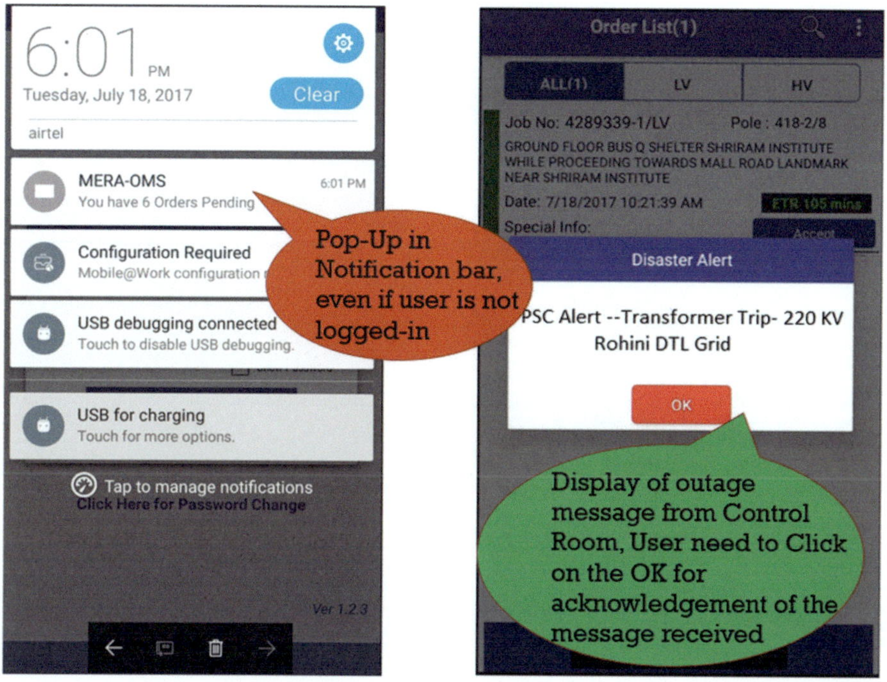

Fig. 12 Notification alert and disaster message pop-up

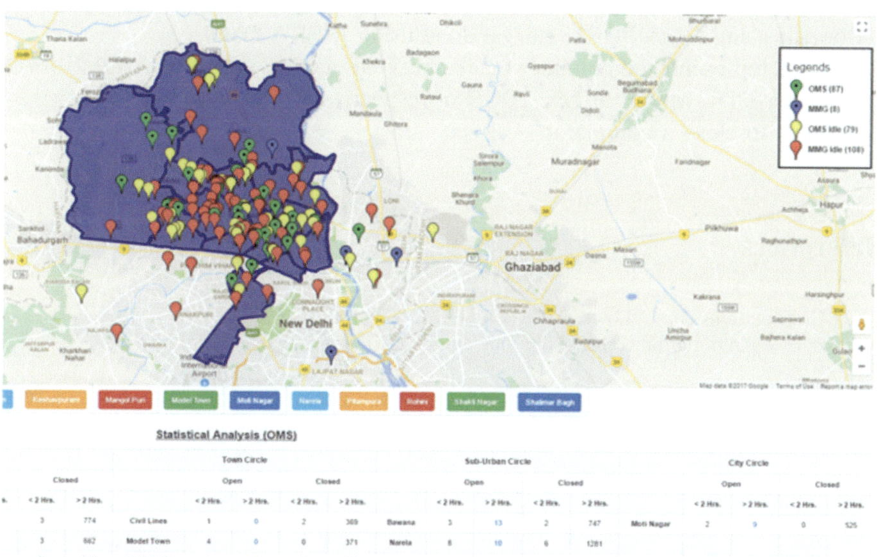

Fig. 13 Crew tracking

Table 1 proposed savings from Feb 18

Man hours saving by substituting zonal TO by MERA-OMS		
	Number*Expenses per month	Cost incurred per month
Zonal TO	100*15,000	1,500,000

Table 2 Quantitative benefits

Year	OMS		MERA-OMS	
	No. of Calls	Avg. Duration of calls (in minutes)	No. of Calls	Avg. Duration of calls (in minutes)
2011	1619	3200.10		
2012	20,428	802.55		
2013	23,124	383.37		
2014	31,656	443.40		
2015	26,041	471.96		
2016	24,710	293.08		
2017	28,173	226.06	19,117	96.30

And the quantitative benefits are:

- Timely closure of complaints resulting in increased time and work efficiency.
- Drastic reduction in the complaints and calls to the back office.

Table 1 shows the cost saving after removing Zonal TO as all the operations performed using OMS PORD can be done using MERA-OMS.

Table 2 represents the number of calls received in OMS and MERA-OMS (PAN Tata Power-DDL rollout in Dec 2017) in respective years along with the average time taken to close a complaint.

References

1. GEGridSolutions.com, "PowerOnRestore_GEA12930_HR,"
2. https://www.mobileiron.com/en/solutions/mobile-device-management-mdm

Self-healing Distribution Grid: Benefits, Challenges and Optimize Way of Implementation

Md Shadab Ahmad

Abstract Tata Power Delhi Distribution Limited (Tata Power-DDL) is a Power Distribution Utility, serving about 1.5 million consumers of North and North-West Delhi. Tata Power-DDL always play a pioneer role in power distribution reforms in its licensee area to reduce AT&C loss and improve reliability and consumer satisfaction through technology adoption. Self-healing Distribution Network is one of the progressive step towards Smart and Digital Utility to provide quality and uninterrupted power to its consumer. As name suggests, Self-healing Network is capable of quick detection of fault, isolation of faulty area and then restoration of healthy section without manual intervention. Although it improve system reliability significantly by reducing restoration time, it require highly mechanized and automated system which need huge investment. But as a part of regulated business model where each investment converts into consumer tariff, it is not advisable to invest such high capital expenditure. This study describe the benefits of Self-healing, challenges to develop it in Tata Power- DDL and the optimized way to implement the same in cost effective manner.

Keywords System reliability · Consumer satisfaction · AT&C loss · Self heal network · Smart and digital utility

Abbreviation

ADMS	Advanced distribution management system
APRS	Automatic power restore system
AT&C	Aggregate technical and commercial loss
DMS	Distribution management system
FPI	Fault passage indicator
FLIR	Fault location identification and restoration
FDIR	Fault detection, isolation and restoration
FFA	Field force automation
FRTU	Field remote control unit

M. S. Ahmad (✉)
Tata Power Delhi Distribution Limited, Delhi, India
e-mail: mdshadab.ahmad@tatapower-ddl.com

R. K. Pillai et al. (eds.), *ISGW 2018 Compendium of Technical Papers*, Lecture Notes in Electrical Engineering 580,
https://doi.org/10.1007/978-981-32-9119-5_26

OMS	Outage management system
GSAS	Grid substation automation system
LBS	Load break switch
NOP	Normally open point
RMU	Ring main unit
RTCFI	Real time communication of fault indication
SAIDI	System average interruption duration index
SAIFI	System average interruption frequency index
SCADA	Supervisory control and data acquisition
VPI	Voltage presence indicator

1 Introduction

It is well known fact that **Reliability** is one of the most important criteria for measuring the performance of power distribution utility. Reliability will improve when consumer outage hours will decrease, that means number of interruption as well as restoration time will decrease. To reduce interruptions, Tata Power-DDL had already taken so many conventional measure like augmentation of sick and under-rated network, segregation of long feeders, providing protective device at optimized location with proper coordination, adding sectionalizing and load break switch at proper location. Besides these conventional method, some cutting-edge technology has also been adopted to reduce restoration time like Auto-reclosers and Sectionalizers, Grid Sub-station Automation System (GSAS), Supervisory Control and Data Acquisition (SCADA), Distribution Management System (DMS), Outage Management System (OMS), Field Force Automation (FFA). Due to all these perennial initiatives, there is significant improvement in Reliability Indices since its inception in 2002. SAIDI decreased from 8.6 to 2.68 h whereas SAIFI decreased from 6.5 to 2.5 in last 10 years (Fig. 1).

Faults are the universal truth for electrical network. It not only reduce consumer satisfaction but also bring good performance of utility in bad light. Although consistent improvement in reliability and performance, expectation of consumer is

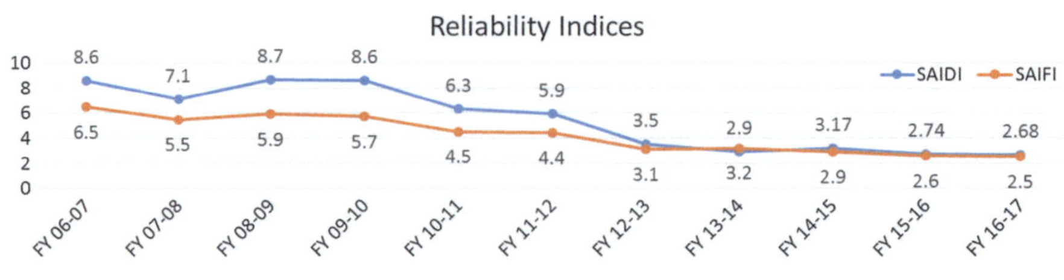

Fig. 1 Reliability Indices (SAIDI and SAIFI) of TPDDL

also going high. Due to increased pressure from consumer as well as regulators to provide uninterrupted power supply, an effective and advanced Fault Management technique is required which can help to reduce outage duration and hence help to over go the bar of expectation. Effective Fault Management means quick detection of fault location, isolation of faulty section and restoration of healthy section. Existing technology (like SCADA, DMS, OMS, RTCFI etc.) provides quick information of fault location. DMS has capability to suggest switching action for isolation of faulty section and outage restoration. Hence these technology only provide sufficient idea on basis of which operator or field crew perform switching action to isolate faulty section and restore rest healthy section.

Self-healing distribution network is basically next generation to the existing technology and a salient characteristics of Smart Grid. As name suggests, it is capable of quick detection of fault, isolation of faulty network and then restoration of healthy section without any manual intervention. That means it is basically to automate the manual procedure of outage restoration. Hence, it will help to convert sustain outages into momentary interruption, which will results in significant improvement in Reliability Indices and ultimately Consumer Satisfaction. But the success of the system depends upon various factor especially quantum of automated and motorized switches and robust & reliable communication network. This study will describe different methodology of self-healing, their advantages and the major challenges to develop and sustain this system. It will also suggest the optimized and cost effective way to implement it.

2 Methodology

Different methods and different technologies are being developed around the world to achieve self-healing distribution network. On the basis of these method, self-healing system can be broadly divided into three categories.

- Distributed Self-healing System
- Semi-distributed Self-healing System
- Centralized Self-healing System.

Distributed Self-healing System

It is a system where automatic switching occur without any information of status of other switches or other sub-station. Switching occur on the basis of algorithm (or logic) based on the status of fault indicator and voltage indicator of the switch itself. There are different technology available under this category.

Conjunction of Auto-recloser and Sectioniser

It is successfully implemented in Tata Power-DDL long year back. **Auto-recloser** is a protective device, design to reclose itself on fault. It is capable of breaking fault current and making on fault current. It is equipped with protection relay and voltage and current sensor. It is generally used in overhead lines where probability of

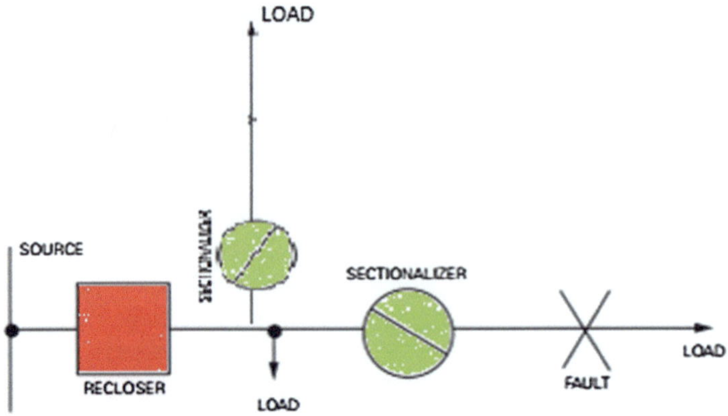

Fig. 2 Conjunction of auto-recloser and Sectioniser

transient fault is high. **Sectioniser** is load break switch, designed to operate at normal load condition. It is not capable to break at fault current. It is also equipped with protection relay, supply interruption counter and voltage and current sensor. It is best used with conjunction of Auto-recloser and coordinated in such a way that it trip during dead time of upstream recloser.

Principle

If any fault occur after Sectioniser (as shown in Fig. 2), Auto- recloser and Sectioniser both sense the fault but only Auto-recloser trip. With this tripping, voltage of network become zero. Sectioniser's Interruption Counter counts when downstream fault is detected, current drop to zero and loss of supply side voltage. When the above condition meet, the counter counts it as first interruption. Auto-recloser will reclose after preset time duration. If fault is transient in nature then the network will restore. But if fault is permanent in nature then the fault condition will again encounter by Auto- Recloser and Sectioniser. Auto-relcoser will trip again and Sectioniser counter count it as second interruption. Sectioniser opens during dead time of upstream recloser when counter will be 2. So, faulty portion will be isolated and when auto-recloser recloses then supply restore the healthy network. Counters reset to zero after complete cycle.

If fault occur between auto-recloser and Sectioniser then its counter will not encounter fault condition and will not trip anyway. Auto-recloser will go in lock-out condition after third tripping. Number of reclosing can be pre-defined.

Advantages

- It is best suited for radial, overhead network having T-off sections. Main line will be protected through Auto-recloser and T-off section will be protected through Sectioniser.
- It is capable of auto-restoration of transient faults of complete feeder and permanent fault in T-off sections.

- It doesn't require any communication medium. Hence probability of failed operation and wrong operation is very less.
- Cost effective solution and best suited for Indian utility having large overhead network with high transient nature of fault.

Challenges

- It cannot be applicable for underground network where try operation is not valid.
- It is not capable of auto-restoration when fault occur between auto-recloser and Sectioniser
- Time coordination of Auto-recloser and Sectioniser should be precise. Sectioniser should open during dead time only.
- Specialized manpower is required for maintenance and troubleshooting of these switchgears.

Semi-distributed Self-healing System

It is also a decentralized system where automatic switching occur on the basis of information shared by neighbor sub-station. Algorithm of automatic switching is based on the status of fault indicator and voltage indicator of switches as well as switches of other substations lie on the same network. Communication medium between substations can be GSM, RF or PLC depending upon communication technology available in utility. There are different method available under this category.

Pause Switching with signal

It is successfully implemented in ENEL, Italy. The **Pause-Switch** is load break switch with direct overcurrent release. Pause switch well-coordinated with Primary Circuit Breaker can be used for fast restoration of supply after fault. It can be applicable for both overhead and underground network with some change in auto-switching logic. Such self-healed secondary substation consist of Ring Main Unit (RMU) having Fault Passage Indicator (FPI), Voltage Presence Indicator (VPI), motorized Load Break Switch (LBS), Circuit Breaker (CB) and Field Remote Terminal Unit (FRTU). This FRTU is responsible to communicate with other substation and execute algorithm to identify fault location and to isolate faulty section automatically.

Principle

Logic of auto switching for overhead network and underground network is different. For overhead line network, existence of a fault is noticed by the VPI as disappearance of supply voltage. All the LBS of the faulty line are opened by the local logic after a certain time delay. Then, circuit breaker at primary substation is reclosed, and if the first line section is healthy, communication up to the first switching sub-station is possible. LBS at first sub-station is then closed as governed by FRTU. If there is no fault in the corresponding line section, then procedure is repeated for the next LBS. When the faulty section is finally encounter, CB will trip again and the voltage disappears again. The loss of voltage, immediately after the reclosing, is detected by a timer circuit and the last switch is opened again. Circuit breaker is reclosed again with a time delay short enough to prevent the reopening of the other line switches.

In case of underground network, the auto-switching logic is somewhat different in order to avoid the reclosing of the faulty section. When a fault happens, the circuit breaker of primary substation get tripped. During the dead time of the line, the remote terminal unit at the substation 1 (FRTU 1), sends a message downstream to check the status of FPI at the second substation. Having received the message, FRTU 2 responds to FRTU 1 and sends a corresponding message to FRTU 3 (Fig. 3b). If there is no response from FRTU 3 then FRTU 2 makes a conclusion that the fault is in the following section and then opens the corresponding switch and sets it in the blocked' mode (Fig. 3c). Then after a certain time delay, the circuit

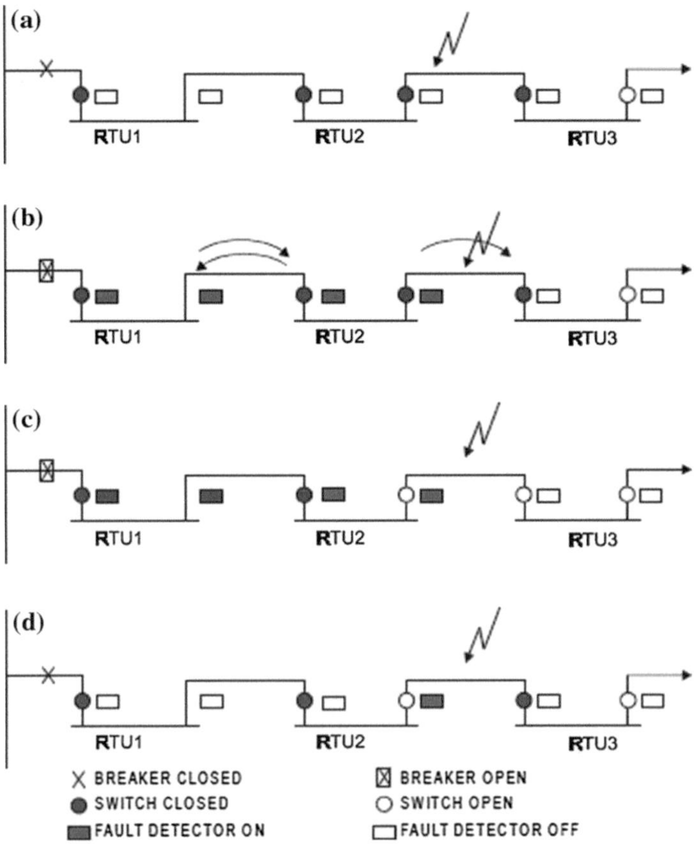

Fig. 3 Pause switching with signal method (in the case of underground systems)

breaker is reclosed, and the substations are reenergized. The return of voltage for a sufficiently long period resets the fault indicators connected to the sound sections (Fig. 3d).

Advantages

- It is equally applicable for both overhead and underground network.
- It is capable of auto-restoration of transient faults as well as permanent fault.
- Better and efficient Fault Management.

Challenges

- It is not capable to auto-restore network which fall downstream to the fault location.
- Healthiness and reliability of communication network is biggest challenge.
- Time coordination of Circuit Breaker and Load Break Switch should be precise. LBS should open during dead time only.

Decentralized FLIR

It is successfully implemented in STEDIN, a Dutch power distribution utility. This scheme is similar to previous one but having capability to handle all type of fault. It is semi-decentralised scheme uses intelligence at master controllers each of which communicates with a limited number of slave devices. The FLIR algorithm uses messages passed between a numbers of Smart Controllers.

Principle

In the restoration process the faulty section is isolated by opening two LBS (Breaking Node) and the healthy sections are re-energized by closing the normally open point (Making Node) or primary circuit breaker. Hence FLIR algorithm execute in two phases:

Upstream Isolation Phase: During this phase, messages are sent downstream from the feeder CB node to NOP (Normal Open Point) via the Breaking Nodes. As each Breaking Node receives a message, it analyses its own FPI to find out if the fault is upstream of itself. If so, it will open its switches to isolate itself from the fault. If a Breaking Node successfully isolates the fault, then it will forward the message to the Making Node (NOP) with the status "Fault Upstream and Isolated". If this status is received by the Making node, it will close the NOP switch. Hence partial network (below fault region) is restored.

Downstream Isolation Phase: During this phase, a second set of messages are sent back to upstream from NOP to the feeder CB via the Breaking Node. During this phase, each Breaking Node will complete its analysis of whether the fault is downstream of itself. If so, then it will open a switch to isolate on the upstream side of the fault. If a Breaking Node successfully isolates the fault, then it will forward the message to the Making Node (CB) with the status "Fault Downstream and Isolated". If this status is received by the CB then it will re-close the breaker. Hence complete network is restored after isolating faulty network (Fig. 4).

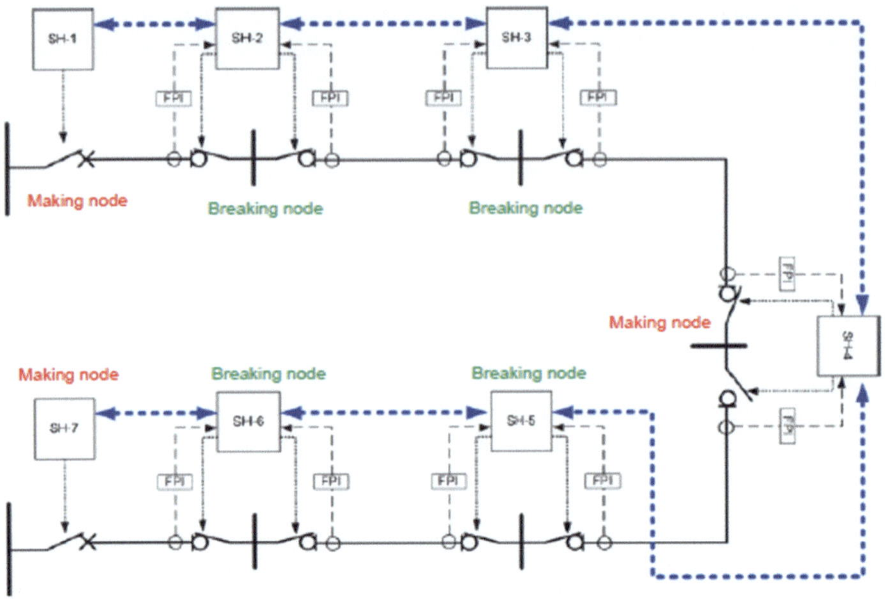

Fig. 4 Decentralized FLIR method (Making and Breaking Nodes)

Advantages

- It is capable of provide complete auto-restoration of network.
- It is capable to handle missing and faulty FPIs.
- Highly robust system as if a switch fails to operate to isolate a fault, then the system will try the next switch.
- Safe operation and minimize chances of wrong operation, as any node is put in local mode, the self-healing scheme is automatically disabled at all the other node also.

Challenges

- It doesn't provide accurate solution in case of more than one NOP that's mean more option to back-feed the load.
- Time coordination of Circuit Breaker and Load Break Switch should be precise. LBS should open during dead time only
- Communication system should be robust and reliable.

Centralized Self-healing System

It is system where automatic switching occur through centralized application (like SCADA, DMS, OMS) used in Control Center for monitoring and control purpose. All these application are efficient to provide fault information and location. SCADA provides information on the basis of telemetered data, DMS provide information on the basis of status of FPI and OMS provide on the basis of prediction logic based on consumer call. ADMS (Advanced Distribution Management System) is recent development in this arena.

Advanced Distribution Management System (ADMS)
ADMS is basically unified platform with one user interface for SCADA, EMS, DMS and OMS equipped with advanced analyzing and planning application. ADMS is ornamented with advanced application which help in fast restoration of distribution network after tripping either in control or advisory mode to squeeze outage duration and ensure safe operation.

Automated Power Restoration System (APRS)
APRS is an algorithmic approach for Fault Detection, Isolation and Restoration (FDIR). APRS uses real time data from network to locate fault. It is capable of executing a sequence of switching actions or can commend the same to isolate the fault and restore power to rest of the healthy network and consumers as many as possible. When a fault occurs on the network and an APRS automation program is invoked, then the following stage occurs:

Fault Information: Information of fault is provided through status of telemetered switches, voltage indicator and analog value of breaker and prediction logic on basis of consumer complaints.

Fault Verification: A downstream trace starts from the tripped device to verify the scope of fault.

Fault Location: The location of the fault is determined based on information provided by telemetered Fault Passing Indicators (FPI). Fault distance is also calculated by ADMS through fault current data received from telemetered feeder breaker and feeder impedance value.

Fault Switching: Switching actions are created or executed to squeeze the extent of the fault in following way:

- Isolate upstream then restore upstream by closing the tripped device.
- Isolate downstream, then attempt to restore each downstream outage by closing best available NOP to shift consumers to neighboring feeder.

3 Tata Power-DDL Challenge

As discussed above, there are substantial improvement in system reliability but there are a lot of challenges also which need to be take care of:

Capital Investment: To set up self-heal network, highly mechanized and automated switches (LBS, Recloser, Sectioniser and Breaker) are required. Also a reliable communication network is needed to sustain the system. All these require huge investment and all we know that capital invested by company will ultimately translate into consumer cost. As a part of regulated business, utilities are neither free to invest at their own wish nor to increase consumer tariff. All investment require prior approval from Regulator.

Communication Reliability: A reliable communication network is required to sustain the system. At present, communication of primary grid substation is connected through fiber optical network but distribution substation is connected through GPRS/ GSM technology. GSM technology is very less reliable and going to be obsolete soon. Laying fiber network to connect distribution substation is very costly job and hence not advisable. So, to improve communication network Tata Power DDL is decided to form their own RF canopy.

Network Diversity: Depending upon situation and site condition, different type and configuration of network had been laid out in Tata Power DDL. In Rural area, maximum circuits are overhead whereas in urban area, maximum circuits are underground. Substantial amount of mixed circuits are also available. Also, some network are radial, but others are in ring and mesh configuration. As we discussed earlier that same method is not applicable for all type of network. So, utmost care should be taken to choose method of self-healing.

Building Confidence: Self-healing will completely change the method to attend fault. It will automate the fault management process. So, Maintenance Engineers and Control Center Engineers will have to adopt the new advanced procedure. But, safety and equipment security are their main concerns. Hence building confidence in these engineers regarding this solution is very important to make it successful. And for developing confidence, regular training and workshop need to be organize for all concern engineers. Moreover, to gain confidence and hands on training of the self-healing grid system a simulator need to be setup in a test environment.

4 Observation

Since it is very expensive system, so it is not advisable to implement it in Pan Network. For optimized and cost-effective implementation, keen attention should be taken to choose the feeder.

High fault prone network: It is suggested to implement Self-heal system in a feeder having high unserved energy due to fault. It will drastically improve reliability of that network as compare to reliable feeder and hence will provide high monetary gain.

High Revenue base network: Network serving industrial area and commercial area is also very much suggested. Since, they are high paid consumers, so same severity of interruption in these network converts into high financial loss as compare to domestic network.

Critical public amenity: It should also be advised for important public amenity like Hospital, Water Works etc. in order to provide uninterrupted power supply to them.

Strategic Location of feeder: It is advisable to automate and motorize LBS at some tactical location only instead of all LBS of complete feeder. As shown in

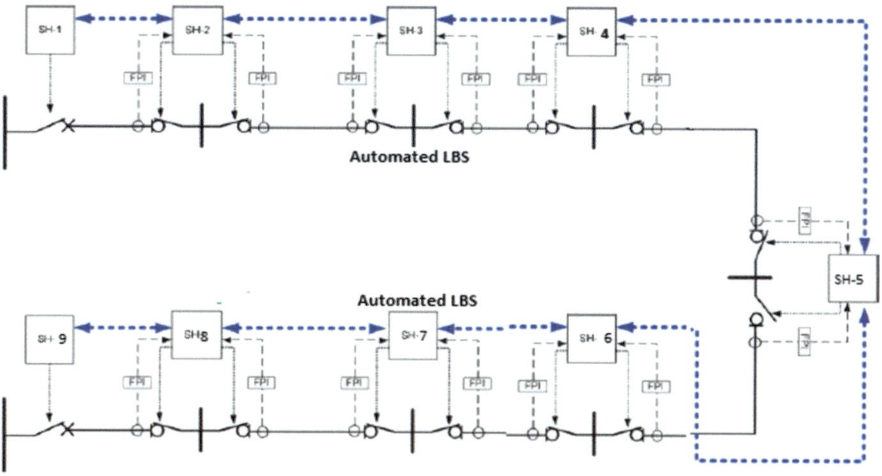

Fig. 5 Strategic location of automated LBS

Fig. 5, there are two alternate feeders having three substations each. And if we automate only three substations i.e. substation having NOP and substation at mid-point of both feeder, then five substation can easily be restored instantaneously through Self-healing. Hence, more than 70% restoration can be achieve only through 30% of automated switches. As substation per feeder will increase, effectiveness will also increase. Hence, it is very much important to choose these strategic points to automate in order to make it cost effective.

Table 1 Comparative study of different methodology of Self-heal distribution network

S. No.	Self-healing method	Network	Fault type	Scope	Communication	Cost
1	Conjunction of auto-recloser and Sectioniser	Overhead	Transient and some extent of permanent fault	Fault isolation and partial supply restoration	No communication	Low cost
2	Pause switching with signal	Overhead and underground	Transient and permanent	Fault isolation and partial supply restoration	Local communication	Medium cost
3	Decentralized FLIR	Underground	Transient and permanent	Complete fault isolation and supply restoration	Local communication	Medium cost
4	ADMS	Overhead and underground	Transient and permanent	Complete fault isolation and supply restoration	Centralized communication	High cost

5 Conclusion

Self-healing distribution network is next step toward Smart Grid and is future of Electrical utility. It will convert sustain outages to momentarily interruption and hence improve reliability and consumer satisfaction tremendously. It will decrease quantum of unserved energy and that is why directly increase fiscal benefits of the utility. There are many technology and methodology available now a day which has their own advantages and disadvantages. (Comparative description of these methods are given in below Table 1). But still there are some major challenges in implementing self-heal grid especially cost of the system which need to be tackle. Some suggestion and observation has been listed above to make it cost effective but still some innovative and economical method need to be evolve, so that it can be implemented in developing nation also.

Smart Metering Solutions for City Gas Distribution Company

Davinder Paul Singh

Abstract Indraprastha Gas Limited (IGL) is supplying piped natural gas to approx. 7.0 lac domestic customers and more than 2800 Industrial and Commercial (I&C) customers in Delhi and NCR. The biggest challenge before any City Gas Distribution (CGD) company is to do the accurate & timely billing of its large customer base which is geographically distributed in wide spread area. Traditionally Bimonthly billing is performed of domestic customer segment. A meter reader visits every domestic customer in its bimonthly billing cycle to get the meter reading. Once the meter reading is collected it is punched in SAP for final billing. All the process is done manually i:e from meter reading collection to Final billing. Manual meter reading collection process is very time consuming and prone to errors. Errors includes reading estimation, wrong reading, manual error and House lock problem. Due to such errors there is monthly reversals done in domestic, I&C segment which is approximately 7 lac SCM/month. Other than that there are issues like not 100% customers get billed. This inaccurate billing have financial impact on Gas reconciliation of any CGD. In order to overcome the Billing related issues in Domestic, Industrial and Commercial segment Automated Meter Reading technology has been implemented in IGL. In this paper in depth study on the various AMR technologies is presented. Author has tried to cover all the pros and Cons of available AMR solutions. Automated Metering Infrastructure (AMI) and AMR solutions presented here are Walk By solution, Drive By solution, Fixed Concentrator type solution, Optical character based solution, And AMR with prepaid solution. Usability, efficacy and requirement of all the above AMR solutions in accord with a CGD is discussed in this paper.

Keywords CGD (City gas distribution) · AMR (Automatic meter Reading) · AMI (Advanced metering Infrastructure) · I&C (Industrial and commercial) · MIU (Meter interface Unit)

D. Paul Singh (✉)
Indraprastha Gas Limited, New Delhi, India
e-mail: davinder.singh@igl.co.in

R. K. Pillai et al. (eds.), *ISGW 2018 Compendium of Technical Papers*,
Lecture Notes in Electrical Engineering 580,
https://doi.org/10.1007/978-981-32-9119-5_27

1 Introduction

In a city gas distribution companies a normal gas meter is used to measure the flow rate and volume of fuel gases consumed by customer. In a CGD sector gas meters are primarily used at residential, Industrial & commercial (I&C) segment.

A smart gas meter in addition to the gas measurement also records & transmits gas consumption as per the programmed time. Time interval can be chosen as per the requirements of CGD, generally it is kept for an hour or daily basis. Smart gas meters communicates gas consumption information at least daily back to the utility for monitoring and billing.

Smart meters enable two-way communication between the meter and the central cloud server. Smart meters are used for remote reporting, daily gas reconciliation & gas leakage detection etc. Smart Automatic meter reading (AMR) system enables two-way communications with the meter. Communications from the meter to the network can be done via fixed connections (such as concentrators) or via wireless (using free radio frequency band).

City gas distribution companies provides gas to large number of residential customers. Normal residential meters comes with the provision of mechanical counter for display. With such meters customer billing is a big issue for any CGD company. Every bi month meter readers as appointed by utility visits each house and collect meter reading for billing. Collected meter reading is then entered in the system for final billing.

Limitations of the traditional method are manifold like.

i. House lock problem during customer visit
ii. Human error in collecting meter reading
iii. Estimate reading for billing.
iv. Delay in billing
v. Time consuming process
vi. High man power required for the billing.

In a smart AMR enabled meter there is a provision of Magnetic, Inductive, Optical or Encoder pulse. A MIU (Meter Interface unit) is attached to the smart meter for transmitting the pulse to cloud server.

2 Smart AMR Solutions

To overcome the limitation of traditional metering system. Smart Automatic meter reading solution emerged with multiple benefits. With the smart metering solutions near real time basis gas reconciliation, checking of gas leakage in the loop, Consumption trends and data analytics become possible. Smart AMR solutions are available in different modes. Type of AMR is decided on the working principle of Meter interface unit fitted in the meter. Following are the broadly classified modes of smart AMR system:

i. Walk by/Drive By (RF based)
ii. Fixed Concentrator (GSM/GPRS based).

3 Walk by Solution/Drive by Solution

Meter Interface unit used in walk by/Drive mode works using Radio frequency for pulse transmission. System requirement is a smart meter with pulse output, a Rf pulse transmitter and a handheld transceiver for communication with the meter. Smart metering solution using this technology required essentially the provision of pulse from the meter. Retroactively AMR device can be installed on the meter index. A hand held device is used to communicate with the meter index using RF signal.

RF signal technology can be based on two way communication or One way communication.

In a two-way communication mode a radio signal is sent to an AMR meter's unique identification number, which wakes the MIU s' transceiver to power-up and transmit its data. The meter transceiver and the reading transceiver both send and receive radio signals.

In one-way communication mode there is continuous radio signal broadcasted. The meter interface unit transmits continuously and data is sent as per programmed time interval. Hand held device in one way communication can be a receiver only, and the meter interface unit works as a transmitter only. Data travels only from the meter interface unit to the hand held device.

The only difference between walk by mode and drive mode is only the programming of hand held unit. In drive by mode there is option for bulk customer data uploading. Meter reader in drive by mode, fix the hand held unit on the vehicle and fetch the data using group data collection module in the hand held device.

Once the data is uploaded in the handheld unit meter reader can push the data on central server using GPRS in the hand held.

In case there is some error in fetching meter reading from MIU. Hand held unit generates alarm for no data collection against the specific MIU in which communication is not established. Meter reader can again retry to fetch and after successful uploading alarm get removed.

Meter reader has two options to transmit meter reading data.

(i) Single point transmission mode: In which after uploading of data from individual MIU to hand held unit, data is transmitted to the server via GPRS from hand held unit.
(ii) Bulk mode: In which data can be transmitted after completing data reading of a wide spread area. Transmission in bulk mode is done once from hand held to central server.

3.1 Advantage of Walk by/Drive by Solution

i. Revenue protection: Accurate and timely billing.
ii. Operational efficiency.
iii. Fast and reliable meter data management.
iv. Easy access to read meters and eliminates home intrusion.
v. Efficient meter reading in high density areas.
vi. Go smart with minimized investment.
vii. Accurate and reliable smart-ready gas metering.
viii. Retrofittable smart radio module.
ix. GPS assisted Mobile meter reading.
x. Secure web based data access from anywhere.
xi. Integrated workforce and asset management.
xii. Sim card required per hand held unit only.
xiii. Meter installation is independent of any factor that can affect its operation.

3.2 Disadvantage of Walk by

(i) Data collection is dependent on manual intervention.
(ii) Daily data cannot be available on server.
(iii) Smart meter with provision of pulse is the essential prerequisite for retrofitting the walk by mode based smart metering solution.

4 Fixed Concentrator Solution

Smart metering solution can be achieved by using fixed concentrator. This system is composed of smart meter, meter interface unit for meter pulse transmission and a fixed concentrator. Meter Interface unit used in fixed network solution operate using radio frequency/LORA WAN for pulse transmission from meter to fixed Concentrator. A fixed concentrator use GPRS/GSM mode to transmit collected data from meters to the central server. System requirement for fixed network is a smart meter with pulse output, Rf pulse transmitting MIU to transmit pulse to fixed concentrator, a fixed concentrator which act as trans-receiver for communication with the meter and transmitting data to the server. Fixed concentrator get installed in public space for covering maximum meters. Separate power supply is required for fixed concentrators.

4.1 Advantage of Fixed Type Solution

(i) Data can be transmitted directly to the server without any manual intervention.
(ii) Daily data can be collected by utility.
(iii) Daily gas reconciliation is possible.
(iv) Entire Network Management is being managed professionally by ISP providing SIM service.

4.2 Disadvantage of Fixed Type Solution

(i) Fixed type solution is useful for High rise buildings only.
(ii) One Concentrator can cater 40–50 smart meters depending on the line of sight.
(iii) It is not economical in case meters are scattered over wide spread geographical area.
(iv) It is not effective solution in low rise residential apartments.
(v) GPRS/GSM communication is dependent on network conditions. In case of network failure data can't be transmitted.
(vi) Extensive radio survey is required per site in order to determine suitable installation position to deploy concentrators.
(vii) Separate power supply is required for concentrator.
(viii) Concentrator is installed in public space.
(ix) Fixed concentrator installation position is dependent on Line of sight from the meter and availability of GPRS signal.

5 Upcoming Technologies

5.1 OCR Based

This technology is based on fixed type Smart metering solution. In this solution meter is retrofitted by Optical character recognition meter interface unit. Depending on the programmed time interval OCR module capture the photo of meter and transmit the meter reading to cloud server. Meter interface units transmits data directly to server using GSM/GPRS network. In this solution per meter a sim card is required which make it a very costly solution.

5.2 *Prepaid Metering*

This is the most advance smart meter solution for utility as it comes with an inbuilt valve. Meter is equipped with Prepaid processor that communicates with the server on GPRS/GSM network. customer can recharge the gas online and when the credit reach zero meter valve gets shut off. Meter display provide the option of flow rate, alarms, available credit on the display.

6 Threats Involved in Smart Metering Solutions

i. Network/Server are vulnerable to potential virus attack & bugs that can manipulate the data.
ii. Increased digital security risk.
iii. Customer usage can be tracked which can cause loss of customer privacy.
iv. Real time data of customer can be tracked. High potential of monitoring by unauthorised entity.

7 Conclusion

Based on the available proven technology, Walk by solution is the most efficient and economical solution for any city gas distribution utility. In CGD sector residential customers are geographically located in a wide spread area therefore walk by solution provides better solution for smart metering as compare to fixed concentrator type solution. Further there is no dependency on network conditions for successful Walk by operations. Using walk by smart metering solution gas distribution companies can do monthly billing of all its customers instead of bimonthly billing.

Acknowledgements Author expresses gratitude to Shri E. S Ranganathan (M.D) IGL and Shri Raman Srivastava (CGM) IGL for their valuable addition during the course of paper writing, although any errors are owned by the author and should not tarnish the reputations of these esteemed professionals.

Electric Vehicle and Energy Storage

Co-creating Responsible Energy Systems

Yvo Thomas Anton Hunink, Linda Manon Kamp and Esther Maria Blom

Abstract Energy system projects in countries like India are often failing. Not only because of technical or economical barriers, but mainly institutional and social issues are at the base of these failures. A co-creation, or participatory, process to align all demands and requirements of the different stakeholders is required. This paper takes evidence from literature on co-creation and energy systems and from case study research in India to help define an approach towards such a co-creation process as a use case for the application of the Responsible Innovation Systems framework. A discussion on co-creation as a solution generates a number of recommendations, after which a set of characteristics is concluded that the co-creation process of energy systems should have towards a responsible approach, so that more robust and sustainable innovations might emerge.

Keywords Co-creation · Participatory process · Responsible innovation systems · Energy systems · Sustainable innovation · Process design

1 Introduction

Worldwide, 1.1 billion people lack access to electricity, of which 276 million are located in India [11]. Kerosene and firewood are among the most used alternatives for energy and can pose serious health threats [3]. The introduction of decentralized

Y. T. A. Hunink (✉)
Responsible Innovation, Energy Bazaar, The Hague, The Netherlands
e-mail: yvo@energybazaar.org

L. M. Kamp
Faculty Technology, Policy and Management, Delft University of Technology, Delft, The Netherlands
e-mail: l.m.kamp@tudelft.nl

E. M. Blom
Delft Centre for Entrepreneurship, Delft University of Technology, Delft, The Netherlands
e-mail: e.m.blom@tudelft.nl

R. K. Pillai et al. (eds.), *ISGW 2018 Compendium of Technical Papers*,
Lecture Notes in Electrical Engineering 580,
https://doi.org/10.1007/978-981-32-9119-5_28

energy resources, such as solar panels, provides a new set of opportunities to improve these numbers and alleviate millions from energy poverty. However, numerous examples are known where energy projects were:

- Not accepted by the communities [17]
- Not providing the necessary improvements in people's lives [7]
- Endangered by expansion of the subsidised, and fossil fuel-based, central public grid [5].

The most apparent reason for failures of energy systems seems to be that the many stakeholders involved in the innovation system, are not sufficiently aligned towards a consensus on how to form the projects. It has become apparent that there are significant influences of social and cultural values involved with energy exchange [19], which are often not accounted for. It stresses that a purely economic model for energy systems does not take into account the demands of a community. Additionally, the research by Comello et al. [5] shows that problems in energy system adoption not only originate from the local level, but also arise from institutional barriers, that can only be solved by policy makers. Also, there are many technologies being proposed for use in energy systems, such as blockchain and artificial intelligence, which are still subject to extensive research and implementation barriers, requiring decision-making activities in both industries and universities.

The main challenge of this research, therefore, is to determine a set of characteristics for a collective innovation process of energy systems, where all these actors are included, shared objectives can be determined, role divisions instigated and collective resources created, so that the activities can be aligned, innovations are more likely accepted and the rate of electrification in the world might go up. The terms co-creation and participatory programs are often found to describe such processes.

2 Methodology

A literature study will expose what is currently known about co-creation, both in the context of India and in relation to energy systems. Thereafter, a relevant theoretical approach is presented that could guide a co-creation process.

A number of field visits to the rural parts of India are used to generate insights on the co-creation processes of energy technologies for these areas. A total of five projects by different organizations with different products are visited and examined on characteristics of the co-creation process in the perspective of the approach explained in the literature study, being:

- Smart meters, Bhodgaya (Bihar), Gram Power
- Solar home systems, Bankey Bazar (Bihar), Rural Spark

- Solar home systems, Badaun (Uttar Pradesh), Rural Spark & Simpa Networks
- Solar irrigation, Daulatpur (Bihar), Claro Energy
- Community microgrid, Buknari (Bihar), Vayam Renewable.

Lastly, a synthesis of the findings is made towards a set of objectives for a co-creation process of energy systems for the context of India.

3 Literature Study

In this section, an overview is given on available research on co-creation processes in India (3.1), co-creation in energy systems (3.2) and the Responsible Innovation Systems approach (3.3).

3.1 Co-creation in India

Across several sectors, participatory programs are found in India. In agriculture, for example, farmer participation in crop variety introduction has proven to successfully increase crop diversity [12]. However, a randomized evaluation on participatory education intervention programs in India sketches a different image. It was exposed that the results of such programs depend on the details of the intervention and the contexts [1]. Three interventions were researched. The results of the interventions, based on the tested level of education of the children, show that the two interventions that were mainly information-based, had negligible effect. The third approach, however, which gave a specific action-based tool, showed significant improvement. Therefore, participatory programs appear more likely to be successful if a clear action plan is provided for participants.

The demand for participatory rural energy planning in India has resulted in a new model for stakeholder's roles, visible in Fig. 1 in the Appendix (Neudoerffer 2001). While since then a slow uptake of energy cooperatives has taken place, such initiatives appear to have helped in reducing distribution losses and electricity theft in India, while also improving billing and revenue collection efficiency [14]. However, while participatory and cooperative initiatives on energy systems in India exist, no exhaustive empirical research could be found to draw extensive conclusions on a best-practice approach, making it valuable to explore other contexts of co-creation in energy systems.

3.2 Co-creation in Energy Systems

Already in 2003, it was found that renewable energy cooperatives in Bangladesh could form the foundation for sustainable rural development, by targeting economic, environmental and community development through the village level supply of energy [18]. A historical example from Canada, however, shows that the successful electrification of the country spanned a period of over 50 years and demanded the active involvement of rural farmers, grouped into cooperatives [20].

Since then, extensive research has been done which has shown that community investment and ownership in energy projects can have far reaching positive impact, but also experiences significant barriers, especially economically and socially [8, 22]. Additionally, a comparative case study of Western energy cooperatives finds that the evolving institutional configuration of an energy sector is a significant factor in the development of community initiatives, where it can both limit and enable community energy projects depending on the configuration [13]. Decentralization of the institutional space and alignment of discourses between stakeholders generally appears to increase the chances for local community players.

In an attempt to create a conceptual framework for understanding the introduction of renewable energy infrastructure in society [23] propose to take a broad perspective, looking at political, market and societal and community acceptance. Devine-Wright et al. [6] later suggest that each of those three need to be expanded and segmented in levels of international, national and local influence and the role of 'middle' actors. Middle actors influence the system both bottom-up, top-down and sideways and often across institutional boundaries [15].

Koirala et al. (2016) take a next step in defining a framework for cooperation in energy systems. After a comparative analysis of different approaches of cooperative energy solutions, they propose the Integrated Community Energy System (ICES), combining the concepts of sustainable energy communities, community energy systems, community microgrids, and peer-to-peer energy. Figure 2 in the Appendix shows what such a system would entail. Central to the ICES is the creation of a local energy market, which would allow solar panel owners to exchange their energy and create notable extra revenues, increasing the attractiveness of buying solar panels instead of traditional grid connection. The ICES model, therefore, appears to be a good foundation that would improve the chances of local adoption of technologies, by creating local energy markets. Still, ICESs are also found to face key barriers in technological, socio-economic, environmental and institutional issues.

It appears that the same barriers keep on returning, but not a single one-size-fits-all solution can be appointed for mitigating such barriers. In complex innovation systems, with many different stakeholders, it can be expected that each situation with different contextual characteristics requires a custom-made process of aligning the activities. This is why a more holistic view needs to be taken.

3.3 *Responsible Innovation Systems*

A new approach towards guiding collective innovation processes is the Responsible Innovation Systems framework, which can help to align stakeholders in an evolving innovation system to come to more responsible and sustainable collective solutions [10]. The framework was tested in the rural energy technology innovation system in India, showing that it can be relevant for this case. The framework is a synthesis of other innovation approaches and consists of 3 elements, the components, relations and functions, which is presented in Fig. 3 of the Appendix.

The components element is composed with the Quadruple Helix [2], which categorizes relevant stakeholders in the four institutions of government, academia, industry and the culture- and media-based public (or civil society). Each institution, defined as a rule-based societal structure and selection mechanism for decision making in innovation processes, is required to be included in the collective innovation process, so that alignment of activities might happen. Each of the institutions creates a specific type of knowledge, respectively, political, human, economic and social 'capital'. The evolutionary characteristic is expressed in the notion that actors can also take over roles of other institutions and create multiple types of capital, effectively creating multi-institutional organizations, which often act as the 'middle' actors explained in Sect. 3.1.

The relations element uses the approach of Open Innovation [4], to describe the knowledge channels within the innovation system and argues that these should be as open as possible, creating two-way communication patterns between actors. It results in a free flow of the different capitals across the institutional boundaries, evidently needed for alignment of activities.

The functions element is largely derived from research by Ranga and Etzkowitz [16], which defined three virtual spaces that are required in a functioning innovation system, namely those of Knowledge, Innovation and Consensus, of which the latter is extended with the Responsible Innovation approach [21]. The Knowledge Space is formed when knowledge is created or shared between actors. The Innovation Space is formed when this knowledge is combined into new products, services or processes. The Consensus Space is formed when the 'dimensions' of Responsible Innovation [21] are found in the activities of actors and their partners. The dimensions can be used to filter innovation processes that might lead to 'collective irresponsibility and are defined as:

- *Anticipation*—Determining future visions, risks, effects, opportunities and situations. Asking the question'What if..?' and setting goals and targets.
- *Reflexivity*—A retrospective view on one's own role and those of others, by holding a mirror up to the activities, commitments and assumptions and how this affects others in society.

- *Inclusion*—Participation of all relevant stakeholders that might influence innovation and are influenced by innovation, in the several phases of the process.
- *Responsiveness*—A capacity to change shape or direction in reaction to changing circumstances.

Only if all dimensions, of which inclusion is already embodied in the components elements, are enhanced in the activities of actors in the system and mutually shared among them, the Consensus Space will form and a responsible collective innovation process is in place. A broad intake of these dimensions is taken, allowing all forms of enhancement to be relevant. Also the dimensions are intertwined and can be in tension with each other, both positively as negatively.

Because the framework was taken to rural India for an initial case study, it has touched upon activities in practice. However, it has not fully left the academic sphere, since it has not yet been used to actually design a collective innovation process and it is subject to a number of recommendations for improvement. Still, a co-creation process as proposed for microgrid design, appears to already be a suitable initial use case to apply the Responsible Innovation Systems framework to, because the process would lend itself for constant iteration of insights along the process.

4 Observations from Fields Visits

In this section, the most notable findings during the field visits are presented. While the interviews described were also used in the original research that instigated the Responsible Innovation Systems framework and a full analysis following the different elements should be sought in the work of Hunink et al. [10], this section takes some observations from those visits that show how the framework can help expose relevant factors for a co-creation process of energy systems.

The smart meters in Bodhgaya, installed by a private DISCOM, had very little co-creation in them. The shop owner that was interviewed, just received this new meter one day. The smart meter had an interface that the shop owner could not access, because he did not possess a smart phone, preventing him from fully exploiting the possibilities of the smart meter platform, such as insights in consumption patterns, of which he was not even aware it existed. It shows that a lack of inclusion of the end user with the activities of the industry partners and closed communication channels, undermines the potential that these smart meters have and resulted in a lack of anticipation, reflexivity and responsiveness.

In the community microgrid, a completely different dynamic was observed. After an initiative of a village member to create a microgrid, 43 of the 200 households expressed their willingness and ability to be part of such a grid. Together with the company building the grid, a custom design was made, taking into account the specific wishes and demands of the community towards a strong Consensus Space. A clear role division was made (reflexivity), communication channels were kept open (inclusion) and an action plan was set up in case of calamities (responsiveness). Looking at the ICES model, explained in Sect. 3.2, this project showed clear community generation and energy management. At the time of interviewing, 1.5 years after installment, the villagers expressed their happiness with the system, the sense of ownership they had in the project and were even actively involved in further innovating the system. They were now trying to persuade the company to expand the system with a running water system powered by solar panels (anticipation and Innovation Space), also showing some human capital creation. Several elements, therefore, were clearly visible. A missing element was the role of the government, showing that this successful project could be vulnerable to changes in the political environment due to a lack of political capital.

The solar home systems in Badaun exposed how a disconnected government, can play a limiting role. While nearby villages had a connection to the central grid, this particular village was not connected. Upon asking why this was, the villagers explained that in the previous election they had voted for the losing party, which has damaged the connections they had to local government. It prevented them from being included in the electrification process and forced them to individually rely on solar home systems instead, preventing the local community from having collective reflexivity and responsiveness, leaving them only to anticipate on the future. With upcoming elections, as they expressed, they would vote for the party in favor of winning, so they would be connected to the grid this time. This would mean the continuity of using and expanding the solar home systems was not ensured, all because of the closed communication and limiting role of the government.

The solar irrigation plant in Daulatpur showed that effective inclusion of government, however, can help in creating a robust project. The irrigation department was explicitly involved with the design of the projects, together with the industry partner and a cooperative of farmers. The project was designed to run for five years and had all this time been going according to the plan that was set up with all those actors. The requirements of each involved institution were included and aligned, showing a strong Consensus Space. Additionally, the effect of inclusion on the other dimensions of Responsible Innovation became apparent. Initially the inclusion was high, resulting in clear role divisions (reflexivity), a 5-year plan (anticipation) and effective billing procedures organized by the farmer community themselves (responsiveness). At the point of interviewing, however, the 5-year contract was almost ending and the farmer was not being included in any conversations regarding the continuity, resulting in doubts about his role, the future and how to respond when the contract ends.

Finally, the solar home systems in Bankey Bazar revealed what the impact can be of taking over the role of human capital creation, originally being a responsibility of the academic institution. By internalizing the skill of entrepreneurship into the product, by adding a set of batteries that can be rented out, the users were found to create new competencies and generate additional income. This characteristic of the product was added after a co-creation process where they observed how people were using lamps and batteries they distributed amongst communities. It shows strong signs of the Knowledge and Innovation Spaces at work and has led to rapid iteration of the innovation process to better suit the needs of the end user, while also increasing the agency to come up with new solutions in using the product at their will. The Consensus Space was also visibly enhanced. The product increased the reflexivity of end users, by adopting a new role in their community of energy supplier, also creating social and economic capital. It enhanced anticipation, because the future changes to the product actively influenced their businesses. Ultimately their responsiveness increased by earning additional income and being able to structure their business model according to local situations. The ICES model is largely visible, since the community takes up generation, management and storage of energy, while also creating a local energy market.

5 Discussion and Recommendations

The observations during the field visits show that the projects that contained more elements of the Responsible Innovation Systems framework in their (co-creation) process, could be seen to harness the potential of the situation much better and generally create solutions that are easier adapted. The sense of ownership and willingness to expand on the existing infrastructure was increased, allowing a faster innovation process to take place, with a better chance of acceptance by the end user. While co-creation was seen to potentially have meaningful benefits, this does not automatically mean that co-creation or participation is a silver bullet, as the research by Banerjee et al. [1] showed in Sect. 3.1. Several things can be noted that might obstruct a smooth process

5.1 Design of the Process

An important part of the co-creation process is taking the time to get to know each other, so that the shared or conflicting values and requirements can be determined and better aligned and trust between the stakeholders can be established. However, the amount of time and resources spent can increase significantly and without results coming in, can actually reduce the responsiveness of actors. It should be recommended that a clear timeline and resource allocation is anticipated, so that negative consequences can be limited to a minimum. This does not mean that each phase and stage of the process should be designed up front and completely followed. A co-creation process rather is a complex set of activities in which each phase and stage is dependent on the previous. The complete co-creation process, therefore, is unclear at the beginning, but two key phases should always be there, in which all institutions are at least included.

First, an exploration should be used to carefully create the specific shared and conflicting demands of the stakeholders in the co-creation process. It will also help define the initial necessary contents of the process. Since it is of outmost importance to know the social and cultural dynamics within communities that participate in a co-creation process, a necessary stage within the exploration phase is the ethnographic study. Here the socio-cultural context should be revealed, where values and objectives of all the different stakeholders can be determined and the right strategy towards the process can be derived. For example [9] conclude that community projects are more likely to succeed when there is already a sense of community present. Conflicting interests within communities can seriously impede the co-creation process in relation to the developers building the energy system. Therefore, whenever such community sense is not there, the goal should be first to create this.

Secondly, a reflection phase must evaluate the process at the end, letting all actors come together and discuss relevant topics, keeping an open knowledge environment. In between these phases, the contextual characteristics of the project will determine what the rest of the process should look like. It should be evaluated whether separate stages are required and if varying involvement of stakeholders is needed. However, after each stage and phase, clear follow-up activities with an action plan should be formed, in line with [1], to ensure that each session is processed and shall be communicated to all the stakeholders, so that they can utilize the outcomes of the sessions to a maximum, also relating to responsiveness. Additionally, it is advised that the continuous design of the process remains subjected to academic research. Since it is new terrain that is explored, research questions should be set up that see to new insights on the constant monitoring and improvement of the process.

5.2 Role of the Facilitator

In the model from Neudoerffer (2001) in Fig. 1, each of the required institutions, according to the Responsible Innovation System, is already visible, however they are only indirectly in contact with each other. This leaves the possibility of demands and requirements of each actor to be missed, overruled or corrupted. In the projects with end users, it was impossible to retrieve the workings of this structure from the interviews done, with only the relation between villages and product manufacturers observed. The model appears static and does not allow for flexible approach towards the role division and the interaction between actors, questioning its usability.

The role of disseminator, or facilitator, is only between NGOs and government in the model. This role, however, could be expanded to act in the middle of all institutions, including industry and academic research institutes. This could be in the form of facilitating the joint co-creation sessions, or simply to take the role of a missing institution whenever decision making on innovations is happening. Therefore, this facilitator effectively becomes the 'middle' actor, as described in Sect. 3.1. It would instigate a different type of process, where the initial role divisions and interactions are not predetermined, but rather evolve from co-creation in an initial stage, as seen in Fig. 4. It would create a process that is able to uptake contextual influences much better and accordingly adjust.

The facilitator should be in the position to help design the process and ensure that the components of the Responsible Innovation Systems framework are included. It should organize the stakeholders, educate them about the importance of co-creation and sketch the path of the process. A large skillset on technical elements of energy systems and the Indian innovation system surrounding the development of the sector and its policies will help to fulfil such a task.

However, communicating with rural end users should always be done by someone that is trusted by the community and has the knowledge on specific socio-cultural contexts. Taking an active role in a co-creation setting with participants that do not trust the facilitator could be more harmful than helpful, due to misjudgment of nuances, and should therefore be minimized. A mediator should be present during the co-creation sessions who is respected and known to all stakeholders, but not feared in such a way that the openness of the conversation is endangered.

It is therefore that the facilitator should allows others to take a leading role during co-creation sessions in the different phases and stages, wherever the contextual characteristics require such. In such cases, the facilitator preferably also designs the session in cooperation with that actor. Similarly, to inclusion of end users, it will increase ownership of the process among all stakeholders. However, the exact role of the facilitator is not completely clear yet and should be subjected to further research.

5.3 Local Energy Markets

The ICES model (Koirala et al. 2016), explained in Sect. 3.2, is seen as the most integrated approach towards cooperative energy grids. However, the model was said to face key barriers in technological, socio-economic, environmental and institutional issues. The Responsible Innovation Framework is meant to prevent such barriers by determining the characteristics of a collective innovation process. From the field visits, it appeared that the projects with the most signs of the elements of the framework (Buknari and Badaun), were also the ones with visible parts of ICES model. It could be a hint that a co-creation process towards an ICES that applies the Responsible Innovation Systems framework has an enlarged change of success, because it mitigates the barriers for ICES adoption. It is recommended to explore such a combination.

6 Conclusion

This research aimed to gain answers on what a co-creation process for the formation of energy systems in India should look like. After a literature study, several characteristics for co-creating energy systems in India were found and it was determined that the Responsible Innovation Systems framework poses as a suitable guidance for the design of a co-creation process with the help of a number of field visits. The characteristics for the co-creation process of energy systems are concluded to be the following:

- Inclusion of all four institutions of government, industry, academia and civil society from the local, national and international level.
- An open knowledge sharing environment.
- A multi-phase process with a clear timeline and resource allocation, always starting with an exploration phase and ending with a reflection phase in which all institutions are present, where each phase is dependent of the previous one and can have multiple stages.
- In the exploration phase, an ethnographic analysis of the community in which the energy system is introduced is required as a separate stage. Here, at least the sense of community within the selected group is determined.
- Knowledge production and sharing is enabled, while innovation of new products, services and processes is fostered.
- Activities for enhancement of anticipation, reflexivity and responsiveness are shared among the stakeholders of the process.
- Creation of a local energy market, following the ICES model, after which further integration in the energy system can take place.
- A flexible role for a facilitator to help design and guide the co-creation process, while also being able to step down from the leading role when the contextual circumstances require such.
- Create clear action plans for stakeholders after each stage and phase, to ensure further development.
- Ending in a consensus forming between all parties on the creation of community energy systems.

Appendix

See Figs. 1, 2, 3 and 4

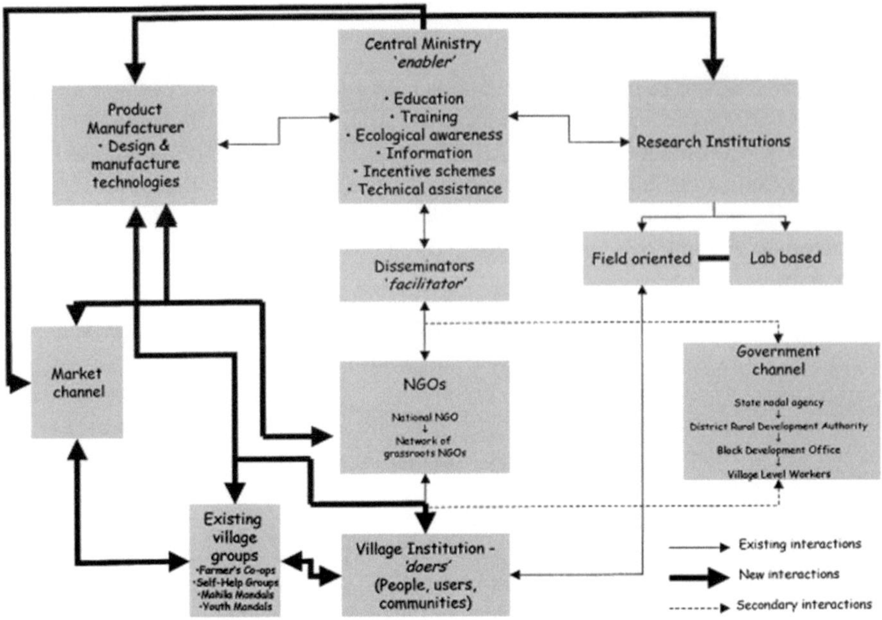

Fig. 1 Model for new stakeholder roles in rural energy planning in India (Neudoerffer 2001)

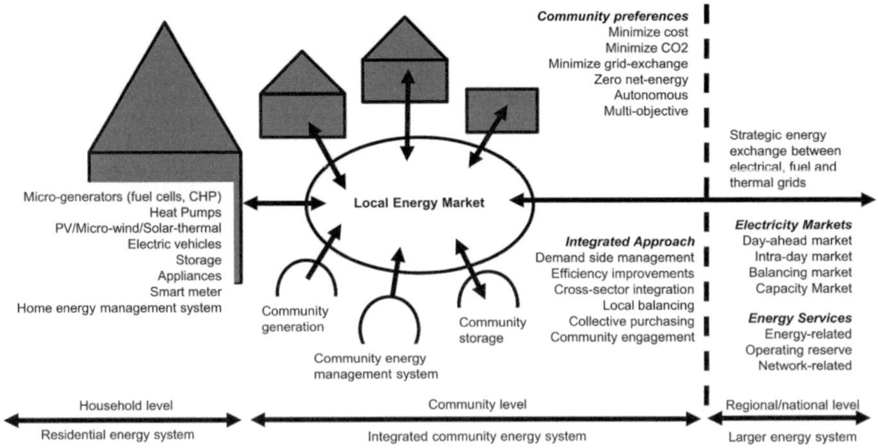

Fig. 2 Functions of an integrated community energy system (ICES) in the larger system (Koirala et al. 2016)

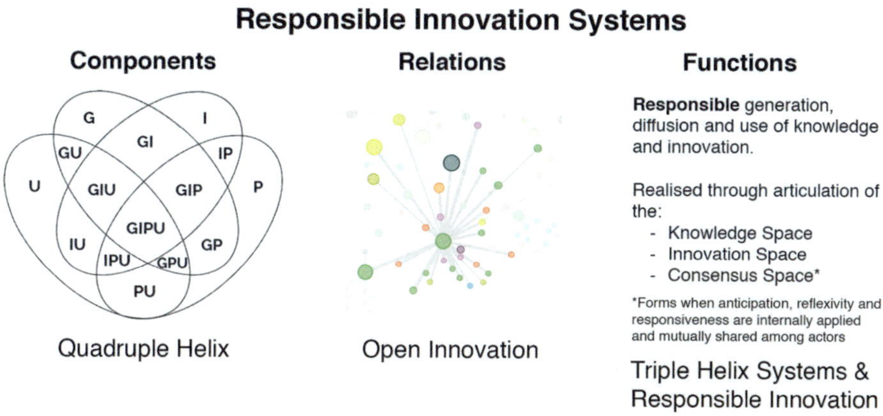

Fig. 3 A synthetic representation of the Responsible Innovation Systems framework, with G, I, P & U, respectively, indicating government, industry, public (civil society) and university (academia) [10]

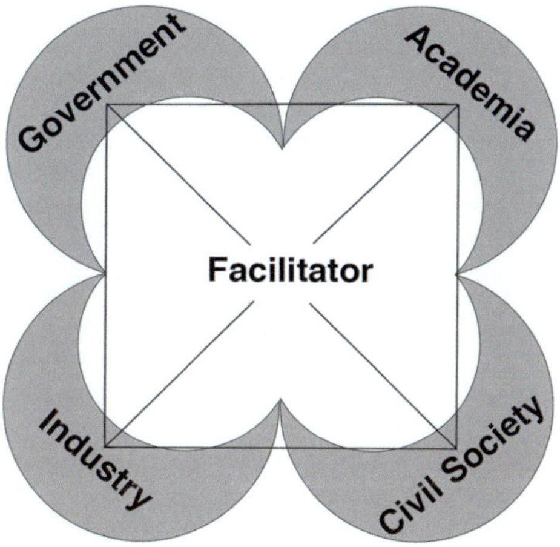

Fig. 4 Proposed institutional division of stakeholders for the initial phase co-creation processes of ICESs, with each line depicting a potential interaction for later phases. Preferably, local, national and international stakeholders should be included

References

1. Banerjee AV, Banerji R, Duflo E, Glennerster R, Khemani S (2010) Pitfalls of participatory programs: Evidence from a randomized evaluation in education in India. Am Econ J: Econ Policy 2(1):1–30
2. Carayannis EG, Campbell DF (2009) 'Mode 3' and 'quadruple helix': toward a 21st century fractal innovation ecosystem. Int J Technol Manage 46(3–4):201–234
3. CEEW (2015) Access to clean cooking energy and electricity. Technical report
4. Chesbrough HW (2006) Open innovation: the new imperative for creating and profiting from technology. Harvard Business Press
5. Comello SD, Reichelstein SJ, Sahoo A, Schmidt TS (2017) Enabling mini-grid development in rural india. World Dev 93:94–107

6. Devine-Wright P, Batel S, Aas O, Sovacool B, Labelle M, Ruud A (2017) A conceptual framework for understanding the social acceptance of energy infrastructure: Insights from energy storage. Energy Policy 107:27–31. https://doi.org/10.1016/j.enpol.2017.04.020
7. Fernández-Baldor Á, Lillo P, Boni A (2015) Gender, energy, and inequalities: a capabilities approach analysis of renewable electrification projects in Peru. In: Sustainable access to energy in the global south. Springer, Cham, pp 193–204
8. Haggett C, Aitken M (2015) Grassroots energy innovations: the role of community ownership and investment. Curr Sustain/Renew Energy Rep 2(3):98–104
9. Haggett C, Creamer E, Harnmeijer J, Parsons M, Bomberg E (2013) Community energy in Scotland: the social factors for success. University of Edinburgh, Edinburgh
10. Hunink YTA, Kamp LM, Blom E (2018) Developing responsible innovation systems: the case of the rural energy sector in India. Working paper awaiting publication
11. International Energy Agency (2017) World energy outlook 2017 data on energy access. Retrieved on 04-12-17 from: http://www.iea.org/
12. Joshi A, Witcombe JR (1996) Farmer participatory crop improvement. II. Participatory varietal selection, a case study in India. Exp Agric 32(4):461–477
13. Oteman M, Wiering M, Helderman JK (2014) The institutional space of community initiatives for renewable energy: a comparative case study of the Netherlands, Germany and Denmark. Energy Sustain Soc 4(1):11
14. Palit D, Chaurey A (2011) Off-grid rural electrification experiences from South Asia: status and best practices. Energy Sustain Dev 15(3):266–276
15. Parag Y, Janda KB (2014) More than filler: Middle actors and socio-technical change in the energy system from the “middle-out”. Energy Res Soc Sci 3:102–112
16. Ranga M, Etzkowitz H (2013) Triple helix systems: an analytical framework for innovation policy and practice in the knowledge society. Ind High Educ 27(4):237–262
17. Ravindra K, Kannan B, Ramappa N (2014) Microgrids: a value-based paradigm: the need for the redefinition of microgrids. IEEE Electrification Magazine 2(1):20–29
18. Siddiqui FA (2003) Linking innovation and local uptake in rural development: potential for renewable energy cooperatives in Bangladesh (Doctoral dissertation, Murdoch University)
19. Singh A, Strating AT, Herrera NR, van Dijk HW, Keyson DV (2017) Towards an ethnography of electrification in rural India: Social relations and values in household energy exchanges. Energy Res Soc Sci 30:103–115
20. Smallridge CG (1999) Rural energy and development: a study of canadian rural electrification. CG Smallridge & Associates, Ottawa, Canada
21. Stilgoe J, Owen R, Macnaghten P (2013) Developing a framework for responsible innovation. Res Policy 42(9):1568–1580
22. Viardot E (2013) The role of cooperatives in overcoming the barriers to adoption of renewable energy. Energy Policy 63:756–764
23. Wüstenhagen R, Wolsink M, Bürer MJ (2007) Social acceptance of renewable energy innovation: an introduction to the concept. Energy Policy 35:2683–2691

Estimating Effect of Electric Vehicles on Indian Grid System and Identifying Consequent Opportunities

Nisargkumar Suthar

Abstract In context to the burning issues of rising fuel demand as well a step towards cleaner energy with lesser carbon emission targets (COP21, Paris Agreement), continuous efforts are made by Indian government for a shift towards E-mobility. Grand targets are set in transport sector, under mobility mission plan, India aims to bring in four lakh passenger electric vehicles (EVs) by 2020 and to go almost 100% electric cars by 2030 is supported by incentive schemes like FAME. The research paper focuses on predicting the growth of electric vehicles in different phases of timeline up to 2030 under various conditions. Further, technical evaluation of the effect of this rise of E-vehicles on Indian grid system at different penetration levels is made in respect to corresponding electricity consumption. Various opportunities to grid and market due to integration of large number of Electric vehicles are discussed in the paper with potential benefits of concept like V2G, and conveniently utilising upcoming quantum of intermittent energy from renewable sources. The challenges due to integration of vehicles and recommendations are also discussed. Possible business models to drive easy growth and smooth integration are outlined.

Keywords Vehicle grid integration · Estimating electrical consumption · Grid impact

1 Introduction

The world is moving towards Electric mobility and electrification of transport plays a large role in all scenarios aiming to achieve the decarbonisation of the electricity sector. Along with ongoing trend of shifting towards Electric Vehicles (EVs) worldwide, the Indian Government has also announced an ambitious goal to switch to electric cars by 2030 and phase out diesel and petrol cars. Significant

---

N. Suthar (✉)
National Power Training Institute, Faridabad, India
e-mail: nsrgsuthar@gmail.com

R. K. Pillai et al. (eds.), *ISGW 2018 Compendium of Technical Papers*,
Lecture Notes in Electrical Engineering 580,
https://doi.org/10.1007/978-981-32-9119-5_29

initiatives are being taken to achieve this objective, including the launch of the National Electric Mobility Mission Plan (NEEMP).

The report in brief discusses about the evolution of Electric Vehicles in country till date and focuses on existing and future technologies related to EVs as well as the Battery technology that accounts as one of the important parameter for growth of EVs in the country.

As India aims to bring in 4,00,000 passenger electric vehicles by year 2020 and to go almost 100% electric cars by 2030 that are supported by incentive schemes like FAME [1, 2]. The report focuses on predicting the growth of electric vehicles in different phases of timeline up to 2030 under various conditions. A parallel projection of the growth of Electricity demand is made and relation and impact of the growing electric vehicles in the grid is established. Further, technical evaluation of the effect of this rise in proliferation of E-vehicles on Indian grid system at different penetration levels is made in respect to corresponding electricity consumption. Various opportunities to grid and market due to integration of large number of Electric vehicles are discussed in the paper with potential benefits of concept like V2G, and conveniently utilising upcoming quantum of intermittent energy from renewable sources. The challenges due to integration of vehicles and recommendations are also discussed.

Furthermore, there is substantial need to promote balanced and sustainable growth of Electric Vehicles in India. Based on the understanding by interactions, surveys and interviews with people part of current value chain of EV market in India, a Pilot Business Model has also been outlined to drive easy growth and smooth integration of upcoming EVs. The business model is based on concept of "Battery swapping" that can prove to be successful as per current market situations prevailing in country.

The study brings out the possible effects of proliferation of Electric Vehicles to the Indian grid and recommendations on different grounds such as Regulations, Vehicle Grid Integration (VGI), Charging Infrastructure etc.

2 The Growth in Deployment of Electric Vehicles

Considering global scenario, the transportation systems are in many cases are mainly dependent on oil. Also, they result in considerable emissions of greenhouse and other harmful gasses. Energy policies in time today are very much inclined towards energy efficiency improvement, reduction of emissions and diversification of transportation fuels. Electric vehicles have good capability to fulfil challenging requirements of the grid. Transportation sector in India is responsible for considerable (around 70%) consumption of petroleum products.

2.1 The Adoption Rate of EV

India is heavily dependent on foreign nations for increasing demand of oil that has direct impact on GDP (Gross Domestic Product) of the country. So, few years down the line it is expected that the adoption and deployment of electric vehicles will considerably increase. As proliferation of EVs will rise, India may face various issues due to charging of vehicle on existing Indian power grid. Effects due to charging these vehicles may lead to compromise in quality of power, harmonics generation, high transformer losses, rising copper losses, issues of equipment heating in power line and high reactive power consumption. As a result of all these effects, potential threat in damage of customer devices and power system equipment may be observed. These issues may lead to severe problems in power system, voltage stability problems, and even collapses.

As per NITI Ayog following are the Elements of India's Mobility Transformation that could help the country in achieving the goals related to electric mobility.

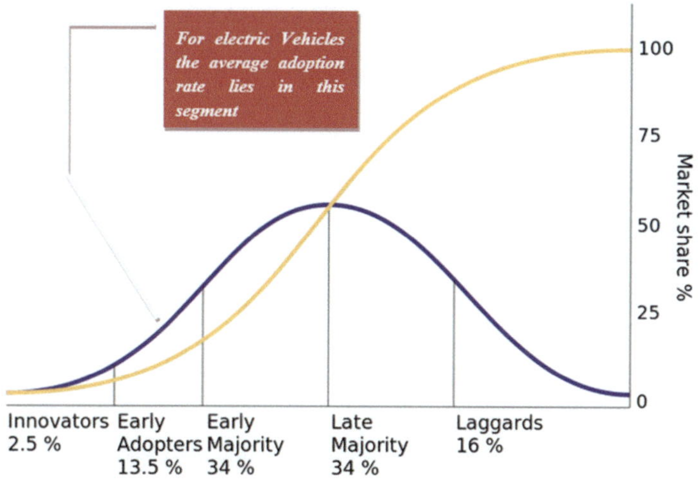

2.2 EV's Penetration and Consume

Year	2008	2009	2010	2011	2012	2013	2014	2015	2016
Stock	370	530	880	1330	2760	2950	3350	4350	4800
New Registrations	370	160	350	450	1430	190	410	1000	450
Market Share	0.02%	0.01%	0.01%	0.02%	0.05%	0.01%	0.02%	0.04%	0.02%

The Table above shows the details of current stock, New Registrations and Market share of Electric Vehicles in India (Source: IEA 2017)

2.3 Total Vehicle Stock in 2030

It is possible to forecast the stocks of total vehicles in coming time with help of mathematical calculations. Initially taking the present number of vehicles registered in country, as per the Ministry of Road Transport and Highways.

[Note: *X*M = *X* millions | *X*B = *X* billions]

	2000	2005	2010	2013
Two-wheelers	34	59	92	133
Cars, Jeeps, and Taxis	6	10	17	25
Buses	0.6	0.9	1.5	1.9
Goods Vehicles	3	4	6	9
Others	5	7	11	15
Total	49	81	128	182

Source: (MORTH, 2015)

Due to unavailability of data of the registered vehicle for later years in the public domain, the cumulative numbers above do not take into consideration the vehicles that have retired. According to (Guttikunda and Mohan 2014), not more than 70% of the given vehicles are operational on roads in reality. Hence, it can be assumed that total active vehicle stock in year 2013 will be approximately 128 M. This count includes about 93 M two-wheeler vehicles and 17 M cars. Then based on historical trends, it can be projected that the sales of vehicle up to 2030. The historical data of vehicle sales is taken from the Society of Indian Automobile Manufacturers (SIAM) as given in table below:

Category	2000	2005	2010	2015
Passenger Vehicles (Cars)	0.6	1.1	2.0	2.6
Two-wheelers	3.8	6.6	9.4	16.0

Source: (SIAM, 2011, 2016)

It can be clearly observed from above tables that vehicle sales have been growing at rapid pace. In between year 2000 and year 2015, the two-wheeler and car sales have increased more than four times. This indicates a 10% compound annual growth rate (CAGR) in particular year. It is also observed that number of registered vehicles have quadrupled.

In order to project the vehicle stock in the future, the use of simplified vehicle stock turnover model is done. It is assumed that sales of cars sales will keep growing and also historical rate, leading to sales nearly quadrupling in 15 years. For two-wheelers, it can be assumed that the rate of sales would slow down due to raised incomes especially in city areas. Two-wheeler sales will double from year 2015 to 2030. The assimilation is inline with industry predictions and some other studies. For e.g. 'Guttikunda and Mohan 2014'. It is assumed that all the cars and two-wheeler vehicles will have a 15 year life span. Below table shows forecast for total sales of vehicle and total active vehicle stock up to year 2030.

		2015	2020	2025	2030
Sales	Two-wheelers	16	20	26	32
	Cars	2.6	4.1	6.5	10
Active stock	Two-wheelers	122	204	289	367
	Cars	22	38	59	89

2.4 *BEV Stock in 2030*

The assumptions are made that by year 2030, all cars and two-wheelers (light duty passenger vehicles) sold in the country are BEVs. In India, sales of BEV in year 2015 were as small as about 20,000. It is assumed that growth of BEV sales on log-linear basis, with rate of growth altering at a term of 5 years each from 2015 to 2030. As per the table given below, it is assumed that the total business of vehicle as well as the stock will be unaltered even if customers select regular internal combustion (IC) engine vehicles or BEVs. Hence, with growing sales of BEV, the sales of IC engine vehicle reduce. So, till year 2030, if BEVs become responsible for almost 100% of vehicle sale in country as per table given below, to achieve the goal of 100%

vehicle electrification, the sales of two-wheeler vehicles BEV should rise against approximately 2000 in year 2015 to approximately 32 M in 2030.

Vehicle Sales (millions)

	Two-wheelers				Cars			
	2015	2020	2025	2030	2015	2020	2025	2030
ICE	16.0	19.7	20.2	0.0	2.6	3.8	4.1	0.0
BEV	0.0	0.5	5.4	32.3	0.0	0.3	2.4	10.3
Total	16.0	20.2	25.6	32.3	2.6	4.1	6.5	10.3

BEV car sales should increase from about 20,000 in year 2015 to approximately 10 M during year 2030.

The table below depicts the stocks of current BEV and IC engine vehicles by the year 2030. BE vehicles will show approximately 44% of the total current car stock and about 29% of the total active two-wheeler stock by the end of year 2030. By mid-2020 full vehicle stock electrification could be expected to realise, if total IC engine vehicles bought prior year 2030 will retire [3].

Active Vehicle Stock (millions)

	Two-wheelers				Cars			
	2015	2020	2025	2030	2015	2020	2025	2030
ICE	121.9	203.5	275.9	262.0	22.2	37.0	51.7	49.9
BEV	0.0	0.7	13.6	105.1	0.0	0.6	7.0	39.0
Total	121.9	204.2	289.5	367.1	22.2	37.7	58.6	88.9

3 Impact to Grid in 2030

On the basis of above calculations, the active stock of Electric vehicles in 2030 is determined. Now, in order to identify the effect of these vehicles and their load impact on grid following calculations can be undergone:

The Average Watt-Hour capacity for Electric Vehicles in Wh/Km for three major categorisations i.e. LCVS, two wheelers and three wheelers are as given in the table below:

WATT-HOUR CAPACITY FOR ELECTRIC VEHICLES Wh/Km

LCVS	Two wheelers	Three wheelers
250	37	135

Considering 40 kms of average travel by each vehicle, moving to further calculation process:

Approximate Travel per Day Kms	40

On basis of Avg distance travelled, and projected number of vehicles, the total travel distance in kilometres can be identified.

No. Of Vehicles X Avg Kms X 365 = Total Travel Kms in Years

Load Consumption (Per Category) = Total Travel Kms X Watt Hour Capacity

Now, considering different penetration levels, the yearly consumption in GWh is calculated in the table below. Considering Peak load of 402 GW in 2030, the percentage effect due to EVs in all penetration cases is given below.

Load Impact on Grid due to Electric Vehicles in 2030 (With different levels of penetration)				
Vehicle Type	LCVS	Two wheelers	Three wheelers	
Watt Hour Capacity of Electric Vehicles (Wh/Km)	250	37	135	
Approximate Travel per Day Kms	40	40	40	
Projected Number of Vehicles in 2030	39000000	105100000	3060000	
Total Travel Kms in Year	14600	14600	14600	
Penetration in 2030	Yearly Consumption GWh			
10%	14235.00	5677.50	603.13	
25%	35587.50	14193.76	1507.82	
50%	71175.00	28387.51	3015.63	
75%	106762.50	42581.27	4523.45	% of Peak Load
100%	142350.00	56775.02	6031.26	
Total Peak Load by 2030 (GW)	402	402	402	
Load in case of 10% penetration (GW)	1.63	0.65	0.07	0.58%
Load in case of 25% penetration (GW)	4.06	1.62	0.17	1.46%
Load in case of 50% penetration (GW)	8.13	3.24	0.34	2.91%
Load in case of 75% penetration (GW)	12.19	4.86	0.52	4.37%
Load in case of 100% penetration (GW)	16.25	6.48	0.69	5.83%
% of Peak Load	4.04%	1.61%	0.17%	5.83%

Hence, Total Load on Grid due to Electric Vehicles in 2030 is

4.04% + 1.61% + 0.17% = 5.83%, i.e. not even exceeding **6%** of total peak (402 GW).

4 Significance of EV Deployment

4.1 Gain to BEV via Cost Benefit Analysis

Cost Benifit Analysis of Mahindra E2O					
Car Model	**Mahindra E20** P4	**Mahindra E20** P6	**Mahindra E20** P8	**Swift** zxi	**Hyundai grand i10** 1.2 CDRI Asta diesel
Car Specifications					
Fuel	Electric	Electric	Electric	Petrol	Diesel
Transmission	Direct drive	Direct drive	Direct drive	Manual	Manual
Length(mm)	3590	3590	3590	3850	3765
Width(mm)	1575	1575	1575	1695	1660
Hieght(mm)	1585	1585	1585	1530	1520
Weight(Kg)	932	940	990	965	935
Power	70 Nm @1000rpm	70 Nm @1000rpm	91 Nm @1000rpm	115 Nm @4000 rpm	190 Nm @ 1750-2250 rpm
Initial Costs and Subsidies					
Total Price including Insurance, Road Tax, Registration	₹ 9,75,868.00	₹ 10,59,195.00	₹ 14,18,742.00	₹ 7,28,790.00	₹ 8,44,604.00
FAME Subsidy	₹ 1,24,000.00	₹ 1,24,000.00	₹ 1,24,000.00	N/A	N/A
Delhi Subsidy	₹ 1,50,000.00	₹ 1,50,000.00	₹ 1,50,000.00	N/A	N/A
On Road Price(New Delhi)	₹ 7,01,868.00	₹ 7,85,195.00	₹ 11,44,742.00	₹ 7,28,790.00	₹ 8,44,604.00
Average Driving Distances					
Kms per day	45				
Kms per month	1350				
Kms per year	16200				
Operating Cost Calculations					
Cost per unit kwh / Liter *(Rs)*	7.44	7.44	7.44	65	55
Mileage per unit *(km-p-kwh / kpl)*	10	10	8.4	15.6	21
On-Road Price *(Lakhs)*	7.01	7.85	11.44	7.28	8.44
Cost per km *(Rs)*	0.74	0.74	0.89	4.17	2.62
Service cost per year *(Rs)*	₹ 2,500.00	₹ 2,500.00	₹ 2,500.00	₹ 3,500.00	₹ 5,000.00
Battery cost for 4th year *(Rs)*	₹ 2,00,000.00	₹ 2,00,000.00	₹ 2,00,000.00	₹ -	₹ -
Total cost of ownership over for 10 year period(Rs) (With one time replacement of battery)	₹ **10,47,396.00**	₹ **11,30,723.00**	₹ **15,13,227.71**	₹ **14,38,790.00**	₹ **13,18,889.71**
Total cost of ownership over for 10 year period(Rs) (With no replacement of battery)	₹ **8,47,396.00**	₹ **9,30,723.00**	₹ **13,13,227.71**	₹ **14,38,790.00**	₹ **13,18,889.71**

Total cost of ownership over for 10 year period(Rs) (With one time replacement of battery)											
	Year 1	Year 2	Year 3	Year 4	Year 5	Year 6	Year 7	Year 8	Year 9	Year 10	Total Cost of Owning
Mahindra E20 p4	₹7,16,420.80	₹14,552.80	₹14,552.80	₹2,14,552.80	₹14,552.80	₹14,552.80	₹14,552.80	₹14,552.80	₹14,552.80	₹14,552.80	₹10,47,396.00
Mahindra E20 p6	₹7,99,747.80	₹14,552.80	₹14,552.80	₹2,14,552.80	₹14,552.80	₹14,552.80	₹14,552.80	₹14,552.80	₹14,552.80	₹14,552.80	₹11,30,723.00
Mahindra E20 p8	₹11,61,590.57	₹16,848.57	₹16,848.57	₹2,16,848.57	₹16,848.57	₹16,848.57	₹16,848.57	₹16,848.57	₹16,848.57	₹16,848.57	₹15,13,227.71
Swift	₹7,99,790.00	₹71,000.00	₹71,000.00	₹71,000.00	₹71,000.00	₹71,000.00	₹71,000.00	₹71,000.00	₹71,000.00	₹71,000.00	₹14,38,790.00
Hyundai grand i10	₹8,92,032.57	₹47,428.57	₹47,428.57	₹47,428.57	₹47,428.57	₹47,428.57	₹47,428.57	₹47,428.57	₹47,428.57	₹47,428.57	₹13,18,889.71

Total cost of ownership over for 10 year period(Rs) (With no replacement of battery)											
	1	2	3	4	5	6	7	8	9	10	Total Cost of Owning
Mahindra E20 p4	₹7,16,420.80	₹14,552.80	₹14,552.80	₹14,552.80	₹14,552.80	₹14,552.80	₹14,552.80	₹14,552.80	₹14,552.80	₹14,552.80	₹8,47,396.00
Mahindra E20 p6	₹7,99,747.80	₹14,552.80	₹14,552.80	₹14,552.80	₹14,552.80	₹14,552.80	₹14,552.80	₹14,552.80	₹14,552.80	₹14,552.80	₹9,30,723.00
Mahindra E20 p8	₹11,61,590.57	₹16,848.57	₹16,848.57	₹16,848.57	₹16,848.57	₹16,848.57	₹16,848.57	₹16,848.57	₹16,848.57	₹16,848.57	₹13,13,227.71
Swift	₹7,99,790.00	₹71,000.00	₹71,000.00	₹71,000.00	₹71,000.00	₹71,000.00	₹71,000.00	₹71,000.00	₹71,000.00	₹71,000.00	₹14,38,790.00
Hyundai grand i10	₹8,92,032.57	₹47,428.57	₹47,428.57	₹47,428.57	₹47,428.57	₹47,428.57	₹47,428.57	₹47,428.57	₹47,428.57	₹47,428.57	₹13,18,889.71

4.2 *Additional Load Due to BEV Charging Is Negligible*

As shown calculation above, in spite of advancing electrification of vehicles at high pace, the additional increase in load due to charging of BEV in 2030 is only 5.8% of country's aggregated electric load (as shown in table above). This is mainly because of the following three reasons:

- Rapid increase in electricity demand from number of other end uses like air conditioner and heaters will be very large over the upcoming 15–20 years
- Vehicle penetration is dominated by lesser consuming two-wheelers than higher consuming cars by 2030.
- There are uncertainties about overall vehicle penetration (expected to be significantly lower than the penetration in other industrialized or emerging economies).

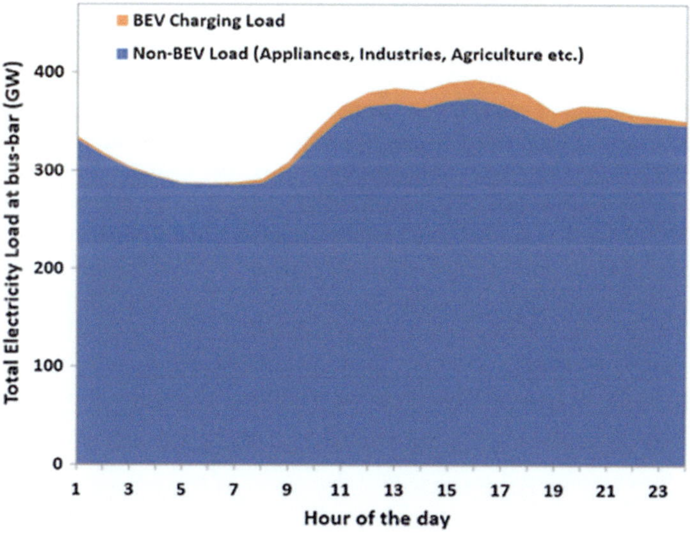

4.3 BEV's Having Positive Impact to Utilities

(1) *Higher load requirement during RE generation peak hours*

In India, the Comptroller and Auditor General (CAG) has estimated that the loss in wind farms of Tamil Nadu (TN) was INR 2.04 billion because of lack of power evacuation during 2007–14. Electric vehicles can be a possible fix to rising requirement of power during low-demand hours by overnight and afternoon charging. Electric vehicles can serve to grid by increasing load requirement during low demand hours, even at times when renewable energy generation peaks. As renewable energy capacity increases in India, Electric vehicles can potentially serve as very good distributed energy storage option, which can be further utilized to smoothen the load curve and provide evacuation of RE power up to sufficient extent.

Renewables when combined with EVs as storage can effectively deliver what can be considered as distributed base-loads. This provides an effective remedy for upholding and managing stability in the grid even with infrequent Renewable energy generation.

Charging of electric vehicles during mid-day or night times can utilize renewable energy peak generation with base-requirement operation to meet the rest of the requirement (load). As a result of this, dip in the total load reduction is observed in the present conditions and also the level of load curve is maintained. The smoothened load curve further can result in saving of cost for end-to-end each component in the value chain by utilizing the base-load capacity i.e. reduced Fixed cost (FC)/unit, also lesser requirement for expensive peaking capacity.

(2) *Revenue diversification for discoms*

Rise in aggregate technical and commercial (AT&C) losses and rise in power generated from captive power plants (both industrial and commercial) results in high cumulative loss financially weak distribution utilities. These utilities are struggling to buy sufficient power to meet demands. The increased electricity consumption because of requirement from Electric vehicles will aid in bridging the Regulatory Assets (RA) on the financial statements of distribution utilities in long run and help reduce the cross-subsidization losses due to industrial capacity (captive). The smart meters can help distribution companies to confirm, support and help accurate billing. Also help to reduce technical and commercial losses with those cases that arise due to imprecise calculations of theft, billing, losses etc.

The distribution utilities are approximated to have suffered from loss of INR 3800B in financial year 2015. Considering that EVs have a battery sizing 100 KWh, the yearly requirement for 6 million Electric vehicles additionally is approximated to be about 93 TWh. This is leads them to an additional revenue of INR 990 billion per annum.

(3) *Optimal utilization of the current coal capacity in the country*

From last years it has been evident that, reduced industrial electricity requirement has lead to constantly diminishing Plant Load Factor (PLF) in domestic coal based power plants. According to Central Electricity Authority (CEA) the Plant load factors lied from 57.60 to 66.80% in financial year 2016.

The requirement from industrial consumers may rise in the coming years. Also, from the regulatory perspective there are various reformations in coal regulations, the introduction of EVs can contribute in optimum utilization of coal. Subsequently after few years EVs can contribute towards demand assurance when demand response and energy efficiency measures are in place.

EY has estimated that an addition of about 6–7 million EVs by year 2020 would require about 7–12 GW of coal generated installations in place. Since, charging of vehicles generally happens during non-peak hours, the consumption of base load installments would rise. As a result of this the extra requirement can be met without adding further new capacity into the system. Additional use of the capacity contributes in reducing the cost of coal-based power plant. Like the Indian Government estimates that the production cost would reduce by 1 rupee per each unit for an increase in PLFs from 60 to 90%.

(4) *Reduction in energy storage cost*

In present scenario, the heavy cost of energy storage is said to be the major weakness for intensive RE growth. Heavy requirement for storage (i.e. major component as battery) would result in growth of well linked battery supply chains which will substantially reduce storage cost in market. Low cost of energy storage would not only aid in reduced landed cost of RE but also increase the efficiency of evacuation. China and US, the proliferation of electric vehicles have started leading in development of battery production houses. In China, about 16.70 GWh of battery capacity is already under commissioning whereas 35.10 GWh is in developing stage. For the case of United Sates, 3.8 GWh of capacity is commissioned and about 36.2 GWh is in development.

(5) *Actively increasing Distributed Energy Resources (DERs) and reduced dependence on utilities for energy*

The power source—Battery, utilized in Electric vehicles generally show life span ranging from 8 years to 10 years, but also, they possess considerable capability to be used for other alternate requirements, like solar backup or similar power storage requirement. There is also a second life of these batteries, which could be a revolution in clean energy drive because of following benefits:

- Help in providing cheap energy storage required because of intermittent and distributed renewable generation
- Help in reduction of effective ownership cost of Electric vehicles.

Cheap energy storage when merged to Net-metering and RE can lead to a high probability in encouraging DER. Also, creating more energy independence with distribution utilities and set-up of viable micro grids in remotely located places with help of Indian power grid which are difficult to reach.

(6) *BEVs can considerably decrease import of oil in country*

At the end of year 2030, electric vehicles have the capacity to cutoff the burning of crude oil by 360 M barrels/year that is almost about 15% of total crude oil consumption by year 2030. In year 2015, the country use to consume almost around 1322 M barrels of crude oil per year. According to 'IEA 2015; Karali et al. 2017', the consumption is predicted to rise to 2246 M barrels/year up to year 2030 and further up to 3199 million barrels/year up to year 2040. It is forecasted that 80% or even higher amount of oil used in country by year 2030 shall be depended on imports. According to 'Planning Commission, 2014', this implies that complete cutback in crude oil usage will probably refrain import of oil. Considering the price of crude oil to be 40 UDS per each barrel in a conservative manner, the electric vehicles can contract oil import by 7B USD per year by 2030 which is almost INR 50,000 Crore each year.

(7) *Integrating intermittent renewable sources to grid can be made more economical and efficient using—Smart Charging*

The load of charging BEVs can be modified in such a manner that system costs remain more economical, like load-shifting to non-peak timings. The concept of such modification (load- shifting) can be termed as **smart charging**. The two graphs below depict two different scenarios for the month of May in year 2030 with Non-BEV load and EV charging load. The first graph shows impact with normal charging option, while the second one shows enhanced changes due to smart-charging. Considering business as usual (BAU).

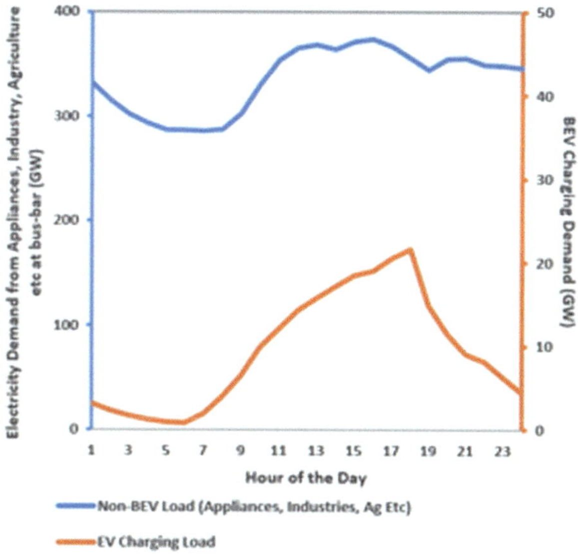
Electricity Demand from Appliances, Industry, Agriculture etc at bus-bar (GW)
BEV Charging Demand (GW)
Hour of the Day
Non-BEV Load (Appliances, Industries, Ag Etc)
EV Charging Load

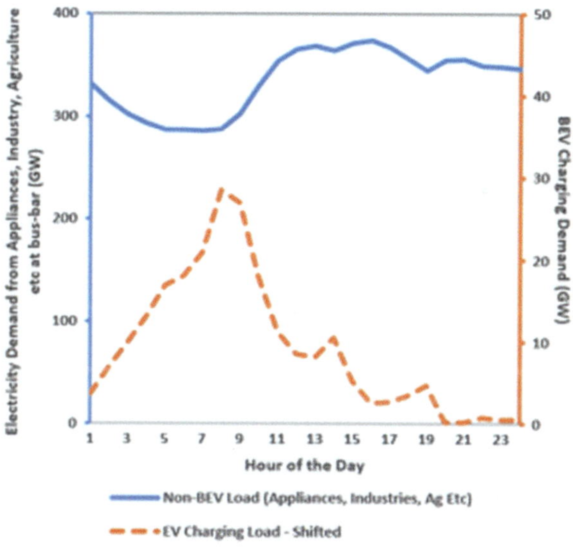
Electricity Demand from Appliances, Industry, Agriculture etc at bus-bar (GW)
BEV Charging Demand (GW)
Hour of the Day
Non-BEV Load (Appliances, Industries, Ag Etc)
EV Charging Load - Shifted

5 Battery Swapping Model

In the conventional system of charging an electric vehicle the electric power is transferred into the vehicle battery while the battery remains fitted inside the vehicle. A charging cable is extended up to the vehicle and the vehicle remains in charging position during entire charging process. The charging time as discussed before in previous chapter, varies right from about 10 h to 30 min for different classes of chargers respectively.

Even the fastest “DC Fast Charger” takes considerable time which leads to non-availability of the vehicle at that time. It creates hindrance during the travel and reduced business opportunities for a fleet EV. Also, the upfront cost of EV due to high battery prices is not convenient.

In the Battery swapping model, the entire empty battery set is replaced by an already charged battery set in the EV. Hence the time take to charge the EV is equal to the time taken by system to replace the charged battery.

The Pilot Model on this battery swapping concept is developed in the diagram given below. The flow of service and revenue between each constituent of the entire value chain is explained with the help of flow lines.

Initially the Electric vehicle needs to be purchased without an inbuilt battery and the contract with Battery Swapping Station is signed. The BSS purchases a battery on behalf of the EV owner and provides charged battery each time he demands against his calculated amount. The amount is decided considering various parameters like Cost of battery, Life cycles, Degradation, Cost of Electricity Etc. In case of multiple franchise based BSS, the user is free to swap the battery from any of the available BSS outlet. The record of driven distance and time is maintained with BSS to keep the track and levy charges accordingly.

5.1 *BSS: Significance, Possible Challenges and Potential Markets*

Major Constituents under BSS value chain	Significance to Constituent	Possible Challenges	Potential target market
Power Utilities	> Higher revenue due to commercial tariff > Time of usage (TOU) bring in shifting and smoothening of load curve	> Infrastructure capable to allow and record bidirectional flows with > TOU meters required > Regulatory inputs	> Utilities in Metro and Mega cities > Franchise models
Battery Manufacturers	> OEM tie ups with BSS developers > Service and replacement easy > Implementation of Cradle to Cradle model	> Continuous Research and Development required > Higher battery costs	> Existing and upcoming Battery Manufacturers
Electric Vehicles Manufacturer	> OEM tie ups with Battery Manufacturer or BSS Services > Higher acceptability and accelerated growth of EVs > Increased convincing power to consumers due to lower upfront costs	> Need to stick to standardised system > Threat due to use of substandard batteries with their EVs	> Existing and upcoming EV manufacturers Segmented: 2W, 3W, LWCV Etc
Consumer (Mobility Service User)	> Excess to increased EV mobility service > Benefits of Clean , Safe and Economic rides in long term	> Absence of evident short term gains	> Commuters in Metro cities
Fleet Owner / EV Owner	> Decreased range anxiety > Reduced upfront cost for battery procurement > Negligible re-fuelling time > Increased vehicle availability > Competitive advantage over market rivals > Remain synchronised to upcoming market technology trends	> Clarity regarding Battery Ownership	> Existing and upcoming EV fleet as well as private EV owners Segmented: 2W, 3W, LWCV Etc
Battery Swapping Station (BSS) Owner	> Maximum utilisation of charging infrastructure > Considerably reduced space requirement > Cost optimisation opportunities > Options for renewable generation > Utilization of storage capacity for participating in power market	> Heavy initial investments > Need of Standardization > Requirement of equipment infrastructure > Need of well developed networked model > Battery degradation	> Electric Vehicle Manufacturers / Showrooms > Charging Stations Other conventional Fuelling stations
Renewable Energy Generators	> Higher capacity utilization > Higher returns over Net Metering > Stronger performance along battery storage systems > DC charging more easier	> Intermittent nature > Reliability exclusively on renewable sources would be difficult > Space requirements	> BSS developers Large rooftops (Domestic, Commercial buildings) near BSS

6 Conclusion

Checker The study attempted to demonstrate growth of Electric mobility in the country supported by various upcoming EV and battery technologies as well as promotional incentive schemes by the Government of India. The assessment of expected growth of 2-Wheelers, 3-Wheelers and light commercial vehicles is done to know the impact on nation's electric grid system. Following conclusions can be drawn from the study due to phase wise penetration of EVs into the grid system:

- **Gain to EV owners**: The cost benefit analysis brings out results that show that 2 Wheelers and 3-Wheeler EVs are already in profit making position, also the four-wheeler EV segment is at the edge of crossing the profit line based on long term Cost of ownership.
- **No considerable load on Grid due to BEV charging**: Total Load on Grid due to Electric Vehicles in 2030 is not even exceeding 6% of total peak (402 GW).
- **Benefits to Utilities**: There can be considerable benefits to Energy utilities in the nation due to EV adoption due to Higher demand during RE generation peak hours, improved utilization of existing domestic coal capacity, Revenue diversification for discoms, Reduction in energy storage cost and propagation of distributed generation and energy independence.
- **Decreased crude imports, Effective RE Integration, Aid as ancillary services and Almost Zero CO_2 Emissions** are other positive impacts due to EV proliferation.

Furthermore, along with above mentioned benefits to the nation, there are some typical areas of concern that need to be focused in order to carry sustainable growth and easy integration of EVs to the Indian grid system:

- Focused efforts and implementation of appropriate Regulation and policies with regard to EVs is required immediately.
- Development of compatible charging infrastructure along with advanced TOU meters, bidirectional energy flow capabilities, easy payment settlement mechanisms etc. is required.
- Appropriate engagement of EVs with forthcoming technologies like **Smart grid**, **Block chain**, **IOT** along with cyber security concern.

Requirement of more realistic Business Models with better practical feasibility

References

1. NITI Aayog Report, May 2017. *INDIA LEAPS AHEAD: TRANSFORMATIVE MOBILITY SOLUTIONS FOR ALL*
2. Electric Vehicles in India And Its Impact on Grid, Compiled By: -NSGM-PM
3. Global EV Outlook (2017) International energy agency IEA

Macro-Level Methodologies to Assess Grid Impacts and Flexibility from Battery Electric Vehicles: Opportunities to Reform India's Energy Sector

Girish Ghatikar, Nikhil Parchure and Rupendra Bhatnagar

Abstract India's plan for electric vehicle (EV) only sales by 2030, which translates to over 206 million EVs, represent energy and power system challenges. While facing intensified renewable integration and continued grid reliability issues, EVs' new demand for power will invariably further strain the power grid. These challenges represent an opportunity for India's goal to reduce greenhouse gas (GHG) emission intensity by 30–35% by 2030 relative to 2005 levels. The electrification of transportation sector represents opportunities for (1) new markets; (2) reduced oil imports; (3) reduced vehicular GHG emissions and air quality improvements, and (4) axiomatic new revenue to distressed electric grid distribution companies (DISCOM). Toward a strategic roadmap to unlock these opportunities, this study proposes macro-level methodologies to assess the grid impacts from EV charging and flexibility using the projections based on historical trends in vehicle sales and driving patterns. The results form the preliminary analysis show new annual revenue opportunities of at least $36 billion for DISCOMs, cost savings of $70 billion from oil imports, and grid impacts that can be mitigated by flexibility services. The policy-makers and grid operators can use the results to optimally plan and operate the energy and power needed for the EVs and develop spatially and temporally flexible EV charging infrastructure.

Keywords Electric vehicles · Grid impacts · Energy sector · Demand response · Revenue models · Methodologies

---

G. Ghatikar (✉)
Electric Power Research Institute, Palo Alto, USA
e-mail: Ghatikar@gmail.com

N. Parchure
Octillion Power Systems, Hayward, USA

R. Bhatnagar
India Smart Grid Forum, New Delhi, India

R. K. Pillai et al. (eds.), *ISGW 2018 Compendium of Technical Papers*, Lecture Notes in Electrical Engineering 580,
https://doi.org/10.1007/978-981-32-9119-5_30

1 Introduction

India has aggressive goals for decarbonizing transportation and electricity generation—toward addressing sustainable growth. With the 2017 ratification of Intended Nationally Determined Contributions (INDC) originating from the Paris Agreement, India's intends to meet 40% of energy needs from non-fossil based resources by 2030 and reduce the emissions intensity of its gross domestic product (GDP) by 33–35% by 2030 from 2005 level [1]. In the post-INDC ratification, the Nationally Determined Contributions (NDC) are post-2020 climate actions that countries plan to implement under the international agreement. India's ongoing national missions for transportation electrification and renewable energy generation play a key role in attaining these international agreements' NDC goals.

India's national- and state-level policies and missions play crucial and symbiotic roles to advance and accelerate high penetration of electric vehicles (EV) and provide local manufacturing, innovation, and energy sector modernization potential. The grid modernization initiatives through National Smart Grid Mission (NSGM) and its counterpart programs such as, Restructured Accelerated Power Development and Reforms Program (R-APDRP) must account for the activities and outcomes from National Solar Mission (NSM) and National Electric Mobility Mission (NEMM) to instill an integrated approach to understand grid impacts and operational preparedness from high penetration of EVs and renewable generation, including enabling flexible EV charging infrastructure. India's national mission for electric mobility, NEMM Plan of 2020 plans to deploy six to seven million EVs (hybrid, fuel-cell, and battery-based vehicles that include two-, three-, and four-wheelers) by 2020 [2]. In a major push for transportation electrification, India recently announced the plans that only pure EVs shall be sold from 2030 [3].

In support of these strategic plans to enable aggressive and accelerated adoption of EVs in India, the study proposes the macro-level methodologies to quantitatively assess grid impacts from battery-based EVs and propose solutions for flexible operations. The focus of this study is an assessment of macro-level grid impacts and the potential of EVs to offer flexibility, as a grid service. Flexibility is temporal and spatial deviations in EV charging and/or discharging with no change in the quality of service. The rippling benefits can play a major role in reducing oil imports and GHG emissions. Albeit environmental factors are out of the scope of this study, early-stage studies are reviewing these benefits [4]. The key objectives this study focuses on are:

- Identify and project the sales trends of all major categories of on-road vehicles in India.
- Analyze and report EV growth projections, their electricity needs, and energy sector opportunities.
- Review demand flexibility needs for EV charging to ensure aggressive EV adoption.

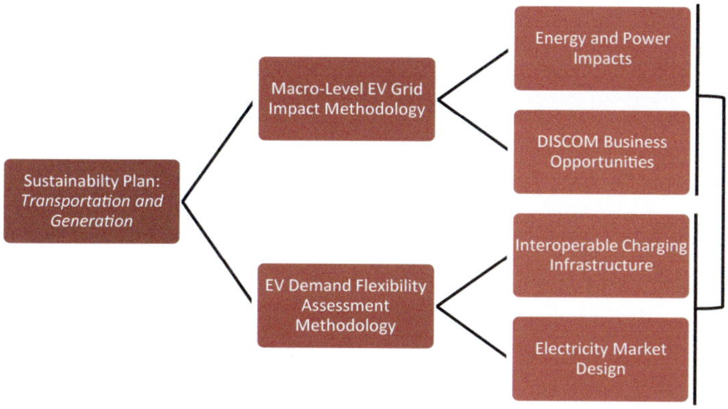

Fig. 1 Study methodology

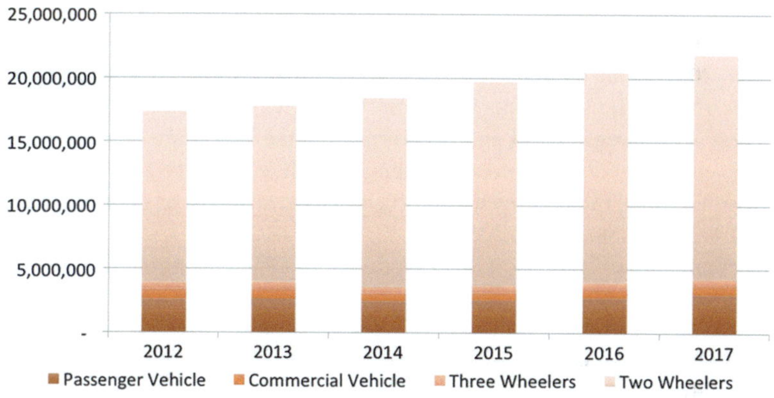

Fig. 2 India's vehicle sales growth trends: 2012–2017

India's plans are leveraged to propose methodologies to assess grid impacts and assess demand flexibility to alleviate any grid impacts from the new EV charging demand. The impacts consider energy and power growth from EVs, and the new revenue opportunities they offer to DISCOMs. Finally, the EV demand flexibility using interoperable charging infrastructure and electricity market design are postulated and presented. Figure 1 shows this methodology.

Figure 2 shows the historic growth trend analysis for domestic vehicle sales in India from 2012 to 2017 using the Society of Indian Automobile Manufacturers' (SIAM) domestic sales data [5]. This include all major on-road vehicle categories—passenger, commercial—comprising of buses, trucks, vans, etc., two-wheelers, and three-wheelers. The combined vehicle categories constitute approximately 22 million in annual vehicle sales at a 6-year average annual growth rate of 5.6%. Full electrification by 2030 is a disruptive transformation considering that India is still in

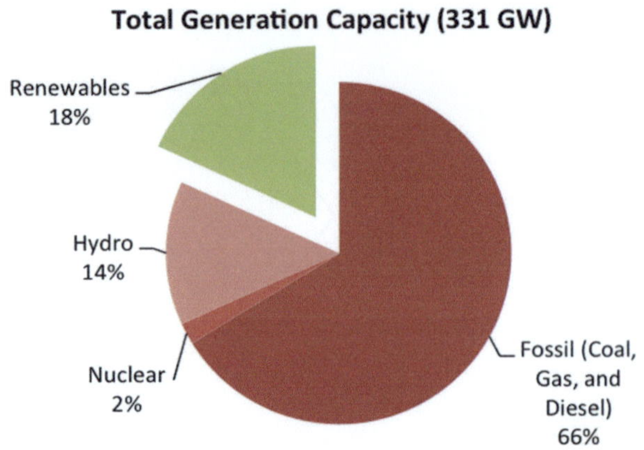

Fig. 3 India's total generation capacity and generation sources

nascent stages of electric mobility. According to India's National Automotive Board, at the end of 2017, there were a total of about 176,000 EVs on the Indian roads [6].

Figure 3 shows India's total power generation capacity of 331 GW, as of November 2017 from all fuel sources [7]. The renewable generation capacity of 60 GW, or 15% of the total capacity, comprises of wind, solar, biomass, hydro, and waste to energy sources. Wind and solar represent the lion share or 80% of the total renewable energy goals and pose challenges with variability [8]. This will only exacerbate, as India moves toward the NDC target of 40% of energy needs from non-fossil based resources by 2030.

India plans large targets of 175 Gigawatts (GW) of total generation from renewable energy sources by 2022. This includes 100 GW of solar, 60 GW of wind, 10 GW of biomass, and 5 GW of hydro. To meet the 2030 NDC targets, the renewable portfolio must exceed 230 GW where over 100 GW may be from solar energy, against the projected total generation capacity of 670 GW [9].

Understanding the grid impacts of EVs, and enabling their demand flexibility are important challenges India must address to accelerate and successfully meet the EV deployment targets while decarbonizing the energy generation through renewables. The demand flexibility also provides an axiomatic benefit to address variability from renewable generation and cost-effective grid integration.

2 Macro-Level Grid Impacts and Flexibility Assessment Methodologies

This section proposes a macro-level methodology to assess national-level grid impacts from EV charging and methodology to assess these impacts and mitigate them cost-effectively and preemptively through flexibility services from EV charging. The section content leverages an earlier study and can be contextually applied to other regions [10].

2.1 Macro-Level Grid Impacts Methodology

The grid impacts from charging the following two key metrics can assess the additional charging needs from the EVs. The energy values indicate total energy consumed by the EVs and the potential for new revenue opportunities for DISCOMs. The power values indicate the additional peak demand increase and the grid infrastructure preparedness to support reliable electricity supply.

1. **Energy**: The total amount of electricity consumed by EVs over time. This assessment can also reveal new revenue opportunities for DISCOMs.
2. **Power**: The total amount of power that the grid must support considering the variable charging needs of the EVs at a specific time.

The study conducted an assessment of both energy and power needs for the Indian grid considering the transition plans for the sale of all EVs by the year 2030. The year 2020 analysis is also considered to align with the NEMM 2020 mission. The analysis for cumulative energy and power numbers show the total grid capacity needed by the year 2030. It should be noted that these numbers are representative based on the preliminary macro-level methodology. The results should be used, as initial numbers for further precision analysis of grid impacts.

Based on the current registered and operational internal combustion engine (ICE) based vehicle sales and growth for all categories, the total number of vehicles and their average driving distances were extrapolated. The vehicle categories include commercial EVs (CEV)—comprising of buses, trucks, vans, etc.—two-wheelers (2-EV), three-wheeler EVs (3-EV), and private passenger EVs (PEV). The linear historical growth numbers were used to project the increase in vehicles for 2020 and 2030. The market share of each of these vehicle categories was used to assign battery capacities (energy) for those driving distances and their planned lifespan of operation. Specific charging infrastructure and charging power levels were assumed and temporally assigned to determine the energy and power levels. These results lead to the identification of new revenue opportunities from EVs for DISCOMs. While the methodology is generic and not country-specific, some of the metrics such as driving distances and battery capacity may vary. For example, Relatively smaller driving distances in India can support smaller battery sizes and lower charging levels. Figure 4 illustrates this macro-level methodology.

The year 2020, the year 2030, and cumulative energy and power results, respectively were used to determine grid impacts. The axiomatic relationships of energy and power growth with the variability in the renewable generation were analyzed to determine the grid impacts and propose flexibility requirements for EV charging.

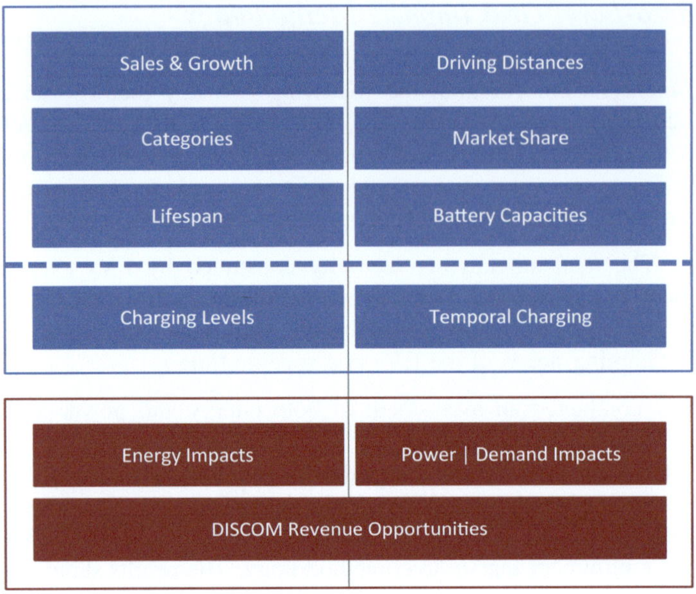

Fig. 4 Macro-level grid impacts methodology for electric vehicles

2.2 *Flexibility Assessment Methodology*

The flexibility metric proposed includes the assessment of EV charging infrastructure and its spatial needs (e.g., the city and highway corridors with a higher concentration of charging stations and charging level types). The early-stage charging initiatives by the Indian DISCOMs are summarized to show the ongoing transition.

The benefits of EV flexibility can be extrapolated from the energy and power, and charging infrastructure and temporal charging data. Depending on the electricity supply and grid infrastructure, the value of the flexibility or managed charging varies. For example, the supply deficit can lead to larger value from flexibility. In this instance, the grid reliability can be maintained without compromising the user's driving patterns or adding any new generation capacity. This provides economic motivation to both customers and DISCOMs to offer flexibility, as a service alongside EV adoption. The flexibility assessment methodology is also generic and not country-specific, some metrics such as electricity market design may vary. For example, In India, the lack of established demand response (DR) markets can lead to underutilization of the flexibility. Figure 5 illustrates this methodology.

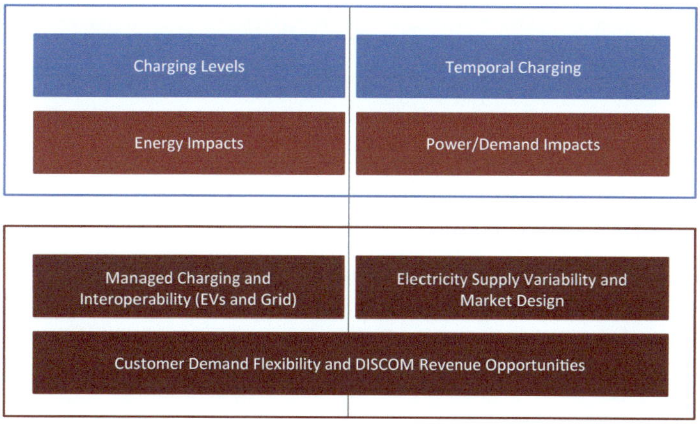

Fig. 5 Flexibility assessment methodology for electric vehicles

In further defining the flexibility services for EVs, two common vehicle-grid integration (VGI) methods are used—V1G (1-way power management between EV and grid, or also called, as managed or smart charging) and V2G (2-way power transfer between EV and grid, or also called, as vehicle-to-grid). An example of V1G is where EV, acts as a DR resource by scheduling the charging to non-peak hours, whereas V2G is where EV acts, as a stationary energy storage system to supply electricity to the grid. The flexibility services that EVs can provide can be valuable to the grid and include DR, voltage and frequency regulation for reliable grid operations [11]. Together with EV owners, the DISCOMs can leverage this flexibility with innovative tariff schemes such as dynamic pricing and peak pricing to incentivize flexibility. This offers new revenue opportunities to both EV owners and grid operators. This will axiomatically also spur technology and business innovation where companies can offer such flexibility, as a service.

The early instances of EV adoption flexibility initiatives for customers that are offered by the Indian DISCOMs such as TPDDL and BSES can be termed, as early adopters; the aggressive growth of EVs necessitates a consideration of charging infrastructure within the low-voltage grid network and regions closer to customers (e.g., public spaces, residential neighborhoods, workplaces). To this end, these initiatives are a blur of the requisite charging infrastructure and support services to meet India's disruptive transformation in electric mobility adoption. These initiatives represent an opportunity to understand the EV charging infrastructure needs and develop the strategic vision that considers the mitigation of grid impacts.

3 Grid Impacts and Charging Flexibility: Preliminary Analysis Results

This section presents the results from a preliminary analysis using the grid impacts and flexibility assessment methodologies. The grid impacts assessment considers 2020 and 2030 scenarios to align with India's NEMM and all EV sales transition plans, respectively and includes:

1. The total energy (GWh) used by the EVs.
2. The total power (GW) or demand by the EVs.
3. The new revenue opportunity for DISCOMs for the total energy sale. The demand charge (for peak power) revenue is not considered in the analysis.

The analysis assumptions are based on the macro-level methodology to assess grid-impacts, as shown in Table 1.

Table 1 Requirements and assumptions to assess grid impacts

Key requirements	Assumptions
1. Energy	Annual travel distance considered for daily energy use—12,800-kilometers (Km)/year for PEV, two and three wheelers [12], 52,000 km/year for CEV
2. Power	EVs that charge over 3 h and 10% of EVs are grid-connected and charge at the same time. The considerations are based on early-stage studies [13]
3. Sales growth	Annual new EV sales growth for PEV, CEV, and three-wheelers and two-wheelers are 3%, 1%, 0.2%, and 5%, respectively. These numbers are deduced from the SIAM historical sales trend data [5]
4. Life span	The lifespan is 10 years and the cumulative total is for the EVs sold between 2020 and 2030 [4]
5. Market share	In 2020 the market share for EVs is—10% for PEVs and CEVs, 50% for three-wheelers, and 25% for two-wheelers. These numbers are deduced from the SIAM historical sales trend data [5]
6. Battery capacity	The EV battery size is 20 kWh, 125 kWh, 3 kWh and 2 kWh for each PEV, CEV, three-wheeler, and two-wheeler, respectively[a]
7. Temporal charging	Commercial EVs (CEV)—buses, trucks, vans, etc.—may charge during the day or night; most of passenger EVs (PEV) and 2-wheeler EVs (2-EV) and 3-wheeler EVs (3-EV) may charge during the night. The PEV charging pattern is based on early-stage studies [14]
8. Charging rates/ levels	The qualitative consideration of effective charging levels of—0.98 kW, 38 kW, 0.58 kW, and 0.27 kW for PEV, CEV, 2-EV, and 3-EV, respectively

[a]The battery capacity for buses and 3-wheelers is in the recommended ranges of the study that considers conditions in Kolkata, India [14]

3.1 *Grid Impacts and Opportunities from Electric Vehicles*

The analyses results and the new revenue opportunities they represent for DISCOMs are described further.

(1) *Preliminary Energy and Power Estimations*

Figures 6, 7, 8 and 9 show the results of the analyses for the power and energy requirements for the year 2020 and cumulative scenarios for all EV categories in the year 2030. While the earlier detailed studies have looked at the energy and power requirements, they focus on PEVs and 2-EVs [12]. These studies have not considered all EV categories.

Figures 6 and 7 show the EV-specific energy needs for 2020, and cumulative scenarios of 2030, respectively. By 2020, India will need an attainable 18.48 GWh of daily energy for EVs. By 2030, India will need a significantly higher 162.73

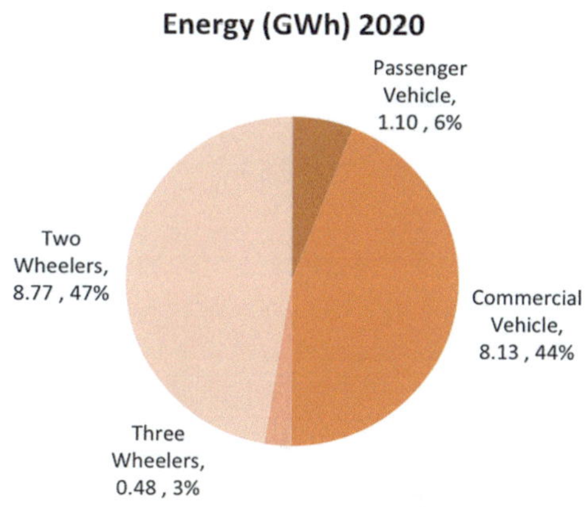

Fig. 6 India's energy use by electric vehicles: 2020

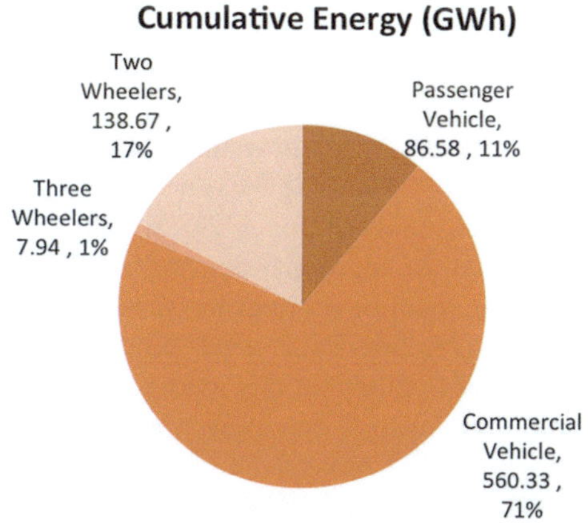

Fig. 7 India's energy use by electric vehicles: cumulative for 2030

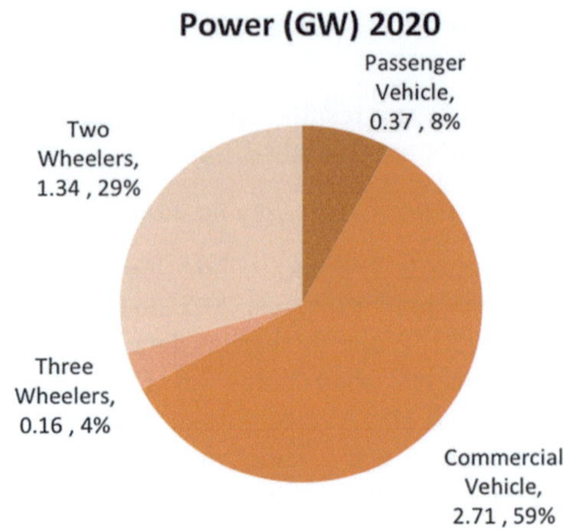

Fig. 8 India's power use by electric vehicles: 2020

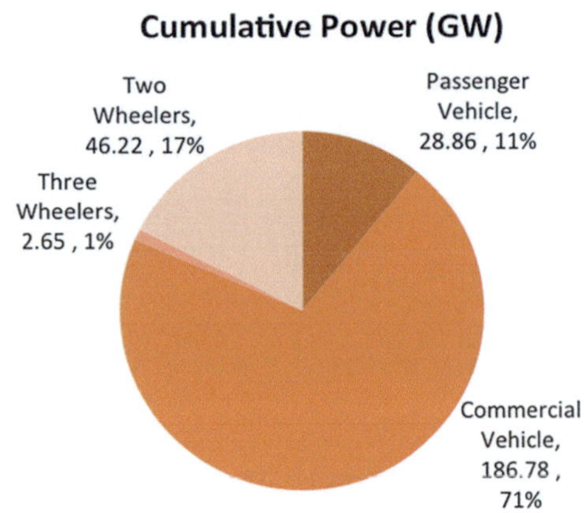

Fig. 9 India's power use by electric vehicles: cumulative for 2030

GWh of additional daily energy and 795 GWh of cumulative energy to support all EV categories on road by 2030. The 2030 cumulative annual energy consumption from EVs will be 290 TWh. This represents about 25% of the total energy consumption of 1142 TWh in India from all the sources in the year 2016 [15].

Figures 8 and 9 show the EV-specific power needs for 2020, and cumulative scenarios of 2030, respectively. By 2020, India will need an attainable 4.58 GW of additional demand for EVs. By 2030, India will need a significantly higher 43.96 GW of additional demand and 264.6 GW of cumulative demand to support all EV categories on road by 2030. This represents about a staggering 80% of the total November 2017 total power generation capacity of 331 GW [7]. It is important to note that this is true when all EVs are charging to the grid at the same time. The energy and power usage from EVs can vary significantly across all categories. For example, a CEV may use 100% of the battery capacity every day, whereas PEV and 2-EV may use 40–50% of battery capacity and CEV may charge at 20 kW whereas, a 2-EV may charge at 0.27 kW.

Table 2 Temporal distribution of electric vehicle charging

Time	PEV (%)	CEV (%)	3-EV (%)	2-EV (%)
7 a.m. to 5 p.m.	20.00	25	0	20
5 p.m. to 11 p.m.	20.00	25	0	20
11 p.m. to 7 a.m.	60.00	50	100	60

Table 3 Temporal distribution of load from electric vehicles

EV category	7 a.m. to 5 p.m.	5 p.m. to 11 p.m.	11 p.m. to 7 a.m.
PV	5.77	5.77	17.32
CV	46.69	46.69	93.39
Three wheeler	NA	NA	2.65
Two wheeler	9.24	9.24	27.73
Total grid load (GW)	61.71	61.71	141.08

The charging time and electricity rate tariff policies will have a significant impact when EVs get charged and their axiomatic grid impacts. Based on the earlier research, and our estimates for electric bus and 3-EV charging patterns, we expect following the distribution of charging events throughout the day [16]. Table 2 shows the temporal distribution of EV charging considered for the study.

Based on the above charging event distribution, the expected utility load for the entire grid is shown in Table 3. This estimate does not consider utility incentives to charge at certain times and other factors such as workplace charging. These policies and factors will significantly affect the charging event pattern.

(2) *Revenue Opportunities to Utilities*

The results provide the Indian power sector an unprecedented access to new markets. In particular, the distressed DISCOMs have a lifetime opportunity to play a critical role in charging infrastructure, the innovation of technologies and electricity rate tariffs, and customer engagement.

Excluding demand charges, the annual EV energy use is a new recurring revenue opportunity of $36 billion (INR 231,705 Crores) at an average $0.11/kWh (INR 8/kWh). Additionally, a good market-design can reduce the peak power demand from EV charging to 61 GW during daytime and 141 GW overnight, as shown earlier in Table 3. While this is still a significant 18.6% of the total power generation capacity during daytime and 42% of the total power generation capacity during nighttime, it is more manageable and benefits customers to lower their EV operation costs.

3.2 *Flexibility Assessment for Electric Vehicles*

The EVs role as a flexible resource is well studied [17–20]. The charging flexibility, as a service, can be used with planned electricity market design and operations and EV infrastructure development that support interoperability with grid-systems. Such flexibility can be used to support to manage India's electricity reliability, resiliency, and better integration of renewable generation. In support of aggressive renewable generation, the VGI methods can be used to by absorbing excess wind or solar generation, which otherwise would be curtailed [10, 11].

However, the owners or the drivers of EV will be able to offer this flexibility under the following key conditions:

1. The established electricity market mechanisms offer value for flexibility services, as new means of electricity grid resiliency and to better address the opportunity of variable renewable generation or peak power management for grid operators.
2. Secure and standards-based interoperable EV technologies among, charging infrastructure, and grid must be well established to foster cost-effective technology innovation to offer flexibility [21].

(1) *Electricity Markets in Support of Flexibility*

Going forward, rising EV and renewable energy penetration could likely create more peaks and valleys in the load curve. The adoption of EV in urban mobility would proliferate electric energy demand. A significant number of EV's will also get charged at home and workplaces. In such a situation, the charging behavior of the residential or commercial consumer impacts the individual- and grid-level load curves. Thus, with the adoption of EVs, demand management would be a challenge for the DISCOMs and grid operators due to the unpredictability of consumer EV charging and their management.

One of the significant programs from the Indian government is the "power for all" initiative, which aims at providing quality, reliable and affordable all-day power to the entire residential, commercial, and industrial customers for a certain period. The demand-side resource flexibility strategies can support such initiatives by optimal use of energy and accelerate reforms and play the role of a virtual power plant, as a supply-side generation option.

Policies in support of EV flexibility will enable cleaner energy resources and also delay adding additional power generators to balance the grid. The EVs should be supported, as a standalone independent resource apart from conventional and renewable energy resources. This requires legislative action to empower state regulatory commission for effective enforcement.

(2) *Electricity Network and Charging Interoperability*

The EV charging infrastructure should be an established system of smart charging infrastructure including the electrical metering point and network connections, The charging stations including the plug, communication and charging, control center for authentication of measured data and billing/invoicing of the energy bill.

The charging infrastructure must be capable of mass market deployments and must include essential features of smart and internationally marketable with various levels of charging capabilities, automatic customer identification, bi-directional communication for optimized friendly charging, management of charging power at station depending on the grid loading and power availability. The public and semi-public charging station areas should have a standard socket with intelligent authentication with suitable communication protocol between EVs and grid across different use cases for security and interoperability. Standardization of charging infrastructure and communication protocols will thus enable the proliferation of electric vehicles and future-proofing them for flexibility services.

4 Conclusions and Future Directions

The study proposed macro-level methodologies to assess the grid impacts and flexibility from aggressive EV adoption to meet India's 2020 and 2030 policy and NDC objectives. The study conducted a preliminary quantitative assessment of grid impacts. Understanding that energy and power use from high penetration of EVs are inextricably linked to India's electricity reliability and renewable generation goals, the study proposes flexibility solutions using EVs.

The quantitative assessment of EV adoption projections based on historical trends in vehicle sales of approximately 22 million annually and 6-year average annual growth rate 5.6%, driving patterns for all EV categories lead to significant energy and power use. The 2030 cumulative energy of 290 TWh that represents about 25% of the total energy use in the year 2016 is a significant need. Also, the 2030 cumulative power demand—of 264.6 GW that represents a staggering 80% of total power generation in the year 2017—can exacerbate India's peak power and grid reliability issues. While the study findings represent grid planning and operation challenges, they also represent opportunities for the Indian power sector. The DISCOMs are on a cusp of new annual revenue opportunity of $36 billion (INR 231,705 Crores).

It would be pernicious for the Indian power sector to not proactively consider flexibility from EV charging, as a prerequisite and plan electricity grid and market design in its support. Grid integration and flexibility reforms are critical for electricity access and to avoid coincidental peak EV charging. The Indian Electricity Act should explicitly name EVs as a DSM resource with an obligation to offer VGI services that benefit both customers and grid. India's grid modernization efforts should mandatorily include interoperability and standards-based interfaces to enable cost-effective use and ease of flexibility offerings from EVs.

Acknowledgements The authors acknowledge the support and review from the India Smart Grid Forum (ISGF) and Working Group 6, Flexibility and Electric Mobility, members.

References

1. United Nations Framework Convention on Climate Change (UNFCCC), India's intended nationally determined contribution: working towards climate justice. Available at: http://www4.unfccc.int/ndcregistry/PublishedDocuments/India%20First/INDIA%20INDC%20TO%20UNFCCC.pdf
2. National Electric Mobility Mission Plan (NEMMP) 2020. Department of heavy industry, The Government of India. Available at: http://dhi.nic.in/NEMMP2020.pdf
3. The Economic Times (2016, Mar) India aims to become 100% e-vehicle nation by 2030: Minister Piyush Goyal. Available at: http://economictimes.indiatimes.com/industry/auto/news/industry/india-aims-to-become-100-e-vehicle-nation-by-2030-piyush-goyal/articleshow/51551706.cms
4. Rai AA, Ghatikar BG, Seethapathy R, Pillai R (2017) Implementation plan for electrification of public transportation in Kolkata (India). Shakti Sustainable Energy Foundation, Oct 2017. https://doi.org/10.13140/rg.2.2.30986.77763
5. Society of Indian Automobile Manufacturers (SIAM). Automobile domestic sales trends. Accessible at: http://www.siamindia.com/statistics.aspx?mpgid=8&pgidtrail=14
6. India National Automotive Board (NAB), Indian Ministry for Heavy Industries. Available at: http://www.fame-india.gov.in/dashboard.aspx
7. Central Electricity Authority (CEA). All India installed capacity of power stations, Nov 2017. Available at: http://www.cea.nic.in/reports/monthly/installedcapacity/2017/installed_capacity-11.pdf
8. Ministry of New and Renewable Energy (MNRE), The Government of India. Available at: http://mnre.gov.in/mission-and-vision-2/achievements/
9. International Renewable Energy Agency (IRENA). Remap Renewable Energy Prospects for India, May 2017. Available at: http://www.irena.org/DocumentDownloads/Publications/IRENA_REmap_India_paper_2017.pdf
10. Ghatikar G, Parchure N, Pillai R (2017) Integration of multifarious electric vehicle charging infrastructure flexibility: applications in India. In: Proceedings of the 1st e-mobility power systems integration symposium, Berlin. Oct 2017
11. Ghatikar G (2016) Decoding power systems' integration for clean transportation and decarbonized electric grid. In: Proceedings of the ISGW 2016. New Delhi, India. https://doi.org/10.13140/rg.2.1.3555.4960
12. Lawrence Berkeley National Laboratory (LBNL). Techno commercial assessment of deep electrification of passenger vehicles in India, May 2017. LBNL-1007121
13. Xcel Energy, Electric vehicle charging station: pilot evaluation report. May 2015
14. Idaho National Laboratory (INL), Plugged in: how Americans charge their electric vehicles. 2015
15. The Government of India, Ministry of Power, All India power sector at a glance, power supply requirement 2016–17. Available at: http://powermin.nic.in/en/content/power-sector-glance-all-india
16. Scoffield D, Smart J (2015) Diversity patterns and coincidence of EV charging with utility systems loads. Idaho National Laboratory (INL), Feb 2015
17. Ghatikar G, Ahuja A, Pillai R (2017) Battery electric vehicle global adoption practices & distribution grid impacts: preliminary case study for Delhi. In: Proceedings of the ISGW, Mar 2017

18. Ghatikar G, Pillai R, Ahuja A (2016) Electric transportation: action plan for India. In: IEEE proceedings for sustainable green buildings and communities, Dec 2016. https://doi.org/10.1109/sgbc.2016.7936085
19. Pacific Gas and Electric Company (PG&E) (2015) Electric vehicle infrastructure and education program application; Before the public utilities commission of the state of California. Feb 2015
20. Greentech Media (2017) BMW and PG&E prove electric vehicles can be a valuable grid resource, June 2017. Available at: https://www.greentechmedia.com/articles/read/bmw-and-pge-prove-electric-vehicles-can-be-a-valuable-grid-resource
21. Ministry of Heavy Industries, GOI; Committee Report on Standardization of Public EV Chargers. Available at: http://dhi.nic.in/writereaddata/UploadFile/Standardization%20of%20protocol.pdf

Smart Charging of Electric Vehicles in Grid-to-Vehicle (G2V) Networks

Harshit Anand, Ayush Agarwal, Pooja Shukla and Suneel Yadav

Abstract The smart charging of electric vehicles (EVs) in grid-to-vehicle (G2V) networks has evolved as a potential solution to enhance the energy efficiency and stability of the grid. However, uncontrolled charging of EVs can lead to the grid instability, which can be avoided if the charging of EVs is appropriately scheduled in the grid. Motivated by this, we herein proposed two algorithms viz., centralized charging algorithm (CCA) and iterative decentralized charging algorithm (IDCA) to optimally schedule the charging of EVs in G2V networks. In CCA, the central aggregator instructs the sub-aggregator (i.e. EVs aggregator) to collect the charging request from the set of EVs. Whereas, in IDCA, the charging request of EVs is controlled only by the central aggregator. To implement this algorithm, we use an additive-increase multiplicative-decrease (AIMD) approach. In both the algorithms, we adopt a time-of-use (TOU) electricity model in order to monitor the energy drawn from the grid in different time instants. Finally, both the algorithms are tested and validated using MATLAB simulations.

Keywords Smart charging · Grid-to-vehicle · Electric vehicles · Scheduling · Centralized and decentralized charging · Time-of-use

---

H. Anand (✉) · A. Agarwal · P. Shukla · S. Yadav
Department of Electronics and Communication, Indian Institute of Information Technology Allahabad, Allahabad, India
e-mail: ibm2014034@iiita.ac.in

A. Agarwal
e-mail: iec2014018@iiita.ac.in

P. Shukla
e-mail: iec2014081@iiita.ac.in

S. Yadav
e-mail: suneel@iiita.ac.in

R. K. Pillai et al. (eds.), *ISGW 2018 Compendium of Technical Papers*,
Lecture Notes in Electrical Engineering 580,
https://doi.org/10.1007/978-981-32-9119-5_31

1 Nomenclature

The main terms used in the paper are presented below and other symbols are explained wherever needed.

N	Total number of EVs
n	Number of EVs that arrive in time slot t; $n \in N := \{1, 2, 3, \ldots, N\}$, $t \in T := \{1, 2, 3, \ldots, T\}$
T	Number of time slots
t	Particular time slot, $t \in T := \{1, 2, 3, \ldots, T\}$
ΔT	Length of each time slot
H_t	Denotes a set, consisting of the number of EVs arriving in the given time slot t; $t \in T$
$SOC_{n,t+1}$	Denotes the updated SOC of the battery of EV n, $n \in N$, at beginning of the next time slot
$SOC_{n,t}$	Denotes the SOC of the battery of EV_n in the previous time slot
$Power_{n,t}$	Denotes the charging rate of the nth EV in the *t*th time slot
$batcap_n$	Denotes the maximum capacity of the battery of nth EV, present in the *t*th time slot
$charge_{effi}$	Denotes the charging efficiency of the charger at every port.

2 Abbreviations Used

EV	Electric vehicle
G2V	Grid-to-vehicle
SOC	State-of-charge
AIMD	Additive increase multiplicative decrease
CCA	Centralized charging algorithm
IDCA	Iterative decentralized charging algorithm.

3 Introduction

Electric vehicles (EVs) are progressively being presented as a means of controlling pollution, battling ecosystem imbalance and climate change. They have immense potential to improve energy security. In the last few years, government guidelines have incentivized sustainable growth and rapid technological advances by the automotive industry. Several nations are focusing on EV penetration in their markets and are dedicating immense resources to accomplish ambitious targets they have presented in the global arena. There are several such countries; the government of Ireland has presented a target of 10% for the dissemination of EVs by 2020

[1]. China is aiming to accomplish 20–30% EV penetration by 2030. Similarly, the European Union has set a target of 48 million EVs by 2015 while the United States of America is aiming to realize a target of 100 million EVs in the same window of time [2].

The Indian governments EV penetration strategy is being shaped on the foundation which has been laid by the Faster Adoption and Manufacturing of (Hybrid and) Electric Vehicles, more commonly known as the FAME India initiative, predominantly in the technology expansion and charging infrastructure sector. There are two principle objectives: a short-term target to get at least 6 million electric vehicles on the road by 2020, and to go entirely electric in terms of new car sales by the year 2030. As per the latest available records, India has a total of 100 charging points, with the majority of these being concentrated in metropolitans such as Bengaluru [3].

There are primarily three variants of electric vehicles: BEV (Battery Electric Vehicles) which rely on a battery as their source of energy. We then have the HEV (Hybrid Electric Vehicles) that utilize batteries in conjunction with more conventional fossil fuels such as petrol, diesel or compressed natural gas as their source of energy. Lastly, we have the PHEV (Plug-in Hybrid Electric Vehicle). PHEV has a distinct trait; the battery can be charged anyplace and anytime. To make this work, RTUs or the Road Side Units must be installed for users who are in possession of these vehicles.

Charging of Electric Vehicles must be optimized because uncontrolled, unmanaged charging would lead to increase in demand for electric power and thereby affect the current electric grid's stability. Increased supply of electricity for charging through the grid may severely damage the grid, leading to fire breakouts. To tackle such issues, researchers are working towards the innovative concept of merging electric vehicles with the smart grid which will be able to meet the demand of the growing power consumption and together with the Internet technology (IoT), one will be able to manage for oneself, the requisite power. The electric vehicle is based on the concept of bi-directional energy flow between the grid and the vehicle, thereby, constituting V2G and G2V networks.

4 Related Work

The load profile will be greatly affected if a large number of EVs plug-in for charging because unmanaged and uncontrolled charging of a large group of EVs will negatively impact the electric grid. The architecture and strategy for centralized and de-centralized charging have been developed in [4]. The concept of real-time pricing, based on TOU (time-of-use) based electricity model for EV charging has been developed from [5, 6]. The concept of battery modelling and battery charging profile has been developed from [7]. The approach for De-centralized charging of EVs using the AIMD algorithm and the capacity event condition is monitored which has been implemented considering [8, 9]. The present targets the G2V based charging of EVs and focusses on the implementation of both Centralized and De-centralized charging scenarios.

The rest of the content has been divided in the following manner. Section 5 describes both the charging architectures-their modelling, the algorithm used and the procedure as to how the charging proceeds in G2V approach. Section 6 describes the battery modelling for EVs, considering the Lithium-ion batteries. Results have been given in Sect. 7. Section 8 discusses the results while we extrapolate our conclusions in Sect. 9. Various assumptions have been used throughout the work and have been stated accordingly.

5 Modelling of Charging Architectures

5.1 Overview

The centralized charging architecture comprises of a central aggregator, which controls a finite number of sub-aggregators. Each sub-aggregator consists of a fixed number of charging stations. Each sub-aggregator collects the charging requests of the EV to make the charging schedules. When an EV is connected to the charger and requests charging, the smart unit uploads the time of arrival of an EV, current SOC and the total capacity of the battery to the aggregator. The EV user should also set the expected SOC and time of departure. Under the premises of satisfying the charging request of EVs, the aggregator determines the charging schedule to maximize total benefits.

In the above-mentioned charging approach, the number of parameters increases drastically with the number of EVs. Therefore, the IDCA has been proposed. So, to address the dimensional problem, we approach the decentralized charging model. The method used to implement this charging is the concept of Additive Increase Multiplicative Decrease (AIMD), which is explained later in the text.

The IDCA does not dwell into too much of complex computations. Each EV owner is postulated to have a smart controller which can receive the price signal issued by the utility. When an EV is connected to the charger the smart controller collects SOC, time of arrival of the EV, battery type and total capacity of the battery. At the same time, the EV user sets the expected SOC and time of departure of the EV.

The following assumptions have been made while implementing the charging scenarios: (1) the total charging duration is limited to six hours from evening up to midnight. (2) For both types charging, the electricity price is constant for a particular time slot and varies for different time slots. (3) The battery model assumed for each EV is the lithium-ion battery characteristics.

5.2 *Centralised Charging Architecture*

1. **Charging Scenario Description**

Figure 1 shows the Centralized Charging model. The Control centre is the electric utility. It involves a two-way information interchange from the central control centre to EVs and from EVs to control centre. The EV aggregator controls the charging. Whenever the electric vehicle arrives at the station for charging, the charging unit uploads the following: time of arrival of an EV, current SOC and the total capacity of the battery to the aggregator. The EV user should also set the desired SOC and time of departure.

2. **Centralized Charging Algorithm (CCA)**

As stated in the above context of designing a conceptual framework for G2V implementation, it is important to perform intelligent scheduling for charging of EVs. For the purpose of simulation, we consider a scenario, where a utility company negotiates with a total of $\boldsymbol{N}$ EVs to schedule their charging profiles over $\boldsymbol{t}$ time slots, of length $\boldsymbol{\Delta T}$. In the CCA, we consider **6** time intervals or time slots, each of one hour duration. $t \in T: = \{1, 2, \ldots, T\}$; $n \in N: = \{1, 2, 3, \ldots, N\}$, see Fig. 2.

The steps involved in CCA are as follows:

Step 1

The charging rate for that time slot $t \in T: = \{1, 2, \ldots, T\}$.

For every EV in the time slot "*t*", the user enters the following: H_t, $t \in T$, denotes a set, consisting of the number of EVs arriving in the given time slot *t*, time of arrival of the EV, current SOC of the arriving EV, maximum battery size of the EV, time of departure of the EV, and charging rate.

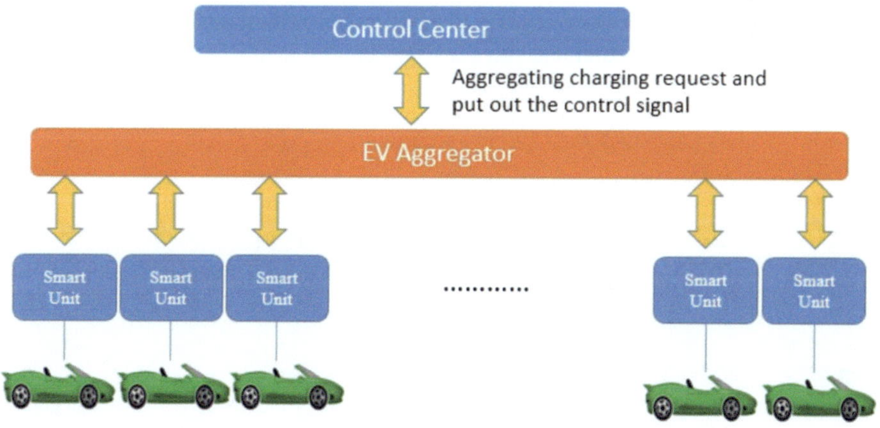

Fig. 1 Centralized charging scenario

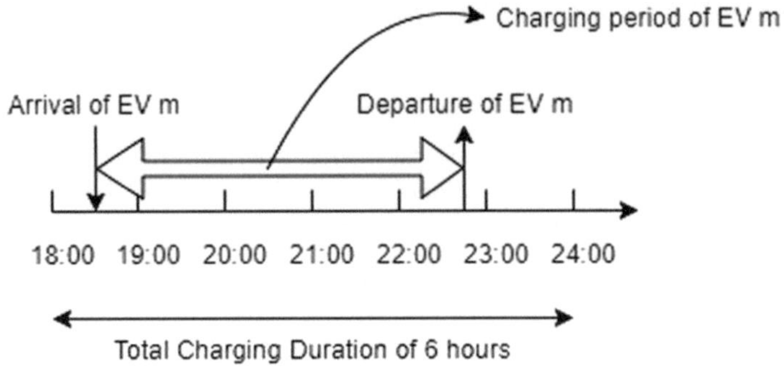

Fig. 2 Charging period and various charging intervals

Step 2
The formula for updated SOC after the end of every time slot *t* is given as:

$$SOC_{n,t+1} = SOC_n + \frac{Power_{n,t}}{batcap_n} \times charge\_effi \tag{1}$$

Step 3
For simulation purpose, we consider a finite set of EVs in each time slot and update the SOC of the EV after each time slot until it departs. The EV is supposed to have been charged to the desired SOC, before or at the given time of departure. If the desired SOC is reached beforehand, it stops charging and leaves; otherwise, it continues charging over next time slots until it's time of departure. If the EV has not charged to its desired SOC even at the time of departure, it is supposed to depart or pay penalty for exceeding the time limit.

Step 4
At the end of each time slot, we calculate the number of EVs that will carry forward their charging. We also calculate the EVs that have left in that interval.

Step 5

$$\begin{aligned} &\text{Total number of EVs in new time slot} \\ &\quad = \text{No. of EVs carried forward for charging from} \\ &\quad\ \ \text{previous times lot} + \text{No. of EVs that arrived} \\ &\quad\ \ \text{for charging in the new time slot.} \end{aligned} \tag{2}$$

Step 6
After the end of the total time T, we find the number of EVs that were charged to their desired SOC and the number of EVs that couldn't be charged to their desired SOC. We also show the cumulative number of EVs present, after a particular time

slot is over. We also present the cumulative number of EVs that departed, when the same time slot is over.

3. The Aggregator Model for CCA

Figure 3 shows how the CCA has been implemented. The aggregator model employs bi-directional communication. The aggregator model is assumed to be like a tree model. The aggregator, which is, the central controller commands a fixed number of sub-aggregators, and each sub-aggregator comprise of a finite number of charging ports. Before proceeding to the next time slot, we add the charging rates of all ports in a particular sub-aggregator and find out which sub-aggregator has the minimum charging rate.

The distribution of EVs in the aggregator is based upon the following three cases: (1) When the number of EVs are less than the total number of charging ports in sub—aggregator 1 (Direct distribution). (2) When the number of EVs are less than the total number of charging ports in the aggregator model. (3) When the number of EVs is more than the total number of charging ports in the aggregator model.

4. Distribution Order of EVs in Aggregator Model and Further Calculations

EVs are distributed based on the availability of charging ports in sub-aggregators. After that, charging begins. At a particular time interval, the total charging rate for each sub-aggregator (which is equal to the sum of charging rates at each charging port), is calculated. After each time slot, the new EVs are distributed in the sub-aggregators in the increasing order of the total charging rate.

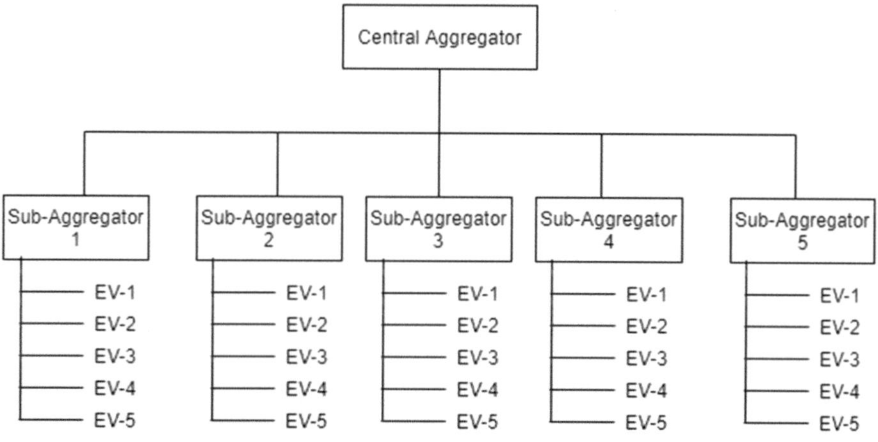

Fig. 3 The aggregator model for CCA

5.3 *Iterative De-centralized Charging Architecture*

1. **Charging Scenario Description**

Figure 4 illustrates the concept of Decentralized charging. In the decentralized approach of charging the EVs, we do not consider the central aggregator. There is one sub-aggregator for one EV, and each of the sub-aggregator is controlled by the power station or utility. Whenever the electric vehicle arrives at the station for charging, and the EV is connected to the charger, the smart controller (present in the sub-aggregator), collects time of arrival of EV, Current SOC, battery type and total battery capacity. At the same time EV user should also set the expected SOC and time of departure.

2. **Iterative Decentralized Charging Algorithm (IDCA)**

In the Decentralized charging algorithm, each EV charge point executes the AIMD algorithm. In the AIMD (Additively Increase Multiplicative Decrease) algorithm, the charging rate of the EVs increases additively with an additive constant and as soon as the total power consumption by all the EVs becomes greater than the total available power, the charging rate decreases multiplicatively with the multiplicative constants decided using a random probable value.

Algorithm 1: Iterative Decentralized Charging Algorithm

do{
 If capacity event **then**
  uniform random number generated, q
  if $q > q_i$ **then**
   $C_i(z+1) = \beta^{(2)} * C_i(z)$
   $SOC_{i,t} = SOC_{i,t} + SOC_{i,t-1} * \exp(1/C_i(z+1))$
  else
   $C_i(z+1) = \beta^{(1)} * C_i(z)$
   $SOC_{i,t} = SOC_{i,t} + SOC_{i,t-1} * \exp(1/C_i(z+1))$
  end if
else
  $C_i(z+1) = C_i(z) + \alpha\,\Delta T$
  $SOC_{i,t} = SOC_{i,t} + SOC_{i,t-1} * \exp(1/C_i(z+1)$
}
While (battery not charged);

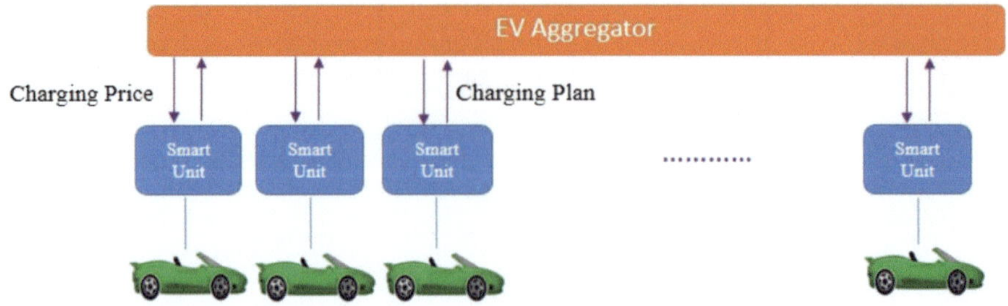

Fig. 4 De-centralized charging scenario

where **Qo** is the maximum available power, **Ci(z)** is the charge rate of **i**th EV, $\boldsymbol{\alpha}$ is the additive constant value in kW/min, $\boldsymbol{\beta}^{(1)}$, $\boldsymbol{\beta}^{(2)}$ are the multiplicative constants selected at random with probability **qi**, $\boldsymbol{\Delta T}$ is the time interval between EV charge rate updates.

A capacity event is deemed to have occurred when the total power **Q(z)** demanded by all active EV charger points at time instance **z** exceeds the maximum available power Q_o at that time instant. Here **N(z)** is computed as:

$$\mathrm{Q(z)} = \sum_{i=1}^{N(z)} \mathbf{Ci(z)} \tag{3}$$

where **N(z)** is the number of active chargers at the zth time instant.

The capacity event condition Q(z) < Qo is monitored by the central monitoring station at the main network distribution substation which alerts the charging points whenever the event occurs in the iterative AIMD algorithm. The main advantage of this approach is that it attains the preferable property while requiring the bare minimum of communication infrastructure and minute computable ability on each EV. Moreover, the basic communication hierarchy and minimum communication bandwidth make it a highly used and cost reducible technique.

6 Battery Model for Electric Vehicle

The most important requirement for an EV is the portable supply of electrical energy, which is converted into mechanical energy in the electric motor for vehicle propulsion. Amongst many portable energy storage devices, Lithium-ion battery is the most popular choice.

In the model of Fig. 5, we use two Lithium-ion batteries. **Battery 1** represents the EV battery and **Battery 2** represents Charging source (or the electric grid in real scenario). Battery 1, i.e.; the EV battery comes for charging with some initial SOC. Battery 2, i.e.; the charging source, acts as the utility or the charging station. In the above model, there are two SOC blocks, i.e.; the **GoTo SOC block** and the **From SOC block.** The **GoTo SOC block** carries the current SOC of the EV when it arrives for charging to **fcn block**. The **fcn block** runs a code, which checks for the current SOC and the desired SOC. From the **fcn block,** two blocks, namely, **the GoTo LoadOn block** and **the GoTo ChargingOn block** emerge. The **GoTo ChargingOn block** becomes the **From ChargingOn block** and is supplied to the two **circuit breakers**, which work only when both the inputs to it is 1. Since the SOC of Battery 2 is greater than the SOC of Battery 1, charge flows from Battery 2 to Battery 1 and charges Battery 1 up.

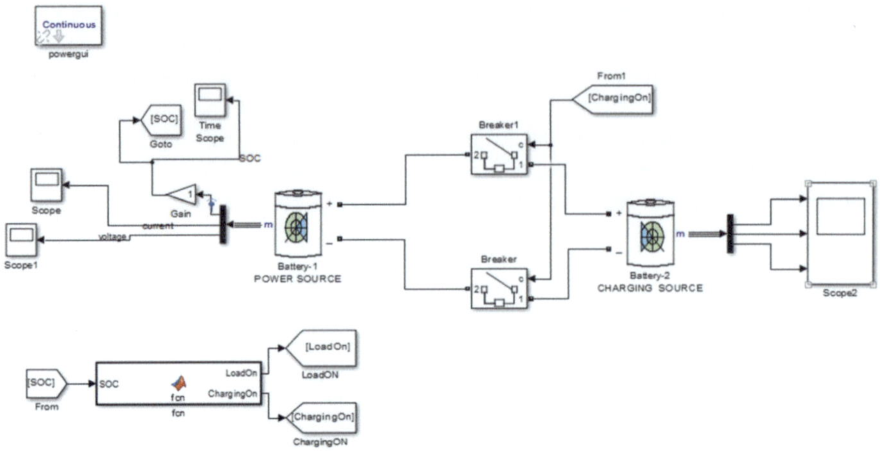

Fig. 5 Simulink model for charging of EV battery

7 Results

The simulations have been obtained using SIMULINK and MATLAB r2015a. For the CCA, we assumed a total duration T = 6 h, starting from 18:00 to 24:00 h, and each time slot of length 1 h. Therefore, we have a total of 6 time slots. The total number of EVs taken for the purpose of the simulation is 10. We also consider rolling price update, i.e. each time slot has a different price of electricity and this price is constant in that time slot, i.e.; the charging rate per time interval is set as 800, 850, 900, 950, 1000, and 1500 respectively. In CCA, we have 6 sub-aggregators and each sub-aggregator consists of 5 charging ports. So, in total, we have 30 charging ports. The number of EVs that arrive in time interval 1, i.e.; t = 18:00 h to t = 19:00 h is taken to be 1. Therefore, in this case, the total number of EVs in time slot 1 is less than the total number of charging ports in sub-aggregator 1. So, the EV occupies the charging port in the first sub-aggregator. The time of arrival of this EV is 18:09 h, the SOC equal to 15% and the time of departure set by EV owner is 21:10 h. The desired SOC is set to 90%. This EV must depart in time slot 4. So, for the first EV, the updated SOC at the end of time slot 1 is 30.2%. Meanwhile, the first EV is charging, we assume that 2 more EVs arrive in time slot 2 at time 19:01 h and 19:06 h respectively. Their current SOC is 16% and 17% respectively. Their desired SOCs are 85% and the time of departures are 21:09 h and 20:30 h respectively. Cumulatively, at the end of time slot 2, EV-1 is 46.35% charged, EV-2 is 31.2% charged and EV-3 is 32.2% charged. While the three EVs are charging, we assume that 3 more EVs arrive at the beginning of time slot 3. The time of arrival for EV-4 is 20:04 h, for EV-5 is 20:05 h and for EV-6 is 20:45 h. Their current SOCs are 16%, 13% and 18% respectively. Their desired SOCs are 90%, 85% and 85% respectively and their time of departures is 22:09 h, 22:17 h and 21:09 h respectively. At the end of time slot 3, EV-1 is 63.45%

charged, EV-2 is 47.35% charged, EV-3 is 48.35% charged, EV-4 is 31.2% charged, EV-5 is 30.1% charged and EV-6 is 35.1% charged. EV-3 departs at end of time slot 3. In time slot 4, we assume 2 more EVs arrive. The time of arrival of EV-7 and EV-8 is 21:05 h and 21:30 h respectively. Their current SOCs are 15% and 18% respectively. Their desired SOCs are 90% respectively and their time of departure is 23:11 h and 22:45 h respectively. At the end of time slot 4, EV-1 is 81.5% charged, EV-2 is 64.45% charged, EV-3 had departed at end of time slot 3, EV-4 is 47.35% charged, EV-5 is 48.15% charged, EV-6 is 47.69% charged, EV-7 is 33.05% charged and EV-8 is 36.05% charged. EV-1, EV-2 and EV-6 depart at end of time slot 4. In time slot 5, we assume 2 more EVs arrive. The time of arrival of EV-9 and EV-10 is 22:08 h and 22:30 h respectively. Their current SOCs are 17% and 16% respectively. Their desired SOCs are 90% and their time of departures is 23:18 h and 23:31 h respectively. At the end of time slot 5, EV-1, EV-2, EV-3 and EV-6 have departed. EV-4 is 64.45% charged, EV-5 is 67.15% charged, EV-7 is 52.05% charged, EV-8 is 55.05% charged, EV-9 is 36% charged and EV-10 is 35% charged. At the end of time slot 5, EV-4, EV-5 and EV-8 depart. In time slot 6, no EVs arrive, so, at the end of time slot 6, EV-7 is 80.55% charged, EV-9 is 64.5% charged and EV-10 is 63.5% charged and they depart.

We have also assumed that all the chargers are not equally efficient. Each has its own charging efficiency in the range from [0, 1]. For each EV, the SOC after every time slot is given by multiplying the obtained SOC with the efficiency of chargers at respective charging ports, and is given as follows:

EV-1: 30.2****0.69***; 46.35****0.74***; 63.45****0.87***; 81.5****0.95***,
EV-2: 31.2****0.74***; 47.35****0.87***; 64.45****0.95***,
EV-3: 32.2****0.74***; 48.35****0.87***,
EV-4: 31.2****0.87***; 47.35****0.95***; 64.45****0.99***,
EV-5: 30.1****0.87***; 48.15****0.95***; 67.15****0.99***,
EV-6: 35.1****0.87***; 50.2****0.95***,
EV-7: 33.05****0.95***; 52.05****0.99***; 80.55****1.0***,
EV-8: 36.05****0.95***, 55.05****0.99***,
EV-9: 36****0.99***; 64.5****1.0***,
EV-10: 35****0.99***; 63.5****1.0***.

Cumulatively,

Number of EVs present at end of time slot 1 is **1**
Number of EVs present at end of time slot 2 is **3**
Number of EVs present at end of time slot 3 is **6**
Number of EVs present at end of time slot 4 is **7**
Number of EVs present at end of time slot 5 is **6**
Number of EVs present at end of time slot 6 is **3**

Number of EVs departed at end of time slot 1 is **0**
Number of EVs departed at end of time slot 2 is **0**
Number of EVs departed at end of time slot 3 is **1**

Number of EVs departed at end of time slot 4 is **3**
Number of EVs departed at end of time slot 5 is **3**
Number of EVs departed at end of time slot 6 is **3**

Figure 6 shows the above data cumulatively, where the thin black line illustrates the number of EVs present at the end of a given time slot and the thick red line illustrates the number of EVs that have departed at the end of a given time slot.

Figure 7 represents the updated SOC of each EV, at end of each time interval. The price of electricity for charging an EV is assumed to be constant during a given time slot and varies over different time slots.

For the IDCA, we have considered input data set for 10 EVs. The total power used is 200 KW. In IDCA, we consider as many time intervals as the total number of EVs under consideration. For example, if we have 10 EVs, then we have 10 time intervals and we assume that one EV arrives in one time interval only. Each time interval is of 1 min duration and the total duration of our simulation is taken to be 120 min.

We first input the maximum duration of simulation as 120 min. Each EV arrives at an interval of one minute with initial SOCs of 21%, 24%, 24%, 24%, 25%, 26%, 20%, 24%, 21% and 23% respectively. The desired SOC for each of the EVs is 60%. The power with which each EV begins to charge, as it arrives at the charging port, is 20 KW, 23 KW, 23 KW, 23 KW, 23 KW, 21 KW, 24 KW, 23 KW, 26 KW and 20 KW respectively. Each power value is subject to an additive increase with additive constant alpha(value = 4), if the total power at a particular time interval, is less than 200 KW, and these power values are subject to multiplicative decrease in the event of the total power exceeding the 200 KW benchmark, at a particular time interval which is decided by the multiplicative constants beta1 and beta2 which are chosen based on the value of random no P_rand which is generated by the MATLAB function rand() and probability Pi with value 0.5 so

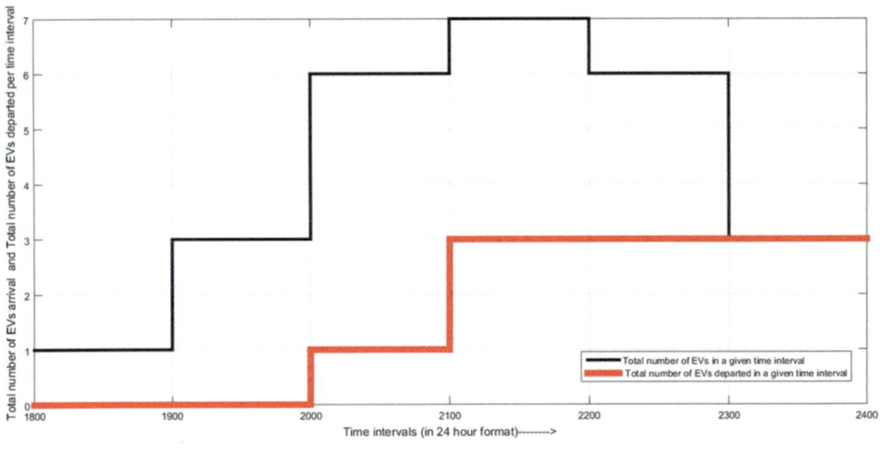

Fig. 6 Graph showing the total number of EVs that arrived and depart in a given time interval

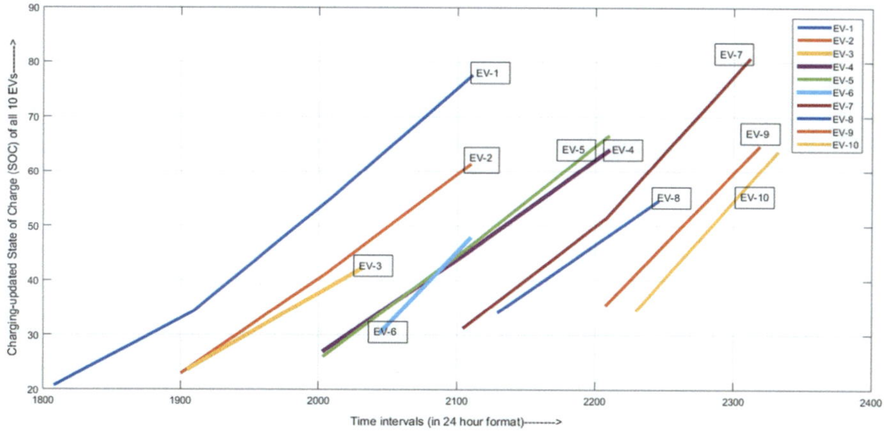

Fig. 7 Graph showing the charging i.e., updated SOC of all 10 EVs

that beta1 and beta2 can be chosen arbitrarily with equal probability. If P_rand < Pi, beta1 is chosen (with value 0.8) and if P_rand > Pi, beta2 is chosen (with value 0.5).

The plot for updated SOC for each of the 10 EVs, using IDCA, is represented in Fig. 8.

The battery model for the EVs has been designed on SIMULINK, MATLAB r2015a. We are using two Lithium-ion batteries. The Battery 1: Represents EV battery and the Battery 2: Represents Charging source (or the Electric Grid in the real world). The parameters for Battery-1 are as follows: battery type is Lithium-ion, nominal voltage is 250 V, rated capacity is 30 Ah (Ampere-hour) and

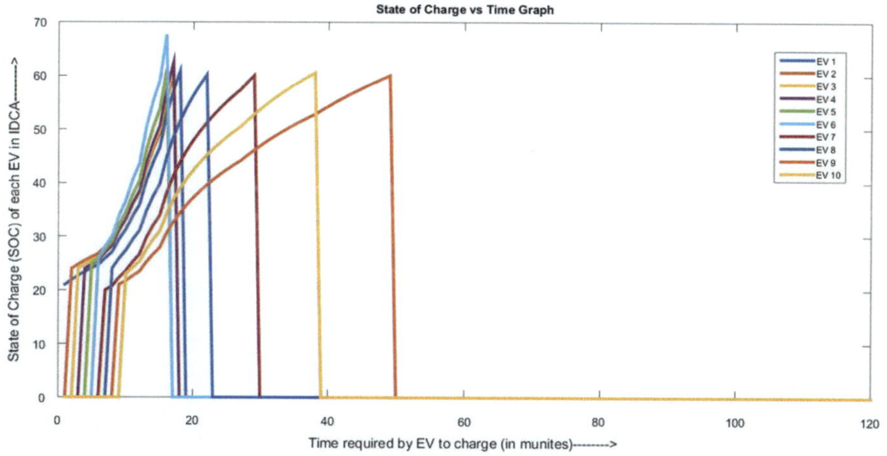

Fig. 8 Plot of updated state of charge (SOC) for 10 EVs

the initial SOC is taken to be 13%. The parameters for the Battery-2 are as follows: battery type is Lithium-ion, nominal voltage is 360 V, rated capacity is 33.1 Ah (Ampere-hour), higher than that of battery-1 and the initial SOC is taken to be 100%.

Figure 9 shows the non-linear model of charging the Battery 1 from its initial SOC of 13–100%. Suppose, we want to charge this Battery-1 to a desired SOC of 95%. We just replace the value “1” in the gain block by “0.95”. This is shown in Fig. 10.

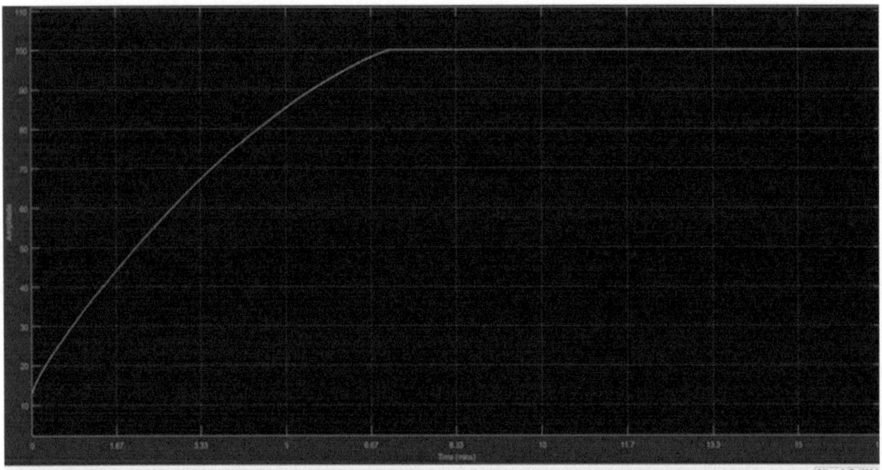

Fig. 9 Plot of updated SOC for battery 1. The maximum updated SOC = 100%

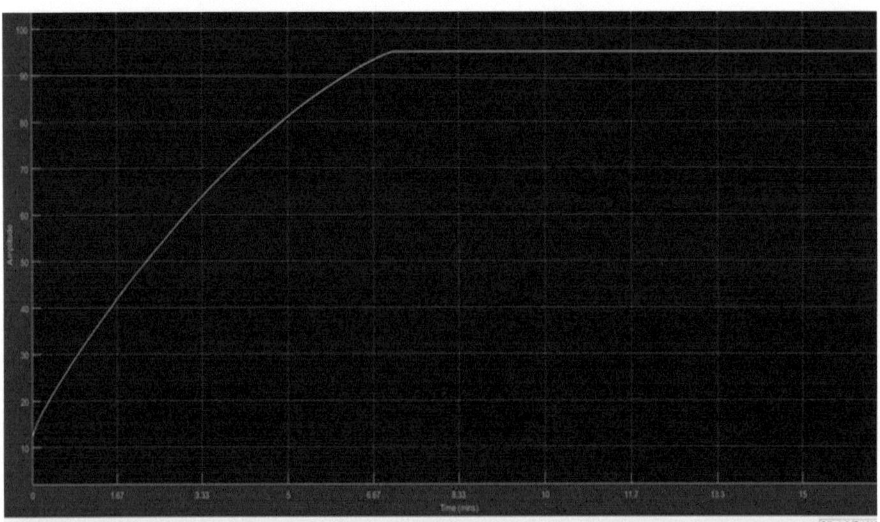

Fig. 10 Plot of updated SOC for battery 1. The maximum updated SOC = 95%

8 Conclusion

In this paper, we have implemented centralised charging architecture and the de-centralised charging architecture for charging the electric vehicles in a grid-to-vehicle network, considering time-of-use pricing model and also developed their algorithms, numerical results and analytical plots for the case of 10 EVs in CCA and IDCA. We also presented the lithium-ion charging model for two cases.

9 Future Work

Our future work includes modelling vehicle-to-grid networks (V2G) and obtaining the performance of each V2G and G2V networks. The authors shall also like to deal with a charging architecture comprising of both CCA and IDCA.

References

1. Foley A, Tyther B, Calnan P et al (2013) Impacts of electric vehicle charging under electricity market operations. Appl Energy 101:93–102
2. IEA (2011) Technology roadmap: electric and plug-in hybrid electric vehicles. In: International Energy Agency. www.iea.org/publications/freepublications/publication/EV_PHEV_Roadmap.pdf
3. Govt push for e-vehicles: No permits, fast-charging stations. The Indian Express. Anil Sasi, 10 July 2017
4. Zhang L, Yan Z, Feng D, Wnag G, Xu S, Li N, Jing L (2014) Centralised and decentralised optimal scheduling for charging electric vehicles. IEEE Trans Smart Grid, Feb 2014
5. He Y, Venkatesh B, Guan L (2012) Optimal scheduling for charging and discharging of electric vehicles. IEEE Trans Smart Grid 3(3), Sept 2012
6. Chung H-M, Alinia B, Crespi N, Wen C-K (2016) An EV charging scheduling mechanism to maximize user convenience and cost efficiency. In: Transactions of Cornell chronicle, June 2016
7. Gan L, Topcu U, Low SH (2013) Optimal decentralised protocol for electric vehicle charging. IEEE Trans Smart grid 28
8. Wang Q, Liu X, Du J, Kong F (2016) Smart charging for electric vehicles: a survey from the algorithmic perspective, July 2016
9. Liu M, Mcloone S (2015) Enhanced AIMD based decentralised residential charging of EVs. Trans Inst Measur Control 37(7):853–867

Harshit Anand was born in Khatauli, Uttar Pradesh, India, in 1995. He is currently a student of Dual Degree B.Tech. in Electronics and Communication Engineering and M.Tech. in Biomedical Engineering at Indian Institute of Information Technology, Allahabad, India. His current research interests include Energy Efficient Systems, IoT, Smart Grids and Electric Vehicles.

Ayush Agarwal was born in Lucknow, Uttar Pradesh, India, in 1995. He is currently a student of B.Tech. in Electronics and Communication Engineering at the Indian Institute of Information Technology, Allahabad, India. His current research interests include Energy Efficient Systems, IoT, Co-operative cognitive relay networks in Wireless Communications.

Pooja Shukla was born in Shaktinagar, Uttar Pradesh, India, in 1995. She is currently a student of B.Tech. in Electronics and Communication Engineering at Indian Institute of Information Technology, Allahabad, India. Her current research interests include Electric Vehicles, Analysis, Algorithm Development and Energy Efficient Systems.

Suneel Yadav received the B.Tech. degree in Electronics and Communication Engineering from Meerut Institute of Engineering and Technology, Meerut, India, in 2008, and the M.Tech. degree in Digital Communications from ABV-Indian Institute of Information Technology and Management, Gwalior, India, in 2012. He has completed his Ph.D. degree in the Discipline of Electrical Engineering at Indian Institute of Technology Indore, Madhya Pradesh, India, in 2016. He is currently with the Department of Electronics and Communication Engineering, Indian Institute of Information and Technology Allahabad, Uttar Pradesh, India, as an Assistant Professor. He has numerous publications in peer-reviewed journals and conferences. He is serving as a reviewer for a number of international journals including IEEE Transactions on Vehicular Technology, IEEE Communications Letters, Elsevier Digital Signal Processing, and KSII Transactions on Internet and Information Systems, IETE Journal of Research. His current research interests are in the areas of Wireless Relaying Techniques, Cooperative Communications, Cognitive Relaying Networks, Device-to-Device Communications, Signal Processing, Physical Layer Security, and MIMO Systems.

Market Diffusion Model of Electric Vehicles for Planning Charging Infrastructure in India

Neeraj Ramchandran, Pradeep Singhvi and Manoj Bansal

Abstract The Indian government has set an ambitious goal of having an all-electric vehicle fleet by 2030. However, limiting factors related to technology, market and policy could impede their adoption. The objective of this research is to forecast how the diffusion of electric vehicles (EVs) will happen in India and the crucial elements that would influence adoption. The research outcomes are expected to help policy makers to optimally phase investments and incentives earmarked for public charging infrastructure. '*Bass diffusion model*' has been used as the base for preparing a system dynamics model using *Vensim* software to forecast the diffusion of EV's from 2017 to 2030. Consumer willingness to purchase EV's has been elicited through a survey conducted among 50 respondents, who drive 4-wheelers. Adoption has been modeled considering the effect of 4 parameters on consumer willingness to purchase EV, namely- price differential between EV's and ICE vehicles, range, recharge time and charging infrastructure density. The model output indicates an S-shaped diffusion curve with saturation near the 50th month. Out of the four parameters, adoption is found to be highly sensitive to charging infrastructure density. The paper concludes that there is a high scope of optimizing government investment in charging infrastructure which would require a detailed view of technical, policy and market related aspects.

Keywords Diffusion · Electric vehicle · Price differential · Recharge time · Consumer willingness

1 Introduction

The transport sector contributes 14% to the total primary energy consumption in India as on 2013. Within the transport sector, road transport is one of the fastest growing end-use sectors contributing 90% of the increase in oil consumption. Road

N. Ramchandran (✉) · P. Singhvi · M. Bansal
GRID-Energy, PwC India, Gurgaon, India
e-mail: neeraj.ramchandran@pwc.com

R. K. Pillai et al. (eds.), *ISGW 2018 Compendium of Technical Papers*,
Lecture Notes in Electrical Engineering 580,
https://doi.org/10.1007/978-981-32-9119-5_32

transport accounts for 86% of the passenger and two-thirds of the freight movement [1]. The road transport sector is predominantly dependent on diesel as the fuel. Dependence on conventional fuels has an adverse implication on the environment, creates uncertainty regarding future energy security and also presents a burden on the foreign exchange exchequer. As on August 2017, India spent USD 35.9 Billion on fuel import [2]. As a mitigation strategy for the above mentioned problems, the Indian Government has set a target for transitioning into an all-electric vehicle fleet by 2030. In the transport sector, the percentage share of 2-wheelers, 3-wheelers and 4 wheeler passenger vehicles among the registered vehicles is 73, 8 and 14% respectively, which present a huge and largely untapped target market for Electric Vehicle and Charging equipment manufacturers [3]. The concept of electric vehicles is at a very nascent stage in India, which has come to the forefront due to Government push rather than market pull. India has traditionally been a country with electricity deficit and only in 2017 has there been power surplus situation at an aggregate level [4]. The barriers to the adoption of electric vehicles are of a multi-faceted nature. As regards the market, there is limited availability of public charging infrastructure, a minimum level of which is viewed as a pre-requisite by customers in making purchase decision [5]. This creates a "chicken and egg" conundrum, in which there will be limited incentive for public investments in electric cars without complementary investments in charging infrastructure and vice versa. Owing to the varying driving conditions across different cities, there is lack of consensus regarding the optimal density of public charging infrastructure among industry players such as fleet management agencies and manufacturers. The availability of indigenous 4 wheeler e-vehicle manufacturers is limited (with only two manufacturers as on 2017) whereas the foreign e-vehicle manufacturers yet to set up shop in India. On the policy front, the Indian Electricity Act prohibits resale of power by any entity which is not a distribution utility or a distribution franchisee. This presents a regulatory hurdle for industry players looking to enter into the energy business. The import duty on electric vehicles is 35.7% [6], which adds to the purchase cost for price sensitive Indian customers.

This modeling exercise carried out in this study concerns itself only with 4-wheeler cars. In India, as far as 4-wheeler cars are concerned, there are three key user groups—(1) Private vehicles, which are registered in the name of a private person and used for private purposes (2) Fleet vehicles, which are registered in the name of a company and used for hire or rent, and (3) Company cars, which are registered in the name of a company and may be used for private purposes by the driver. The uptake in private passenger vehicles is expected to be driven by the market attractiveness, unlike fleet vehicles which can be driven by bulk procurement programs. Therefore, it is important to understand and consider their preferences while designing strategies for phasing government investment in public charging infrastructure network. This study considers only private vehicles in modeling the customer adoption. Also, in this study, we consider only Battery electric vehicles, which are driven purely by electrical energy and not plug-in hybrid electric vehicles (PHEVs) or Range Extended Electric vehicles (REEVs), both of which can be refueled through conventional fuel.

The structure of this paper is as follows: The section on Background provides an overview of the existing methods of modeling diffusion process and literature reviewed as a part of this study. The section on Research methodology explains the approach followed for this study. The model and its key variables are explained in the section named Model. Thereafter the assumptions, analysis and scenarios of the model have been provided. The key findings of the study have been discussed in the section titled 'Summary and Conclusion'. The novelty of this research is that it discusses the diffusion of charging infrastructure considering the interaction with the electric vehicle adoption. This study in its original or extended form can also be replicated in other emerging market countries.

2 Background

There has been considerable research in modelling the market diffusion of new technologies in general and Electric Vehicles in particular. However, the research has largely ignored the role of charging infrastructure in promoting diffusion of vehicles. The type of models have included Agent Based Models and Market Models which are described below:

Consumer Choice Model/Agent Based model: This technique of modeling is essentially a bottom up modeling technique that is applied in situations where the purchase decision is complicated and technology is expensive. A few examples of recent research using Consumer Choice models include Bockarjova et al. [7], Brownstone et al. [8], Potoglou and Kanaroglou [9], Glerum et al. [10].

Market models/Population models: Population models are essentially top down models that predict aggregate outcomes at market level. A few examples of research using Population models includes Keles et al. [11], Lamberson [12] and Meyer and Winebrake [13].

One of the earliest research in market diffusion was the Bass Diffusion model. The Bass model is an economic model that describes the diffusion of a new product due to interaction between a user and a potential user of that product. The differential equation used in the Bass Model is as follows:

$$A'(t) = (p + q^{*}A(t)/M)^{*}(M - (A(t))$$

Where

i. M is the total population of potential adopters which is the fixed saturation level;
ii. p is the coefficient of innovation;
iii. q is the coefficient of imitation;
iv. t is time;
v. A(t) is the adoption at time t.

The solution of the Bass equation is as follows:

$$A(t) = M(1 - \exp(-t(p+q)))/(1 + (q/p)\exp(-t(p+q))))$$

The Bass diffusion model depicts a process in which there are a few early adopters. The population of adopters increases slowly in the beginning. With time, there is increased information sharing between existing and potential users, which creates a reinforcing feedback loop leading to increased rate of adoption.

The literature reviewed as a part of this study has involved interaction between the adoption of alternate fuel vehicles and its diffusion infrastructure. Janssen et al. [14] discussed the market introduction of natural gas cars in Switzerland. They used a combination of stakeholder analysis and system dynamics modeling to analyze the market penetration of natural gas cars. An important observation of this study was that a critical balance is required between investments in fueling stations and natural gas car sales.

A similar study was carried out in context of Hydrogen fuel cell vehicles and refueling infrastructure by Meyer and Whinebrake [13]. The approach involved preparation of a system dynamics model for investigating the complementarity of vehicle and refueling infrastructure. The study concluded that any policy approach for market development of alternative fuel vehicles would be most effective when it simultaneously encouraged the vehicle as well as the refueling infrastructure.

Bongard [15] analyzed the adoption behavior of Dutch population related to electric vehicles using four key elements as defined in Gärling and Thøgersen [16] and using Bass diffusion model as the base model. The adoption model showed exponential growth in the population of adopters, with slow and steady growth in the beginning and a large growth towards the end of the modeling timeline.

3 Research Methodology

The methodology followed for this research included the following steps post the literature review:

1. ***Secondary research on country level transport data***: This data is intended to capture the key parameters related to existing internal combustion engine vehicles such as population of 4 wheelers, price points, fuel costs and associated trends. This data is used to set values for parameters in the model.
2. ***Design of Survey***: Stated preference surveys were carried out on a sample of 50 respondents. Survey tool *Qualtrics* was used to circulate the survey to the intended respondents for eliciting a response online. The objective of the survey was to propose a hypothetical product (EV and charging infrastructure to consumers with the objective of gathering data around the consumer preferences and sensitivities to different features of the product. The data gathered through such surveys was used as the input for the system dynamics model for simulating the future behavior of the consumer.

3. ***Preparation of the system dynamics model***: The model was prepared using a software tool Vensim PLE which allows for modeling quantities that change over time. The steps involved in the preparation of the model are as follows:

   a. *Building the model*
   b. *Specifying input* parameters
   c. *Running the model*
   d. *Checking units*
   e. *Examining the output.*

4 Model

Considering the advantages of using Agent based models over Market models, the model chosen in this study is an agent based system dynamics model. The adoption of charging infrastructure **for Tier 1 cities in India** has been simulated using the system dynamics model and has been formulated to have logistic growth (s-shaped curve). The Bass diffusion model is a fixed saturation model, meaning the saturation point of the adoption is fixed from the start till the end of the timeline. 20% of the total population of 4 wheeler passenger vehicles has been assumed to be in Tier 1 cities. Out of this, 40% has been assumed to be in the private passenger car segment to estimate the saturation level at 22 Lakh vehicles. The timeline considered for the model is from calendar year 2017–2030, which is the targeted year for 100% EV transition. The model has a base part which is based on the Bass diffusion model. The variable where an arrow originates is the causal variable and the variable where it terminates is the effect variable.

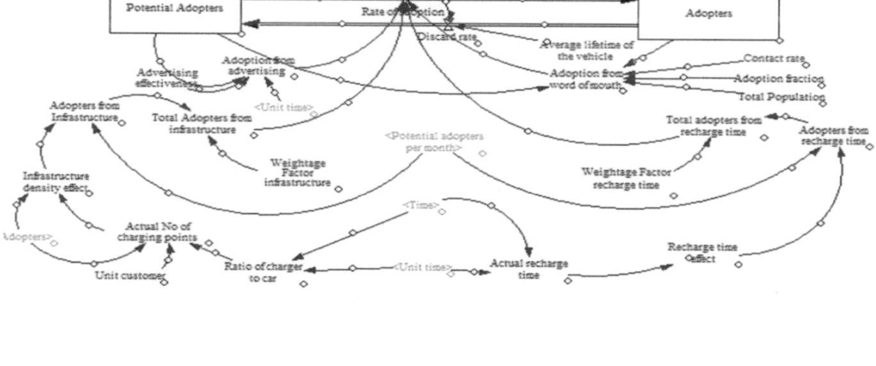

The key variables in the base model and their explanation is provided below:

1. Potential adopters (Unit-Customers): These are 4 wheeler passenger vehicle owners who may potentially convert to Electric vehicles. This is modelled as a stock variable.
2. Adopters (Unit-Customers): The consumers who switch to electric vehicles from traditional internal combustion engine vehicles. This is modelled as a stock variable.
3. Rate of adoption (Unit-Customers per month): The rate at which the potential adopters get converted into adopters. This rate is influenced by the adoption due to word of mouth, adoption from advertising and 4 other variables which have been added to the base model. This is modeled as a flow variable.
4. Discard rate (Unit-Customers per month): The rate at which adopters are converted into potential adopters due to end of life of electric vehicle. In this model the life of an electric vehicle is considered to be 7 years at the end of which the vehicle is retired.
5. Adoption due to word of mouth (Unit-Customers per month): As the name suggests, this variable decides the adoption due to the interaction between adopters and potential adopters. It is in turn influenced by 3 variables namely Contact rate, adoption fraction and total population.
6. Adoption from advertising (Unit-Customers per month): Number of potential adopters who become adopters due to advertising.
7. Time variable (Unit-Dmnl): Simulates time steps of 1 month for 156 months i.e. from 2017 to 2030.

The base model has one reinforcing loop and 2 balancing loops. The reinforcing loop generates the exponential growth and the balancing loop saturates the market and limits growth.

In addition there are four additional variables which have been used based on Bongard [15]. These variables are described below:

1. Actual Driving range (Unit-km): This is the range covered by the vehicle per charge. Driving range has been modeled as a lookup variable, with future values based on forecasts of vehicle range.
2. Actual Fixed Cost difference (Unit-Dmnl): Difference between the fixed cost (ex-showroom price + vehicle registration fee) for an IC Engine vehicle and an electric vehicle.
3. Variable Cost reduction percentage (Unit-Dmnl): Percentage reduction in the cost per month of recharging an electric vehicle with respect to month cost of refueling an IC Engine vehicle.
4. Infrastructure density effect (Unit-Dmnl): Actual Number of chargers installed/ Adopters (Number of EVs).
5. Actual Recharge time (Unit-Minute): Time for recharging an electric vehicle. This variable models fast charging times as based on previous research, Bongard [15], reduction in slow charging times don't have a significant effect on the adoption behavior of consumers.

Similar to Bongard [15], we introduce effect variables for each of the 4 additional variables. The effect variables use the survey outputs to provide percentage adoption for different levels of corresponding variable, e.g. level of adoption when recharge time is between 80 and 130 min. In addition, weightage factors have been introduced for each of the 4 additional variables, which are also influenced by the consumer preference captured through the survey. Overall, the weightages allocated to each of the 6 variables add up to 100%.

5 Assumptions

1. The model assumes all installation of public charging infrastructure to be done by Government and doesn't consider the involvement of private entities, which would evaluate the business case before entry.
2. All IC engine vehicles are assumed to have the same price and all electric vehicles are assumed to be have the same price for the sake of simplicity. Ex-showroom Price of E-car at 0th time step is assumed as INR 11, 99,470 and that of a comparable IC engine vehicle in the 0th time step is assumed as INR 7, 50,000.
3. Registration charges applicable in Delhi NCR region (12.5% of ex-showroom price for price >10 lakh and 8.75% of ex showroom price for price between 6 and 10 lakh) have been considered for simplicity.
4. Variable cost difference has been calculated assuming 80 km/day commute, 26 days of use per month, electricity cost in 0th time step as INR 5.5/kWh (with YoY escalation of 5%), petrol price in 0th time step as INR 70/l (with YoY escalation of 4%).
5. All the tier 1 cities are assumed to have no problems with electricity supply (in terms of quality and quantum).

6 Data Analysis

Based on the response to the survey:

1. The level of adoption pertaining to proposed vehicle range was as provided in Table 1:
2. The level of price premium that the respondents were willing to pay was as follows (Table 2):
3. The preference for charging time is as follows (Table 3):

Table 1 Survey results—range

Range (km)	% adopters
>200	61.90
130–150	4.76
150–180	9.52
180–200	9.52
80–130	14.29

Table 2 Survey results—price premium

Price premium over ICE vehicles (lakhs)	% adopters
0–2	66.67
2–4	14.29
4–5	9.52
5–7	9.52

Table 3 Survey results—charging time

Charging time (min)	% adopters
0–30	68.18
30–60	9.09
60–90	18.18
90–120	4.55

The tables shown above have been used to set the values for the effect variables pertaining to range, price difference and charging time respectively. In addition response obtained regarding reason for not using EV presently have been used to set the values for weightage factors in the model. The fixed and variable cost difference have been added to the model as lookup tables (Figs. 1, 2 and 3).

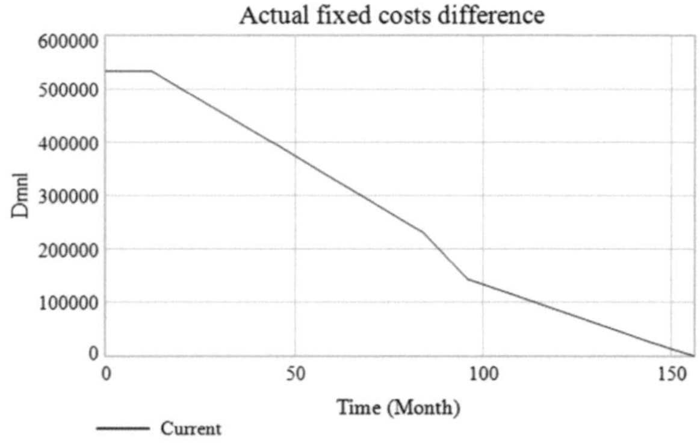

Fig. 1 Actual fixed cost difference

Fig. 2 Actual driving range

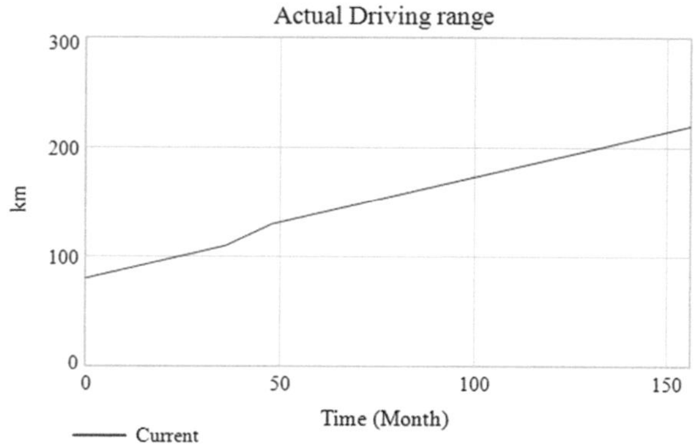

Fig. 3 Actual recharge time

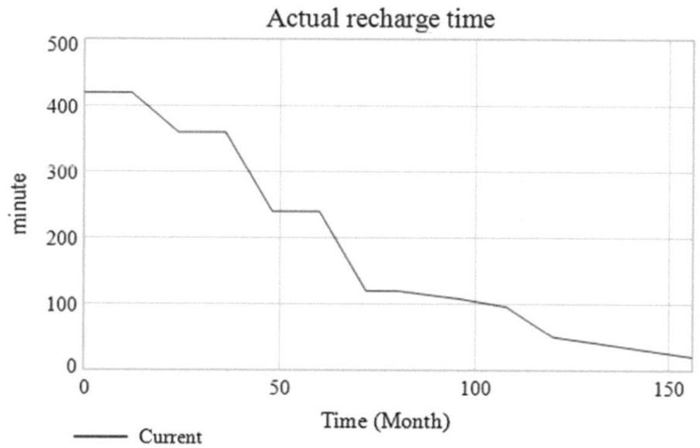

7 Results and Policy Options

Planning for publicly owned bulk procured e-car fleets can be done by applying benchmark ratios (of number of chargers per car) on the number of cars the government intends to procure. However, planning for privately owned passenger vehicle fleets would involve development of forecast regarding the future uptake of the e-cars.

In this study, the chicken-egg conundrum for e-cars and chargers has been modelled by making the rate of adoption of e-cars dependent on the number of charging stations and the number of charging stations dependent on the number of adopters.

Access to workplace charging will bring down the ratio further (Fig. 4).

The model shows two distinct adoption upcycles at the 41st month and 130th month. This is because the life of the e-car modelled is 7 years, post which the vehicle is retired and the customer would come back to repurchase the vehicle. It is also observed that the second curve is broader than the first curve indicating that the first wave will be relatively faster (Fig. 5).

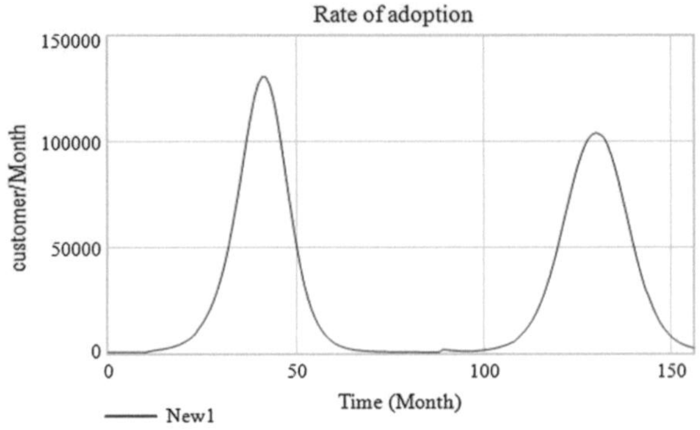

Fig. 4 Rate of adoption

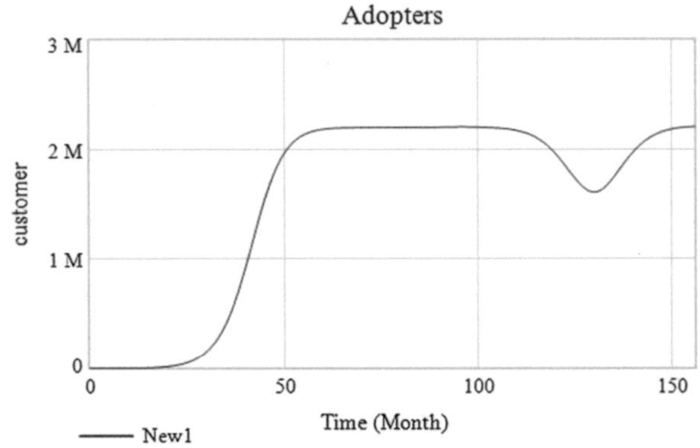

Fig. 5 Cumulative adoption

At a cumulative level the adoption perfectly imitates the classic s-shaped logistic curve, which later shows a dip as there is attrition of adopters due to end of life. Between the 40th and the 50th month, the diffusion reaches critical mass. When critical mass is reached policy incentives become less necessary. Therefore, all policy incentives for accelerating adoption are to be provided before the 40th month (Fig. 6).

It can be seen from the cause strip above that adoption in early stages is more sensitive to infrastructure development than any other variable. Driving range and price are found to matter more towards the end of the timeline. Some policy options to shift them earlier are one time capital subsidy on purchase, lowered interest rate for purchasing EV and tax exemption (Fig. 7).

The figure above shows the adoption of charging points required which has been modelled taking the 'charging point to car' ratio as 1.6 during the 0th time step, which reduces to 0.33 towards the end of the timeline due to transition to fast charging and increased range of e-cars. This presents an interesting trade-off before the Government-whether to invest a lot in not so advanced infrastructure early on (near the peak) and risk it becoming obsolete later on or invest sub-optimally and wait for new technology (faster charging) to invest in.

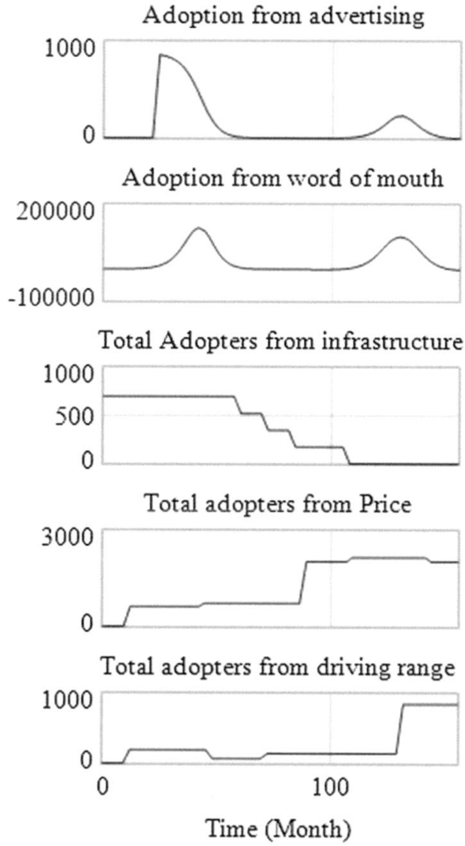

Fig. 6 Adoption rate contributors (customer/ month)

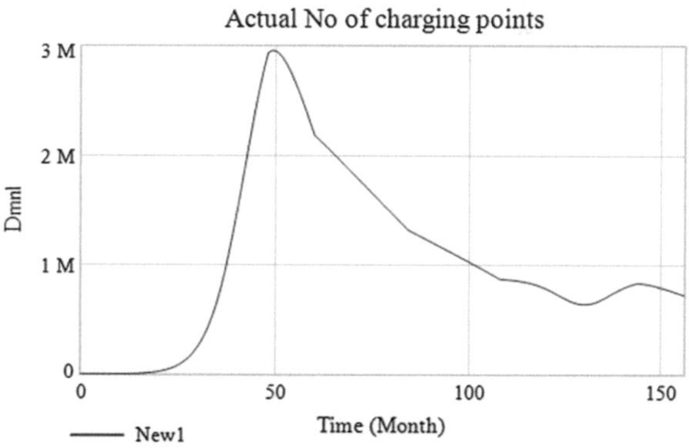

Fig. 7 Requirement of charging points to cater to the e-cars

8 Summary and Conclusion

The actual number of charging points required during e-car upcycle is close to 3 million which gives an average ratio of charging points to cars as 1.34. The paper concludes that there is a high scope of optimizing government investment in charging infrastructure. This will require a detailed view of technical, policy and market related aspects.

9 Limitations

1. The study does not consider the impact of market entry alternate technologies such as fuel cell based vehicle.
2. Bass diffusion model is essentially a model with fixed saturation level which may result in small market prediction.
3. For the sake of simplicity this model does not distinguish between different vehicle classes which will be available in the market and uses one class only. The model does not consider the profitability of the business of operating charging station to have any effect on the government investment.

Acknowledgements The authors would like to thank PwC staff for participating in the survey and for useful discussions. The views represented are those of the authors and only they are responsible for any acts of omission or commission.

References

1. India Energy Outlook (2015) Available: https://www.iea.org/publications/freepublications/publication/IndiaEnergyOutlook_WEO2015.pdf
2. https://energy.economictimes.indiatimes.com/news/oil-and-gas/indias-crude-oil-production-falls-2-per-cent-in-august-import-bill-rises-12-per-cent/60796316
3. http://www.knowindia.net/auto.html
4. http://pib.nic.in/newsite/PrintRelease.aspx?relid=161811
5. Gnann T, Plötz P (2015) A review of combined models for market diffusion of alternative fuel vehicles and their refueling infrastructure. Fraunhofer Institute for Systems and Innovation Research
6. http://www.exportgenius.in/india-import-duexety/electric-car
7. Bockarjova M, Rietveld P, Knockaert J, Steg L (2014) Dynamic consumer heterogeneity in electric vehicle adoption. Innovation
8. Brownstone D, Bunch DS, Train K (2000) Joint mixed logit models of stated and revealed preferences for alternative-fuel vehicles. Transp Res Part B 34(5):315–338
9. Potoglou D, Kanaroglou PS (2007) Household demand and willingness to pay for clean vehicles. Transp Res Part D Transp Environ 12(4):264–274
10. Glerum A, Stankovikj L, Thémans M, Bierlaire M (2013) Forecasting the demand for electric vehicles: accounting for attitudes and perceptions. Transp Sci 48(4):483–499

11. Keles D, Wietschel M, Möst D, Rentz O (2008) Market penetration of fuel cell vehicles—analysis based on agent behaviour. Int J Hydro Energy 33(16):4444–4455
12. Lamberson PJ (2008) The diffusion of hybrid electric vehicles. Technical report, Center for the Study of Complex Systems, University of Michigan
13. Meyer PE, Winebrake JJ (2009) Modeling technology diffusion of complementary goods: the case of hydrogen vehicles and refueling infrastructure. Department of STS/Public Policy, Rochester Institute of Technology
14. Janssen A, Lienin SF, Gassmann F, Wokaun A (2006) Model aided policy development for the market penetration of natural gas vehicles in Switzerland. Transp Res Part A Policy Pract 40:316–333
15. Bongard T (2011) A system dynamics model for the adoption process of electric vehicles from a consumer perspective. Eindhoven University of Technology
16. Gärling A, Thøgersen J (2001) Marketing of electric vehicles. Bus Strat Env 10:53–65

Application of Battery Energy Storage System in Industrial Plants

Anupam Roy

Abstract With the soaring demand for cleaner, safer, high quality and more reliable power, the necessity for an alternative energy sources and smart electrical grid increases which can respond to the growing needs of system operators and energy consumers. Power systems in utility grids and industries also undergoing major structural transition with reducing cost of renewable and energy storage system. In addition there is a strong global pressure to mitigate greenhouse gas (GHG) emission. Battery energy storage system (BESS) emerges to play an important role in stabilizing power supply to industrial plants with improved power quality as well as reducing carbon footprint. BESS performs the tasks of load leveling/peak load shaving, voltage and frequency regulation and maintaining the power supply to critical loads in case of grid outage, virtually removing the requirement of emergency DG power sources. The article discusses about the feasibility of maintaining electrical power to critical loads pertaining to a continuous process plant by integrating grid islanding and load shedding to ensure safe shutdown of the plant. This will save the plant from major capital damage.

Keywords Battery energy storage system · BESS · Peak load shaving · Voltage and frequency regulation · GHG · Smart electrical grid

1 Introduction

As the environmental norms are becoming stricter day by day with reduced carbon footprint, implementation of new fossil fuel based power plants has become a major concern nowadays. But power demand is soaring with rapid urbanization and industrialization. Hence there is a demand for cleaner and safer power which will provide high reliability and quality to the modern power system. Renewable and energy storage technologies present an ideal opportunity to meet the demand supply

A. Roy (✉)
Discipline Head—Electrical, Tata Consulting Engineers Limited, Jamshedpur, India
e-mail: roya@tce.co.in

R. K. Pillai et al. (eds.), *ISGW 2018 Compendium of Technical Papers*,
Lecture Notes in Electrical Engineering 580,
https://doi.org/10.1007/978-981-32-9119-5_33

gap. Most renewable energy sources are intermittent in nature, which poses a great challenge in an effort to balance energy generation and load affecting power system stability and reliability [1]. Various options were evaluated including electrical energy storage (EES) system, load shifting through demand side management and interconnection with external grid etc. Amongst all possible solutions, EES has been recognized the most acceptable approach [1]. Battery Energy storage system (BESS) is considered one of the matured and widely accepted technologies in EES. Advanced lead acid battery, lithium ion battery, sodium sulfur battery and flow battery being the most matured and proven technologies form part of the BESS.

In continuous process industries there is a combination of critical, non-critical and emergency loads. In view of criticality of loads, maintaining the continuity of industrial production and optimization of energy cost and reliability, fossil fuel based (including coal/pit, natural gas and oil) captive power plants are considered essential power sources in addition to utility grid connectivity. However in an effort to reduce GHG emission new industries are facing stiff challenges from statutory authorities in obtaining necessary clearance for setting up such fossil fuel based power plants. Many of these industries use by product gas of higher calorific value to produce power through heat recovery steam generation (HRSG) process. Magnitude of power through HRSG process fully depends on quantity of by product gas. On occurrence of grid power failure operation of process plants and generation of by product gases are badly affected resulting in poor or almost no generation of power. This leads to power blackout condition for the process plants. As there is no power source to meet critical power demand of the plants, possibility of major capital damage of the process plants can't be ruled out. Under such circumstances BESS brings in reliability and stability to the power system of the process plants. Some advantages of BESS are its modular design, short construction time and small environment impact.

2 Benefits of BESS

Distributed BESS strategically placed at a geographical area and connected with the electrical grid can increase the operational performance and reliability of the utility network, helping in peak load shaving/load leveling, alleviating the intermittence of renewable source power generation. BESS supports the realization of smart grids and ensure that energy is readily available when grid power and renewable energy sources are interrupted. In addition voltage and frequency regulation, power smoothing (both active and reactive) and power factor control are other benefits achievable through BESS.

Deployment of BESS for peak shaving application plays an important role for industrial consumers. Today large industrial plants can install the BESS capable of discharging for short periods of time during the peak hours and charging during the low demand periods at night hours, hence reducing the peak demand charge [2].

3 Integration of BESS in Industrial Plant Power System

Typical power scheme with integration of BESS in industrial plant power system as depicted in Fig. 1 may be envisaged to maintain critical process load during disturbance/outages of grid power source. To accomplish this, BESS is connected at 11 kV level of the plant power distribution system, in parallel with the plant loads [3]. If the capacity of BESS and its associated equipment is not adequate to meet the power demand of the plant, on occurrence of grid outage, BESS is exposed to the load surges of the plant bus resulting in discharging of batteries in a shorter period of time. Under such circumstances grid islanding will take place to isolate the grid from the plant power system through active detection method, such as rate of change of frequency (df/dt). Load management system will come into operation to shed the non-critical loads to ensure continuity of battery power supply to critical and essential loads for a time period necessary for the plant to safely shut down its

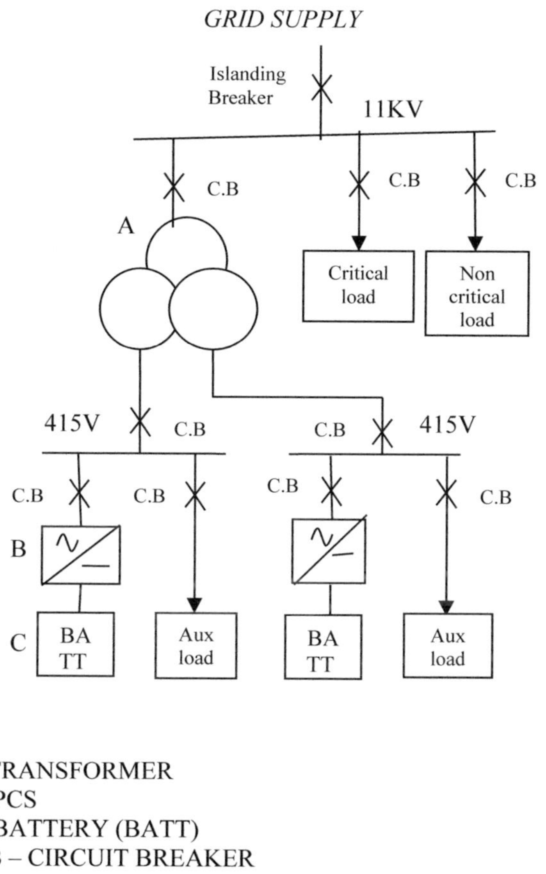

Fig. 1 Typical power scheme with integration of BESS in industrial plants

critical processes. Provision can be made to restore the grid power supply if the utility power is back before the end of the critical processes [3]. The LMS consists of PLC based automation system and a HMI-user interface for operation and parameterisation of the load management system functions.

4 Major Equipment Associated with BESS

Major power and control devices and components involved in BESS are as follows:

(i) Stationary batteries (advanced lead acid, lithium ion, sodium sulfur and flow batteries)
(ii) Battery management system (BMS)
(iii) Power conditioning system (PCS)
(iv) Station control system
(v) Load management system (LMS)
(vi) Network and communication devices
(vii) Harmonic filters
(viii) Relay panel.

Among the battery technologies advanced lead acid, lithium ion (Li–ion) and sodium sulfur (NaS) batteries are quite matured and commercially available. To qualify for utility grid connection batteries should possess the features of high power density, high depth of discharge and higher life cycle efficiency. Li–ion and NaS batteries score higher than advanced lead acid batteries in these respects, although cost wise advanced lead acid battery is cheaper. Vanadium redox also a matured technology is coming in a big way. Battery will be normally connected series-parallel combination to achieve desired voltage and power rating of the installation. One or more battery strings are connected either in parallel to one central inverter or each battery string has its own inverter. Concept of centralized/de-centralized system will be based on the basic tasks intended for BESS to perform, battery capacity and cost.

Battery management system (BMS) is an integrated module of the battery system which is used to control battery charging, discharging operation and monitor each vital element of the battery such as voltage, cell temperature, current, state of charge (SOC), state of health (SOH) etc. The module monitors individual cell voltage. Temperature can be monitored at multiple locations in the module, if not at every cell. It records battery operation for future optimization and warranty management [3]. One more important function of BMS is cell balancing which is used to equalize cell voltage and capacity. All functions are controlled by a smart algorithm and implemented in a PLC. BMS is equipped with built-in multi-protocol communication module to share all operating information for external monitoring and integration with other system. Operator interface consists of industrial grade PC running a graphical interface program for data storage and display [3].

Power Conditioning system (PCS) is a four-quadrant direct current/alternating current (DC/AC) converter connecting the DC system to grid via transformer as shown in Fig. 1 [4]. PCS consists of voltage source inverter and pulse width modulation (PWM) technology. It is a bi-directional equipment which operates either as an inverter or rectifier while discharging or charging the battery respectively. Power converter/inverter is a modular design with reliability and flexibility of enhancing energy storage capacity to be coupled to the grid. Voltage and frequency control and monitoring feature enable the converter to create a micro or islanded grid if a problem occurs with the utility. When utility grid is disconnected, converter should be capable of supplying local load with minimum disturbance. After recovery of grid, converter can resynchronize the island and allow seamless reconnection to the grid. Energy saving feature enables the converter to reduce losses by detecting low power standby condition. As the PCS is a modular design, uninterrupted operation is ensured with N-1 reduced capacity in case of a module failure. Active power (P) and reactive power (Q) are controllable independently. By regulating active and reactive power frequency and voltage support can be achieved respectively. Power factor compensation can be achieved when the active power demand is low. Harmonic filter limits the individual voltage harmonic, total harmonic distortion (THD) and individual current distribution distortion (CDD) as per IEEE 519.

Grid monitoring feature is an integral facility in PCS. Grid voltage and frequency abnormalities are detected by using passive detection methods such as over voltage (59), under voltage (27), over frequency (81O) and under frequency (81U) and active detection methods such as rate of change of frequency (df/dt) and reactive power variation. On delay timer is provided to allow variable parameter to settle to prevent nuisance tripping.

Microprocessor based intelligent controller which is considered an essential component of BESS is responsible for the entire operation of the battery storage solution. While the operation of individual battery block is coordinated by BMS, the operation of all battery blocks is coordinated by the system controller, which manages the total system power and the allotment of that power among the power blocks [4].

DC/AC switchgear is used to control DC and AC power of the BESS. Battery power is transmitted by cable.

Operation, control, monitoring and parameterization of the complete system is envisaged through supervisory control and data accusation (SCADA) system via Ethernet connection. All components such as BMS, PCS/controller, AC/ventilation, fire detection system, DC/AC switchgears are interfaced with Ethernet switch for local/remote control and monitoring of the BESS operation.

5 Conclusion

To overcome intermittency of renewable power sources and over dependency on fossil fuel/industry by product gas based power plants, BESS can be considered as one of the most reliable power sources in order to maintain critical power demand of any continuous process plant. It has the potential to effectively contribute to reduce carbon foot print. BESS can provide important services such as integration of renewable energy, grid stabilization, flexibility, energy security and independence. In spite of battery storage having so many benefits, the cost of the storage unit is a challenge. As environmental concerns becoming prevalent, the potential market for BESS in the future will be much larger than the existing market. New technologies have helped batteries to have higher cycle life, energy density and depth of discharge. With increase in demand and usage, battery storage price will continue to fall. All these aspects will make battery storage more viable for wide deployment of BESS in India and across globe.

References

1. Luo Xing, Wang Jihong, Dooner Mark, Clarke Jonathan (2015) Overview of current development in electrical energy storage technologies and the application potential in power system operation. Appl Energy 137:511–536
2. Oudalov A, Cherkaoui R, Beguin A Sizing and optimal operation of battery energy storage system for peak shaving application
3. Miller (SM) NW, Zrebiec (NM) RS, Delmerico (M) RW, Hunt (NM) G, Design and commissioning of a 5 MVA, 2.5 MWH battery energy storage system
4. Lauder MT, Suthar BK, Northrop PWC, De S, Michael Hopp C, Leitermann O, Crow ML, Santhanagopalan S, Subramanian VR Battery energy storage system (BESS) and battery management system (BMS) for grid-scale applications

Second Life of Energy Storage Battery: Promising Sustainable Growth for Grid and Related Applications

Akshay Ahuja, Hem Thukral and Amol Sawant

Abstract Electric vehicle (EVs) cost is coming down and EVs are getting the top priority in the government's sustainable development agenda worldwide. Usage of fossil fuel will be declining and future of transportation will be electric. As manufacturing of EVs increases, there is tremendous growth in demand for energy storage batteries. Electric vehicle demands high performance from their batteries, so once battery capacity declines to 70–80%, it has to be replaced with a new one. During that point, batteries can still handle a good amount of charge and discharge and thus, there is a second life of a battery which can be deployed at static energy storage applications such as grid storage, renewable energy power plants, ancillary service market, residential usage, data center back-up applications, etc. This paper studies the role of second life of a battery in the electric grid and its contribution towards the growth of renewable energy penetration in India. This paper will look at the options to tackle the different dimensions, pros and cons of the second life of batteries, from the vehicle life to the stationary life; life cycle assessment of second life battery applications and alternatives for ageing vehicles batteries besides the grid applications.

Keywords Second life of batteries · Electric vehicles · Renewable energy · Smart grids · Lithium-Ion batteries · Energy storage

---

A. Ahuja · H. Thukral · A. Sawant (✉)
Pune, India
e-mail: amolrs.49@gmail.com

A. Ahuja
e-mail: akshay.ahuja@hotmail.com

H. Thukral
e-mail: hemthukral17@gmail.com

R. K. Pillai et al. (eds.), *ISGW 2018 Compendium of Technical Papers*,
Lecture Notes in Electrical Engineering 580,
https://doi.org/10.1007/978-981-32-9119-5_34

1 Introduction

India may face long-term economic and environmental challenges relying on these intermittent resources without deploying more energy storage, such as batteries, compressed air, and pumped hydro facilities. Large-scale energy storage technologies specially, batteries can capture surplus renewable energy during times of low demand for later dispatch at the scale needed to decarbonize our electricity supply over the coming decades. India with its aggressive targets for renewable energy would need the additional support from energy storage technologies in order to deal with the intermittent nature of the renewable energy. Presently the cost of Lithium-Ion batteries is high with an average cost of $209 per kWh which comes down to lot more in India as at present, Lithium-Ion cells are not produced in India and are imported which increases the cost of Lithium-Ion battery packs [1].

With the Government of India's push for Electric Vehicle and as electric vehicles proliferate, so too will used EV batteries. Repurposing retired electric vehicle (EV) batteries for stationary applications provides a potential way to reduce first-cost hurdle of lithium-ion batteries used for EVs.

Presently, higher price of EVs is impending the public acceptance as compared to ICE vehicles which cost comparatively less. Another additional cost which comes with the EV is battery replacement cost which needs to be replaced after the life of the battery. Typically, an EV battery needs to be swapped out once the remaining capacity is 70–80% of the new battery of similar size which is called end of life of battery for mobility applications. At that point, when a battery fails to serve for mobility application, it can still handle sufficient charge to serve less demanding stationary applications like renewable energy storage and integration, UPS etc. This use of battery for the secondary applications will help in delaying the process of recycling the battery or disposing of the resourceful product used in batteries [2]. Giving these batteries a second life will help in bringing down the cost of the EVs and storage for the stationary applications.

2 Battery Chemistry for Mobility

Multiple chemistries are going to exist as the technology evolve and multiple technologies are going to be used for different applications due to their specific requirement and advantage which a particular technology can provide. No one technology is complete in all the dimensions but have its own advantages and disadvantages. Table 1 shows 5 parameters for various chemistries which are available in Lithium-Ion [3].

Table 1 Lithium-ion chemistries

	LCO	NMC	LMO	LTO	LFP	NCA
Specific energy (W h/kg)	150–200	150–220	100–150	50–80	90–120	200–260
Cycle life	500–1000	1000–2000	300–700	3000–7000	1000–2000	500
Charge C rate	0.7–1C	0.7–1C	0.7–1C	1C	1C	0.7C
Discharge C rate	1C	1C	1C	10C	1C	1C
Cost	2	2	2	4	2	3

Cost 1 represents lowest and 5 represents highest
LCO Lithium Cobalt Oxide
NMC Lithium Nickel Manganese Cobalt Oxide
LMO Lithium Manganese Oxide
LTO Lithium Titanate
LFP Lithium Iron Phosphate
NCA Lithium Nickel Cobalt Aluminum Oxide

3 From Internal Combustion Engine to Batteries

Charging of electric vehicles takes longer than pumping fuel to an ICE vehicle. Though the Lithium-ion battery has less energy per weight as compared to the fossil fuel, it takes longer to recharge compared to the refueling of fuel. A modern Li-ion for EVs produces up to 250 W h per kg; energy from fossil fuel is 13,000 W h/kg, 50 times higher but the advantages of the electric drive are high energy efficiency and clean power [4].

A 50 L of fuel in tank holds a calorific value of 600 kWh with a quick refill whereas, battery stores only 20–100 kWh of energy (even more for buses) and takes more time to charge/refill. The two types of charging which are used for the charging of electric vehicles are alternate current (AC) and direct current (DC) charging systems. Table 2 shows the level of charging which are available globally.

The battery is an electrochemical device that can only absorb a given amount of energy and in order to do the ultra-fast charging of the battery it must meet few conditions mentioned below:

1. Chemistry should support ultra-fast charging
2. Battery condition should be good as old batteries have slow charging acceptance

Table 2 Different charging levels

Power levels	Grid voltage/current (in amperes) input	DC power to battery (kW)
AC level 1	108–120/15–20	$\sim$1.4– 2.4
AC level 2	208–240// $\geq$ 30	$\sim$7.2–19
DC level 3 (DCFC)	400–800/ $\geq$ 120	$\geq$ 50

3. Ultra-fast charging works only till 70% state of charge of the battery and post that the charging is comparatively slow.

4 Applications of Second Life of Battery

As manufacturing of electric vehicles increase, there is a tremendous growth in demand of energy storage batteries. As electric vehicles demand a very high performance from their batteries, once the capacity of the battery falls to about 70–80%, they have to be replaced with a new battery. At that point in the lifetime of the battery, it can still handle a good amount of cycles of charging and discharging, and thus there exists a potential second life of a battery which can be deployed for numerous applications, including but not limited to, residential batteries, grid-scale storage, solar PV firming, frequency regulation and data centre powering.

4.1 Residential Batteries

One of the biggest applications of used batteries of electric vehicles will be batteries for residential purposes. The residential energy storage market is expected to rise to about US$ 5.4 billion annually by 2025 and used batteries are expected to contribute significantly [5]. To name a few, automakers such as BMW, Mercedes and Nissan have already started offering residential batteries that are no longer fit to server their purpose in electric vehicles.

4.2 Grid Storage

With increasing amount of renewable energy resources being used, used batteries of electric vehicles can be used as grid-scale (MW-scale) batteries for supplying power to the grid in case of insufficient power in the grid or a fault. Hence utilities will be able to balance the grid in near-real time and without releasing any greenhouse gas emissions.

4.3 Solar PV Firming

With the Governments around the world increasing the usage of solar PV which is intermittent, used batteries of electric vehicles can be used for solar PV firming. As a result, the a.c. side of the grid meter will see a continuous power output, thereby increasing the reliability and stability of renewable energy.

4.4 Frequency Regulation

In order to regulate the grid frequency, gas turbines supported by used batteries of electric vehicles can be used as batteries will absorb the sudden changes from the electricity grid operator while smoothening the working conditions of gas turbines. As a result, there will be a significant improvement in the quality of power supplied.

4.5 Date Centre Powering

As data centres are increasingly being powered by renewable energy resources, used batteries of electric vehicles can be used to alleviate the unpredictability of renewables. Thus a combination of used batteries of electric vehicles and renewable energy resources will be able to power a data centre both, while connected to and independent of the grid.

5 Challenges

More and more use of Li-Ion batteries will bring down the cost of batteries but the question arises as to what should be done with the retired batteries after they are swapped out from the EVs. The lead acid batteries have been recycled in the past but recycling the lithium-ion batteries needs heavy investments which foster reusing these retired lithium-ion batteries for other stationary applications which would help in delaying the process of recycling the battery. At present, less than 5% of spent li-ion batteries are currently recycled globally. Moreover, existing li-ion battery recycling method unit economics are often unprofitable and strategic components such as lithium are not recovered [6]. While this is true in some applications, there are several reasons why reusing EV batteries is not ideal for most stationary energy storage applications.

Challenges which we would face while re-using batteries in second life:

1. Motivating vehicle owners to dump their vehicle batteries at the provided market place.
2. Availability of enough incentives for vehicle owners for participating in second life of battery program.

3. As we have car selling dealers and networks, we need to set up battery selling dealers and networks who should collect those batteries and make them available for re-use.
4. Calibration of remaining capacity of second life of batteries is a big challenge. We suggest that not all batteries can be suitable for stationary applications. Capacity check and safety test should be conducted before re-use.
5. Lack of data availability on battery performance in first and second life applications.
6. Creation of safety and performance standards for second life of battery applications.
7. Awareness creation and knowledge dissemination among the markets and regulatory bodies about second life of batteries.
8. Utility and grid operators should allow industry players to test and run pilots with second life of batteries on grid applications
9. Most importantly, vehicle manufacturers should design the BMS in such a way that, batteries will be prepared for first and second life uses.

5.1 *Complexities of Second-Life Use*

The process of making a battery ready for the reuse is a resource intensive process and not as simple as it may sound to be. Getting a battery ready for reuse requires the packs to be dismantled into cells and these cells needs to be tested to determine the state of health as low capacity batteries would not be solving the desired purpose of stationary applications. The cells must be sorted according to similar remaining capacity to make the performance better. These batteries then need to be fixed with the new or modified battery management system to suit for the required applications. These all process are resource and cost intensive process which may discourage the manufacturers or recycling companies to invest in the refurbishment of lithium-ion batteries for their second innings [7, 8].

Cell packaging type	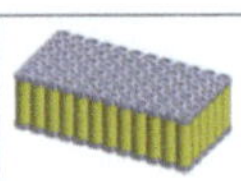		
	Cylindrical	Pouch	Prismatic
Advantage	• High specific energy • Long calendar life and low cost	• Light and cost effective	• Space efficient
Disadvantage	• Less ideal packaging density	• Exposure to humidity and high temp can shorten life	• Costlier to manufacturer
Major manufacturers	Panasonic	LG Chem, AESC	Samsung SDI, BYD
Models	Tesla	Renault Zoe, Nissan Leaf	BYD

Apart from the processing of these batteries, some other issues which will impact the second life of the battery is variation in the chemistries of the lithium-ion which is used in the electric vehicles. For example, the chemistries like nickel cobalt aluminum oxide (NCA) used by some of the EV manufactures to provide higher torque and acceleration are not suited for the stationary applications due to its low cycle life whereas, other chemistries like lithium iron phosphate (LFP) and nickel manganese cobalt oxide (NMC) are suitable for the stationary applications which can be used in the second-life application as the battery coupled with a photovoltaic installation can have a lower energy density, but it needs a longer cycle life, it needs to be able to be charged and discharged many more times [9].

Some other challenges which could act as a barrier in second life of the battery would be lack of regulatory policies and lack of standardization of remanufacturing process which could lead to lack of investment in this domain.

6 Conclusion

As the electric vehicle production and use increase in the market, business models for second life of batteries will start taking different folds. Roles of government, manufacturers, dealers and end consumers are very vital in this EV era. The amount of EVs going to come into the market and limited natural resources going to get consumed are worth considering segments in this paradigm shift of road transportation. While overcoming these challenges of second life of battery applications, it is very important to understand the battery from its first life usage. As the number of EVs increases, role of regulator becomes very vital to keep a cap on too many variables in development of market for second life of batteries.

References

1. The latest bull case for electric cars: the cheapest batteries ever. Bloomberg New Energy Finance, Dec 2017 https://about.bnef.com/blog/latest-bull-case-electric-cars-cheapest-batteries-ever/
2. Spector J (2016) The road to thriving second life EV battery market. Greentech Media, Sept 2016 https://www.greentechmedia.com/articles/read/the-road-to-a-thriving-second-life-ev-battery-market#gs.ykxqez
3. BU-205: types of lithium-ion. Battery University https://batteryuniversity.com/learn/article/types_of_lithium_ion
4. Buchmann I, Ultra-fast charging. Battery University https://batteryuniversity.com/learn/article/ultra_fast_charging
5. Residential energy storage report. Navigant Research https://www.navigantresearch.com/reports/residential-energy-storage
6. Kochhar A, Lithium battery recycling—a global opportunity for positive impact. Li-Cycle https://li-cycle.com/2017/04/12/lithium-battery-recycling-a-global-opportunity-for-positive-impact/

7. Reid G, Julve J (2016) Second life batteries as flexible storage for renewable energy, Apr 2016 https://www.bee-ev.de/fileadmin/Publikationen/Studien/201604_Second_Life-Batterien_als_flexible_Speicher.pdf
8. BU-301a: types of battery cells. Battery University https://batteryuniversity.com/learn/article/types_of_battery_cells
9. Robinson C (2017) Recycle vs reuse: why EV batteries may not often get a second life as stationary applications, energy-storage. News, Jan 2017 https://www.energy-storage.news/blogs/recycle-vs-reuse-why-ev-batteries-may-not-often-get-a-second-life-as-statio

Virtual Energy Storage from Air Conditioning Loads

Prabir Barooah

Abstract It is often said that energy storage holds the key to the power grid of the future due to the highs and lows of uncontrollable generation caused by intermittent renewable sources such as solar and wind. We argue that many loads can offer the same service as batteries. With appropriate intelligence, loads can manipulate demand around a nominal baseline consumption so that the increase and decrease of demand around that baseline appears like charging and discharging of a battery. Loads thereby can act as virtual energy storage (VES) systems. A frequency domain approach is useful to describing the tradeoffs between VES capacity and consumers' quality of service (QoS). We describe a way to characterize the VES capacity of loads in terms of its power spectral density and analyze the case for air conditioning loads in particular.

Keywords Demand response · Grid integration · Renewable energy · Buildings to grid · Demand side control · Energy storage

1 Introduction

With increase in the use of green energy sources such as solar and wind, the power grid is likely to experience significant intermittency in generation. For an example of such intermittency, see Fig. 1; the data comes from BPA (www.bpa.org), a balancing authority (BA) in the Pacific Northwest of the United States. The figure also shows the net demand, which is the difference between demand for power and (uncontrollable) renewable power generated. The net demand must be supplied by conventional, i.e., controllable, generation resources such as coal and nuclear generators. The sharp ramps and fast variations in the net demand are a cause of concern for conventional generators. They are designed for supplying slow-varying

P. Barooah (✉)
Department of Mechanical and Aerospace Engineering,
University of Florida, Gainesville, FL 32611, USA
e-mail: pbarooah@ufl.edu

R. K. Pillai et al. (eds.), *ISGW 2018 Compendium of Technical Papers*,
Lecture Notes in Electrical Engineering 580,
https://doi.org/10.1007/978-981-32-9119-5_35

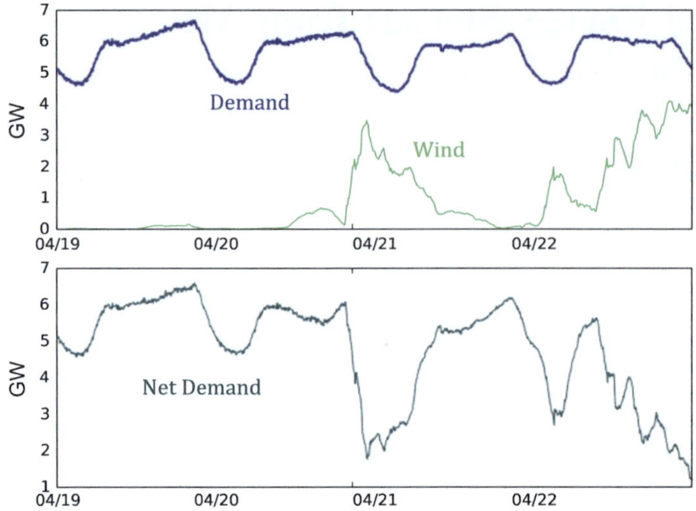

Fig. 1 (Top) Total demand and renewable generation, and (bottom) net demand in BPA (Bonneville Power Administration: www.bpa.gov), April 19–22, 2016

baseload, and not are not designed to track such a fast varying signal. Inability to track the net demand can seriously degrade reliability of the power grid and the service it provides. Demand-supply imbalance causes the grid-frequency to deviate from the nominal value, which adversely affects certain consumers that require a constant frequency. A sufficiently large frequency deviation can cause cascading blackouts, such as the one India experienced in 2012.

Clearly, additional resources are needed to mitigate the volatility created by solar and wind. One possibility is installation of additional fast-ramping conventional generation, such as hydro and gas, and to keep them on standby. Hydro is limited by geography, while use of additional fossil-fuel based power plants as backup will negate the environmental benefits of renewables, apart from increasing overall cost of energy. Another possibility is the use of energy storage resources such as batteries, flywheels, pumped hydro, and compressed air systems. Among these, electrochemical batteries seem to be the only scalable solution. It is often said that energy storage, and in particular, battery technology, holds the key to a sustainable energy future. At present, however, using batteries to store renewable generation at the scale of a national power grid is a prohibitively expensive option.

Another possibility is to employ demand side resources. Loads have been used for a long time to provide demand response (DR) across the world. DR usually involves reducing peak demand in response to contingency events. However, loads can provide much more than only contingency reserves. Many loads have flexibility in power consumption, and this flexibility can be exploited to vary their power consumption without adversely affecting the consumers' quality of service (QoS). With appropriate intelligence, loads' demand can be varied up and down around

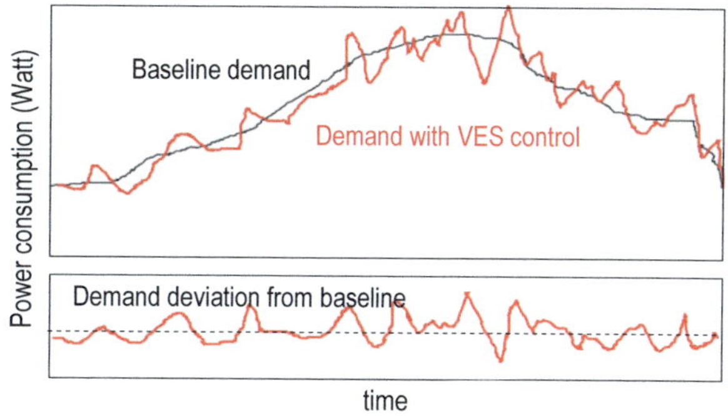

Fig. 2 Virtual Energy Storage (VES) from loads: demand is varied around the load's baseline so that the demand deviation is akin to the charging and discharging of a battery

their nominal demand—the baseline—in such a way that mismatch between demand and generation is reduced. The loads in effect provide the same service as that of a battery. We call this virtual energy storage (VES) from flexible loads; see Fig. 2 for a schematic illustration of the VES concept. The key constraint in the use of loads to provide VES is to ensure that consumers' QoS is maintained within acceptable bounds.

There is an extensive and growing literature on the control of loads to provide various grid services. Among various ancillary services, use of loads to provide frequency regulation has drawn the most attention in recent times; see [1–3] and references therein. It is argued in [4, 5] that loads can be used to provide services at various time-scales, not just the seconds-to-minutes time scale of frequency regulation. A Fourier (frequency-domain) decomposition provides a convenient framework to assign grid's needs to all supply side resources, including traditional generators, loads providing VES service, and batteries providing real energy storage services. Base-load power generation is scheduled based on predictions of the net demand at the slowest time scale (lowest frequency), load following and frequency regulation at intermediate and fast time scales is performed by automatic generation control that adjusts generation set points [6].

In this paper we expand upon the frequency domain characterization espoused in [4] for capacity characterization of VES systems. We argue that VES capacity is most conveniently expressed as a function of frequency, and in the general case, as a power spectral density of the power variation (from the baseline) for the load.

Among all loads, HVAC (heating, ventilation, and air conditioning) loads provide perhaps the best resource for providing VES. Air conditioning accounts for a large fraction of the demand, especially in advanced economies. In addition, the thermal inertia of a building acts as a low pass filter, so that fast variation in power consumption does not lead to large change in the indoor climate. That makes it easy to maintain consumers' QoS. In this paper we focus on air conditioning loads for that reason. However, the definitions and method of capacity characterization discussed here are applicable to any type of load in which power consumption can be varied continuously. For a load that can only be turned on or off, when considers a collection of them, again these concepts become applicable.

The rest of the paper is structured as follows. Section 2 describes the VES idea in detail. Section 3 describes VES capacity characterization for air conditioning loads by using a simple model of the thermal dynamics of a building with air conditioning. The paper concludes with a summary in Sect. 4.

2 Virtual Energy Storage (VES) from Loads

A load's power consumption can be varied around a baseline to provide a battery like service. Let $p_b(t)$ be the baseline power demand of a load (or a collection of loads). Suppose its (their) demand is varied through the use of appropriate control software to be $p(t)$ so that the demand deviation from the baseline:

$$p_{\text{ves}}(t) := p(t) - p_b(t) \tag{1}$$

is zero mean: $\lim_{T\to 0}\frac{1}{T}\int_0^T p_{\text{ves}}(t)dt = 0$; cf. Fig. 2. It can be said that the load is now providing VES, or, that it is acting like a "virtual battery". The demand deviation $p_{\text{ves}}(t)$ is the charging power consumption of the virtual battery: positive $p_{\text{ves}}(t)$ means the load is drawing more power from the grid than the baseline value; so the virtual battery is charging. Conversely, negative $p_{\text{ves}}(t)$ means the virtual battery is discharging. The zero mean nature of the demand deviation means the net energy consumption/generation of the virtual battery is 0, just like a real battery.

The two key questions that arise are:

- For a specific load and a bound on the acceptable change in the QoS, what kind of demand deviation ("virtual charge/discharge signal") $p_{\text{ves}}(t)$ is allowable that ensures the QoS bound is satisfied? In effect, what is the VES capacity of a load? And, how does this capacity vary from load to load?
- How is the net demand signal to be apportioned among the loads so that together they can supply it, while each load maintains its QoS bound?

In this paper we examine the first question, and specialize to air conditioning loads. The second question has been the topic of a number of works; see [2, 7, 8] and references therein. We therefore do not address the second question here.

2.1 Constraining Loss of QoS Via Constraining Bandwidth

QoS measures are a function of load type. There are a large variety of flexible loads, such as refrigeration systems, electric vehicles, pool pumps, water heaters, data centers, municipal pumping systems, HVAC systems, etc. that can be used for VES. Each has their own QoS metrics, and a distinct degree of flexibility. Measures of QoS that are specific to HVAC systems include indoor temperature and ventilation rate, the latter being a surrogate for indoor air quality. Hot water heaters—and pool pumps in some areas—are also large sources of demand. A QoS measure for a pool pump is the average number of hours the pump is on (as a surrogate for water cleanliness) [9, 10]. For hot water heaters, it is the availability of hot water that is critical. For an aluminum plant, a measure of QoS is the temperature of the smelter [11]. For all types of loads the cost of energy used and equipment lifetime are also QoS measures.

The diversity of QoS metrics among distinct load types is a challenge in developing control algorithms to exploit their demand flexibility. It is argued in [4] that in fact an unifying framework can be developed based on the spectral content of the demand variation. For every load type, maintaining a specific bound on the QoS can be translated to maintaining a bound on the *bandwidth* of its demand deviation. For instance, a small and fast variation of power consumption of a commercial HVAC system can be obtained by a small and fast variation of airflow. The resulting temperature deviations will be small since the large thermal inertia of the building will act like a low pass filter to such airflow variations. This has been experimentally verified [3]. However, even a small amplitude airflow variation can lead to large deviation in indoor temperature if the variation persists for a long time, i.e., the frequency is small enough. For a given amplitude, higher the frequency of airflow variation, the smaller the effect on QoS metrics of indoor temperature and average ventilation rate. However, above a certain frequency, QoS will reduce since equipment life will degrade. Figure 3 illustrates this idea. For loads that can only be turned on or off, such as hot-water heaters, again limiting the frequency of turning on and off is needed to reduce short cycling and ensure delivery of hot water.

When the demand variation, $p_{\text{ves}}(t)$, is a pure sinusoid, the VES capacity is the maximum amplitude of the sinusoidal variation that still satisfies the QoS constraints. This amplitude has the unit of power, and therefore represents the power

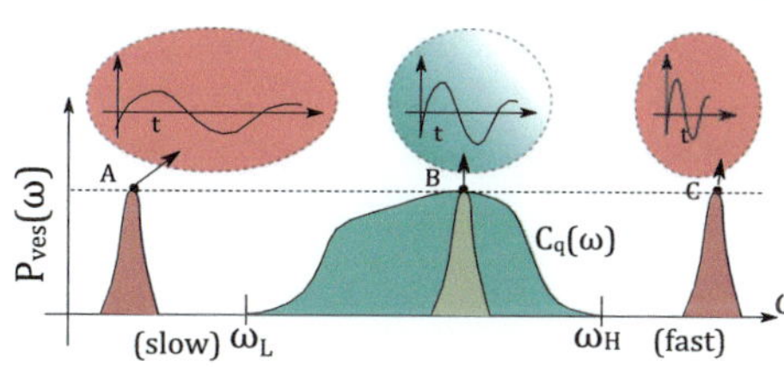

Fig. 3 Constraint on QoS is a constraint on bandwidth of demand variation. The x-axis is frequency and the y-axis is the PSD of demand variation. The PSD must lie in the region under the curve $c_q(\omega)$ to meet the QoS measure q. For a different value of q, this curve would change. The low and high limits of the frequency in which this particular load class can provide VES service are denoted as ω_L and ω_H, respectively. The three signals shown in A, B, and C, have PSDs that have the same total power (i.e., the integral of their PSDs are the same), but distinct bandwidths. The signals A and C violate the QoS metric q, because their bandwidths are too low and too high, respectively. The signal B satisfies the bandwidth requirement

capacity of the virtual battery. The energy capacity of the virtual battery can then be obtained by integrating the sinusoidal power deviation with amplitude equal to the power capacity.

For more general power deviation signal, the VES capacity of a load can be characterized in terms of the power spectral density (PSD) $P_{ves}(\omega)$ of the demand variation, $p_{\text{ves}}(t)$. The PSD must lie in a specific region to meet a given QoS constraint; the latter parameterized by, say, a scalar q. For every value of q there is a curve $c_q(\omega)$ so that QoS will be respected only if the PSD of p_{ves} lies under the curve $c_q(\omega)$. *The curve corresponding to the minimum acceptable QoS* $q*$ *determines the load's VES capacity. We call* c_q* (ω) *the load's VES power capacity density.* Since the unit of the PSD of power deviation is Watt^2/Hz, the energy capacity of the virtual battery can be determined from the power capacity density curve by integrating its square root with respect to frequency: $E_{q^*} = \int \sqrt{c_{q^*}(\omega)} d\omega$.

An illustration of the curve $c_q(\omega)$, for some q, is shown in Fig. 3. For a specific load, or load-class, the curve $c_q(\omega)$ can be determined either through modeling or experimental evaluation [3]. The next section describes how to do this for air conditioning loads.

3 Determining VES Capacity of Air Conditioning Loads

To determine the VES capacity of a particular load, one needs the relationship between how the power deviation from the baseline affects the QoS. This relationship may vary greatly from one load class to another, and on the QoS metrics. In this paper we focus on HVAC loads whose power consumption can be continuously varied between 0 and a high value. Most modern commercial buildings are equipped with Variable Air Volume (VAV) systems; these fall into the category of "loads with continuously variable demand". These are distinct from most residential loads whose demand can be varied usually between two distinct values corresponding to on and off. The QoS constraints in this case include temperature, humidity, and weekly operating cost. In this paper we consider only one QoS metric: indoor temperature.

3.1 Model Relating Power Deviation to QoS for HVAC Systems

We use the standard resistor-capacitor (RC) network model for the dynamics of indoor temperature T in a building with an HVAC system [12]. In particular, we use the following first-order RC model:

$$C\frac{d}{dt}T(t) = -\frac{1}{R}(T(t) - T_A(t)) + Q_{gs}(t) - Q_{A/C}(t)$$

in which C (J/K) is the total thermal capacity of the building, R (K/W) is the effective thermal resistance of the building's envelope to heat exchange between the interior space and the ambient. The variable $T(t)$ is the indoor temperature, $T_A(t)$ is the ambient temperature, $Q_{gs}(t)$ (W) is the total rate of heat generated inside the building including solar heat gain, and $Q_{A/C}$ (W) is the rate of cooling provided by the HVAC system.

The electric power consumption P is approximated by $Q_{A/C}$/COP, where COP is the coefficient of performance of the HVAC system. This is only an approximation since there are efficiency losses, and phase lag between electric power consumed and thermal (cooling) power provided.

The baseline behavior of the building is taken to be static. Using an asterix "*" to denote steady-state quantities, the steady state cooling $Q^*_{A/C}$ provided by the A/C and the steady state indoor temperature T^* satisfy the linear equation,

$$0 = -\frac{1}{R}(T^* - T^*_A) + Q^*_{gs} - Q^*_{A/C}$$

The deviation about these steady state values satisfy,

$$\frac{d}{dt}y(t) = -\frac{1}{\tau}y(t) - \frac{1}{C}v(t),$$

in which $y(t) = T(t) - T^*$, $v(t) = Q_{A/C}(t) - Q^*_{A/C}$, and $\tau = RC$ is the *time constant* of this ODE. The transfer function from cooling power deviation v to indoor temperature deviation y is $\frac{R/\tau}{s+1/\tau}$. Therefore, the transfer function from deviation of the electrical power consumption from baseline, $p_{ves} = v$/COP, and the deviation of indoor temperature from its baseline value, y, is

$$H(s) = \frac{\frac{R}{\tau}\mathrm{COP}}{s + 1/\tau} \tag{2}$$

Using an HVAC system as a VES resource will need a control system that will vary some controllable variable, such as airflow rate, so that the power consumption deviation tracks the VES reference signal $r(t)$. The VES reference $r(t)$ for a building will be most likely computed on-site based on some global reference transmitted from the grid operator. For instance, a regulation reference from the grid operator can be filtered appropriately on-site to compute the reference power deviation [3, 5]. In any case, we ignore the details of the control system, and imagine the controller tracks the reference perfectly: $p_{ves}(t) \equiv r(t)$. The input-output model $H(s)$ between $u(t)$ and $y(t)$ can then be used for analysis on the QoS output y as a function of r.

3.2 Ves Capacity with Pure Sinusoidal Power Deviation

It is instructive first to examine a simple case first, in which the reference for the power consumption deviation is a pure sinusoid: $r(t) = A \sin \omega t$, and a strict bound on indoor temperature is to be maintained, $|y(t)| \leq \Delta_T$ for each t, where Δ_T is a pre-specified constant. *For a given HVAC system and a given frequency* ω, *the maximum A that can be used while still maintaining the QoS constraint is the power capacity of the VES system (at that frequency).*

Since energy is the integral of power, it turns out that the energy capacity is $\frac{A}{\omega}$. Clearly the power capacity of the VES system is a function of frequency of the charge-discharge signal, and for the same energy capacity, peak power delivered by the VES system (i.e., the power capacity) is higher if the charge-discharge signal is of higher frequency. It should be noted that this is exactly analogous to that in case of a real battery.

If the QoS requirement is only imposed in steady-state, then it can be re-interpreted as a bound on the transfer function. Since for an input $u(t) = A \sin \omega t$ to a stable system $H(s)$, in steady state the output amplitude is $A|H(j\omega)|$, the inequality $|y(t)| \leq \Delta_T$ means $A|H(j\omega)| \leq \Delta_T$. Equivalently,

$$A \leq \frac{\Delta_T \tau}{R\,\mathrm{COP}} \sqrt{\omega^2 + \left(\frac{1}{\tau}\right)^2} \tag{3}$$

In practice, there is an upper bound on the frequency ω that can be used, which depends on characteristics of the building hardware rather than building dynamics. This comes from the cost in terms of wear and tear that increases with frequency.

Commercial buildings typically have large values of R and C values, and hence a large time-constant τ. This makes them attractive for VES. The range of possible values for R and C is large, depending on climate and building age. The values estimated for $\sim 540\ \mathrm{m}^2$ small commercial building in [13] are $C \sim 20$ kWh/K and $R \sim 0.5$ K/kW. The corresponding time constant τ is 10 h. We use these values to estimate the VES capacity for such a building.

Most building occupants are unlikely to feel any difference if the temperature deviations for less than 0.1 °C, so $\Delta_T = 0.1$ °C is used. Figure 4 shows the capacity as a function of frequency of the regulation signal it is asked to track, with a COP of 3. The figure also shows the baseline demand, which for that building was around 7 kW (fan + cooling). As we can see, even at the slowest frequency we examine—1/(10 h)—in which the VES capacity is smallest, the capacity is still 0.42 kW, about 6% of the baseline power. At frequency around 1/(30 min), the VES capacity is 100% of the baseline power. Of course, in practice power variation higher than a certain fraction of the baseline power is not advisable since airflow rates and ventilation constraints will be violated even if the temperature deviation remains small.

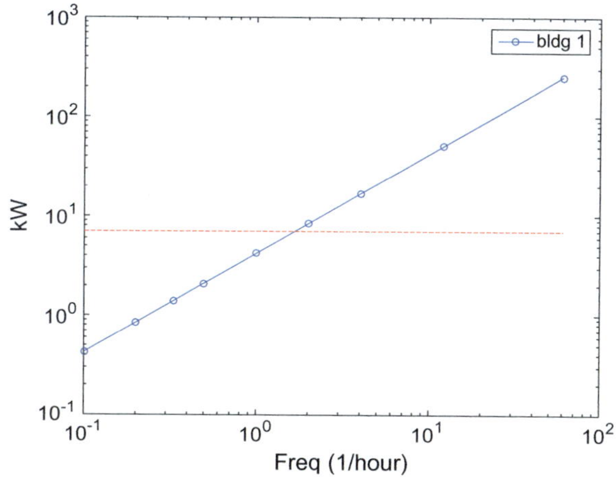

Fig. 4 VES capacity (kW) as a function of frequency (1/h) when power deviation reference is a pure sine, for three different building HVAC systems

Note also that if $\Delta_T = 1°$ is used, which most occupants are likely to tolerate, VES capacity is ten times higher.

Although the analysis here is based on a simplified model, the conclusions point to useful trends. For commercial buildings, large power deviations can lead to very small change in QoS metrics, leading to high VES capacity. This is particularly so at higher frequencies of power variation, since the thermal inertia of the building acts as a low-pass filter to reduce its effect on indoor temperature. At such high frequencies, additional considerations are more likely to determine the peak power deviations allowed (i.e., the VES capacity) rather than indoor temperature.

3.3 Determining VES Capacity Density Curve for HVAC Systems

Now we go back to the question raised at the end of Sect. 2.1: how to determine the curve $c_q(\omega)$ for continuously variable air conditioning loads? As in the previous section, let $y(t)$ be the deviation of the temperature from its baseline (which measures QoS), while $r(t) \equiv p_{\text{ves}}(t)$ is the power deviation of the load from its baseline (the VES signal). Recall that we say the VES system has a capacity density $c(\omega)$ if $P_{rr}(\omega) \leq c(\omega)$ ensures that $y(t) < q$ for each t, where q is the QoS bound that is predetermined.

For ease of analysis, we use a slightly weaker condition, that the QoS bound is respected with high probability. That is, *we say that the QoS constraint is satisfied if*

$$\mathrm{P}(y(t) > q) < \varepsilon \tag{4}$$

for some small predetermined number ε*, where* P *denotes probability.*

Assuming the VES signal $p_{\text{ves}}(t)$ is a wide sense stationary (WSS) signal, the output $y(t)$ will also be WSS since y is the output of a stable linear system $H(s)$ with r as the input, where $H(s)$ is given in (2). From Chebychev inequality, $\text{P}(|y(t)| > q) \leq \frac{\sigma_y^2}{q^2}$ Thus, the condition (4) is satisfied if

$$\frac{\sigma_y^2}{q^2} \leq \varepsilon \tag{5}$$

holds. Since y is the output of H with $p_{\text{ves}} = r$ as the input, we have [14].

$$P_{yy}(\omega) = |H(j\omega)|^2 P_{rr}(\omega)$$

Since p_{ves} is zero mean, y is also zero mean WSS, and for such as process, we have $\sigma_y^2 = \frac{1}{2\pi}\int_{-\infty}^{\infty} P_{yy}(\omega)d\omega$ [14].

Therefore,

$$\sigma_y^2 = \frac{1}{2\pi}\int_{-\infty}^{\infty} |H(j\omega)|^2 P_{rr}(\omega)d\omega$$

From this equation and (5) we need the following for the bound (4) to be satisfied:

$$\frac{1}{2\pi}\int_{-\infty}^{\infty} |H(j\omega)|^2 P_{rr}(\omega)d\omega \leq q^2\varepsilon$$

Any $P_{rr}(\omega)$ that satisfies this inequality will ensure that the output process y will satisfy the QoS constraints. To determine the VES capacity, we look for a PSD $c(\omega)$ that achieves equality in the above inequality:

$$\frac{1}{2\pi}\int_{-\infty}^{\infty} |H(j\omega)|^2 c(\omega)d\omega = q^2\varepsilon \tag{6}$$

There are still many possibilities since more than one function $c(\omega)$ will satisfy (6). Additional constraints will need to be used to obtain the capacity density curve. For instance, recall the discussion in Sect. 2.1: every load has a low and high value of the frequency, ω_L, ω_H to which power deviation signals must be restricted. As an illustration, suppose we look for a capacity curve $c_q(\omega)$ that is constant (say with value b) for $\omega \in [\omega_L, \omega_H]$ and 0 at all other frequencies. In that case (6) translates to

$$\frac{b}{2\pi}\int_{\omega_L}^{\omega_H} |H(j\omega)|^2 d\omega = q^2\varepsilon$$

The number b can be now be uniquely determined from this equation, which uniquely determines the curve $c(\omega)$.

4 Summary

Loads can vary their power around a baseline in a zeromean fashion to effectively act like batteries. With the aid of appropriate control algorithms, loads can thus provide virtual energy storage (VES) to help the grid. A frequency domain framework for characterizing loads flexibility, i.e., VES capacity, vis-a-vis consumer's QoS is advocated, following [4]. The framework is powerful enough to handle not just flexible loads but also conventional generators and batteries. We describe how to characterize the VES capacity of loads in general using amplitude-frequency curves for pure sinusoidal power variations, or in PSDs in the general signal case. Although we have used air conditioning loads for the specific examples in this paper, the methods described for VES capacity estimation are applicable to any load whose power consumption can be continuously varied.

An weakness of the frequency domain characterization of VES capacity is that time-variations due to exogenous factors (such as weather) are not conveniently captured. For instance, an HVAC system may have to operate at peak power at certain times, and a zero-mean deviation from the baseline is not possible at those times. An alternate way of quantifying capacity that has been explored in the literature is a time-varying upper and lower bound of total power consumption that ensures QoS [7, 15]. These approaches necessarily lead to conservative estimates since a constant power deviation from a baseline must be allowed in this framework. A general framework is needed that retains the advantage of frequency-based characterization but is capable of modeling the effect of exogenous factors. This is a subject of future work.

The dependence of VES capacity definition on the baseline is another weakness. The baseline is not possible to measure if a load is providing VES services, only the total power is. This makes estimating the baseline challenging. In [5], the baseline is not estimated but specified as part of the control computations involved in obtaining VES. This "bring-your-own-baseline" approach may offer an attractive alternative.

Acknowledgements Financial support from NSF (through grant 1646229) and DOE (through a BTO-GMLC project titled *virtual batteries*) is gratefully acknowledged.

References

1. Walawalkar R, Fernands S, Thakur N, Chevva K (2010) Evolution and current status of demand response (DR) in electricity markets: insights from PJM and NYISO. Energy 35 (4):1553–1560
2. Mathieu J (2012) Modeling, analysis, and control of demand response resources. Ph.D. dissertation. University of California at Berkeley
3. Lin Y, Barooah P, Meyn S, Middelkoop T (2015) Experimental evaluation of frequency regulation from commercial building HVAC systems. IEEE Trans Smart Grid 6:776–783
4. Barooah P, Bušić A, Meyn S (2015) Spectral decomposition of demand side flexibility for reliable ancillary service in a smart grid. In: 48th Hawaii international conference on systems science, invited paper, Jan 2015
5. Lin Y, Barooah P, Mathieu J (2017) Ancillary services through demand scheduling and control of commercial buildings. IEEE Trans Power Syst, Jan 2017
6. Kirby B (2007) Ancillary services: technical and commercial insights. Prepared for Wärtsilä North America Inc
7. Vrettos E, Oldewurtel F, Andersson G (2016) Robust energy-constrained frequency reserves from aggregations of commercial buildings. IEEE Trans Power Syst 31(6):4272–4285
8. Brooks J, Barooah P (2017) Virtual energy storage through decentralized load control with quality of service bounds. In: American control conference, pp 735–740
9. Nayyar A, Negrete-Pincetic M, Poolla K, Varaiya P (2016) Duration- differentiated energy services with a continuum of loads. IEEE Trans Control Netw Syst 3(2):182–191
10. Meyn S, Barooah P, Bušić A, Chen Y, Ehren J (2015) Ancillary service to the grid from intelligent deferrable loads. IEEE Trans Autom Control 60:2847–2862
11. Todd D, Caufield M, Helms B, Generating A, Starke I, Kirby B, Kueck J (2008) Providing reliability services through demand response: a preliminary evaluation of the demand response capabilities of Alcoa Inc. ORNL/TM, vol 233
12. Kramer R, van Schijndel J, Schellen H (2012) Simplified thermal and hygric building models: a literature review. Front Arch Res 1(4):318–325 [Online]. Available: http://tinyurl.com/kbtak79
13. Coffman A, Barooah P (2017) Simultaneous identification of dynamic model and occupant-induced disturbance for commercial buildings. Build Env (in press) [Online]. Available: https://doi.org/10.1016/j.buildenv.2017.10.020
14. Papoulis A (1984) Probability, random variables, and stochastic processes. McGraw-Hill, New York
15. Hao H, Wu D, Lian J, Yang T (2017) Optimal coordination of building loads and energy storage for power grid and end user services. IEEE Trans Smart Grid 99:1

MIX
Papier aus verantwortungsvollen Quellen
Paper from responsible sources
FSC® C105338

If you have any concerns about our products,
you can contact us on
ProductSafety@springernature.com

In case Publisher is established outside the EU,
the EU authorized representative is:
Springer Nature Customer Service Center GmbH
Europaplatz 3, 69115 Heidelberg, Germany

Printed by Libri Plureos GmbH
in Hamburg, Germany